Universitext

T0210986

Springer
Berlin
Heidelberg
New York
Hong Kong
London
Milan
Paris
Tokyo

Riccardo Benedetti · Carlo Petronio

Lectures on Hyperbolic Geometry

With 175 Figures

 Springer

Riccardo Benedetti
Carlo Petronio
Università degli Studi di Pisa
Dipartimento di Matematica
Via F. Buonarroti 2
56127 Pisa
Italy
e-mail: benedetti/petronio@dm.unipi.it

Cataloging-in-Publication Data applied for
A catalog record for this book is available from the Library of Congress.

Bibliographic information published by Die Deutsche Bibliothek
Die Deutsche Bibliothek lists this publication in the Deutsche Nationalbibliografie;
detailed bibliographic data is available in the Internet at <http://dnb.ddb.de>.

ISBN 3-540-55534-X Springer-Verlag Berlin Heidelberg New York

Mathematics Subject Classification (1991): Primary: 51M10, 32G15, 58B20
Secondary: 51M05, 55NXX, 57M25, 55S40, 57R20

Springer-Verlag Berlin Heidelberg New York
a member of BertelsmannSpringer Science+Business Media GmbH

http://www.springer.de

© Springer-Verlag Berlin Heidelberg 1992
Printed in Germany

Cover design: *design & production*, Heidelberg
Typesetting by the author
Printed on acid-free paper   41/3142ck-5 4 3 2 1

A Martina, Paolo, Clara e Stefania
R. B.

Ai miei genitori, Vitta e Luciano
C. P.

Table of Contents

Preface

In recent years hyperbolic geometry was the object and the inspiration for
an extensive study which produced important and often amazing results and
open questions: it suffices to recall W. P. Thurston's works about the topology
and geometry of three-manifolds and the theory of the so-called hyperbolic (or
negatively curved) groups. However, it is still difficult to find graduate-level
textbooks in the theory of hyperbolic manifolds, starting "from the beginning"
and giving a rather complete and reasonably accessible treatise of some recent
results. The authors became aware of this difficulty while preparing a course
to be held at the Università degli Studi di Pisa, and this book originated from
the first notes sketched on that occasion, widely modified and expanded by a
profitable collaboration during several months. The aim of this text is to give
a modest contribution to the filling of the gap mentioned above, and to the
knowledge of such a fascinating field of mathematics.

⋆   ⋆   ⋆

One of the main themes of this book is the conflict between the "flexibility" and the "rigidity" properties of the hyperbolic manifolds: the first radical difference arises between the case of dimension 2 and the case of higher
dimensions (as proved in chapters B and C), an elementary feature of this
phenomenon being the difference between the Riemann mapping theorem and
Liouville's theorem, as pointed out in chapter A. This chapter is rather elementary and most of its material may be the object of an undergraduate
course.

In chapter B we prove the existence of continuous moduli of hyperbolic
structures on a compact surface via a parametrization of the so-called Teichmüller space: we chose the Fenchel-Nielsen parametrization as it is fully
placed in the realm of hyperbolic geometry.

Chapter C is devoted to the proof of G. D. Mostow's rigidity theorem (in
the compact case). We say two words about this theorem (and the proof we
present in the book) in order to point out the difficulties we mentioned above:
we chose the Gromov-Thurston proof of the rigidity theorem, as it makes use
of a machinery coming essentially from hyperbolic geometry (while Mostow's
original proof made extensive use of analysis). However if you follow the chief
references [Gro2] and [Th1] a problem arises: the core of the proof (where
the differences with Mostow's methods are sharper) consists in establishing a

formula relating the Gromov norm of a compact hyperbolic manifold to its volume; following [Th1] and [Gro2] you first consider a larger class of chains than the singular one and then either you extend the norm to this enlarged class or you even change the definition of the norm (and it is not evident that you get a norm equivalent to the original one). The use of these techniques would suit a more advanced course (as it allows one to make the proof shorter and more conceptual) while it is too demanding for a graduate course. We decided to carry out the proof with the usual singular chains and with the natural and elementary definition of the Gromov norm, following a suggestion M. Gromov ascribes to N. H. Kuiper ([Gro2]); we have filled in the necessary details and we have tried to make the proof as transparent as possible.

Together with the rigidity theorem, a basic tool for the study of hyperbolic manifolds is Margulis' lemma, a detailed proof of which we give in chapter D; as a consequence of this result in the same chapter we also give a rather accurate description, in all dimensions, of the thin-thick decomposition of a hyperbolic manifold (especially in case of finite volume).

Chapter E is devoted to the space of hyperbolic manifolds and to the volume function. We start with the introduction of a natural topology (the so-called geometric topology) on the space of all hyperbolic manifolds having fixed dimension $n \geq 2$, and we discuss different characterizations of such a topology (we shall be interested in particular in the notion of convergence of a sequence). As a corollary of this discussion we shall obtain quite easily the fact that for $n \geq 3$ the volume function (defined on the space of finite-volume hyperbolic n-manifolds) is continuous and proper. This result, together with an extensive use of the fundamental tools developed in the previous chapters (the rigidity theorem and the study of the thin-thick decomposition), will allow us to prove Wang's theorem for hyperbolic manifolds of dimension $n \geq 4$ and most of the so-called Jorgensen-Thurston theory for $n = 3$ (the case of dimension 2 will be treated independently.) The radical difference of behaviour between the case $n \geq 4$ and the case $n = 3$ can be seen as another very important example of the conflict rigidity-flexibility we mentioned above. We shall point out the way the exception of the case $n = 3$ depends essentially on the purely topological fact that all closed 3-manifolds can be obtained via Dehn surgery along links (for instance, in S^3), and on the crucial remark that "almost all" these surgeries can be "made hyperbolic": this is Thurston's hyperbolic Dehn surgery theorem, of which we give a detailed proof based on the possibility of expressing a non-compact hyperbolic three-manifold as ideal tetrahedra with glued faces; this proof is as far as possible elementary and constructive.

Though we confine ourselves to the major aspects of this theory, we are confident that the results explicitly proved in this book are enough to appreciate the sharp difference between the cases $n = 3$ and $n \geq 4$. On the other hand Jorgensen-Thurston theory provides some information whose proof requires the extension of the definition of the Gromov norm and of the techniques used in the proof of the rigidity theorem: we shall give sketches of these results and quote the references for the proofs. We shall also mention other features of flexibility in dimension 3.

We want to emphasize that our discussion of the volume function is mostly deduced from the properties of the natural topology (whose existence is proved *a priori*) on the set of all hyperbolic n-manifolds. As for Jorgensen-Thurston theory the line we shall follow presents some remarkable differences from the chief reference [Th1, ch. 5,6]: a reason is that in these notes we met a difficulty we were not able to overcome, so we needed to re-organize many proofs (we refer to section E.4 for a discussion of these facts).

The notion of Gromov norm introduced in chapter C for the proof of the rigidity theorem can be naturally placed in the general theory of bounded co-homology (developed in [Gro3]); indeed, we can say that proportionality between the Gromov norm of the fundamental class and the volume of a hyperbolic manifold provides the first natural example showing that the theory of bounded co-homology is non-trivial. In chapter F, very far from being complete, we shall just briefly sketch a few other viewpoints of this theory and try to provide some more motivations for it. In particular we shall discuss another interesting example of non-trivial bounded class coming from the study of the Euler class of a flat fiber bundle (due to Milnor, Wood, Sullivan, Gromov and others). In this context we will meet the notion of amenable group, to which we shall devote some space.

The list of references has no pretensions of completeness: it represents the texts we actually used during our work.

$$\star \qquad \star \qquad \star$$

While drawing up this text we have tried to be as self-contained as possible, though we are conscious that this aim remains often closer to an aspiration than to an actual realization: for instance the knowledge of the very basic notions of Riemannian geometry and algebraic topology is essential for a complete understanding of the most part of the book. On the other hand almost all the results mentioned are explicitly proved, and for those which are not easily accessible bibliographical references are given. We hope our aim to be self-contained has been realized at least in the following weak form: the reader can follow the topics and the techniques of this text without needing to stop too often and fill some gap in the pre-requisites, and, in the meanwhile, without feeling he is being asked to accept too many acts of faith.

Finally, a few acknowledgements: it is quite evident from the present preface that this text was largely influenced by the works of W. P. Thurston and M. Gromov; indeed we could say its aim is to divulge in an accessible way a (very little) part of their work. Personally, we are keen on saying that the present text owes much to M. Boileau, G. Levitt, J. C. Sikovav and to the course they organized at Orsay in 1987/88: the first author had the good fortune to attend it, and this fact certainly influenced the choice of the topics and sometimes the details of the proofs. This is true in particular for the section concerning amenable groups, which is largely inspired by some notes Sikovav wrote on that occasion. The second author would like to thank the University of Warwick; he was a visitor there when the final version of the book was completed and he used the University's computer facilities extensively. We also acknowledge some valuable suggestions concerning chapter E made by C. C. Adams. Lastly, we warmly thank Andrea Petronio for the very accurate illustrations and David Trotman for his help in checking our English.

Riccardo Benedetti
Carlo Petronio

Pisa, May 1992

Chapter A.
Hyperbolic Space

This chapter is devoted to the definition of a Riemannian n-manifold $\mathbf{H}^n$ called hyperbolic n-space and to the determination of its geometric properties (isometries, geodesics, curvature, etc.). This space is the local model for the class of manifolds we shall deal with in the whole book. The results we are going to prove may be found in several texts (e.g. [Bea], [Co], [Ep2], [Fe], [Fo], [Greenb2], [Mag], [Mask2], [Th1, ch. 3] and [Wol]) so we shall omit precise references. The line of the present chapter is partially inspired by [Ep2], though we shall be dealing with a less general situation. For a wide list of references about hyperbolic geometry from ancient times to 1980 we address the reader to [Mi3].

A.1 Models for Hyperbolic Space

Let n be a fixed natural number. In order to avoid trivialities we shall always assume $n \geq 2$. We shall give different models for a real Riemannian n-manifold denoted by $\mathbf{H}^n$, which we shall call <u>hyperbolic n-space</u>; these models will be by construction isometrically diffeomorphic to each other. We shall introduce different symbols for them, and we shall use these symbols in order to emphasize a concrete representation of the manifold, while the symbol $\mathbf{H}^n$ will be used for the abstract manifold. We shall not get involved in categorial definitions: every Riemannian manifold isometrically diffeomorphic to $\mathbf{H}^n$ will be identified with $\mathbf{H}^n$.

HYPERBOLOID MODEL. In $\mathbb{R}^{n+1}$ let us consider the standard symmetric bi-linear form of signature $(n, 1)$:

$$\langle x|y\rangle_{(n,1)} = \sum_{i=1}^{n} x_i \cdot y_i - x_{n+1} \cdot y_{n+1}$$

and let us consider the upper fold of the hyperboloid naturally associated to $\langle \cdot|\cdot\rangle_{(n,1)}$:

$$I_n = \left\{ x \in \mathbb{R}^{n+1} : \langle x|x\rangle_{(n,1)} = -1, \ x_{n+1} > 0 \right\}.$$

Since I_n is the pre-image of a regular value of a differentiable function, it is a differentiable oriented hypersurface in $\mathbb{R}^{n+1}$; in particular it is endowed with a differentiable structure of dimension n. For $x \in I_n$ the tangent space to I_n in x is given by

$$T_x I_n = \{y \in \mathbb{R}^{n+1} : \langle x|y \rangle_{(n,1)} = 0\} = \{x\}^{\perp}.$$

Since $\langle x|x \rangle_{(n,1)} = -1$, the restriction of $\langle .|. \rangle_{(n,1)}$ to $\{x\}^{\perp}$ is positive-definite, *i.e.* it is a scalar product on $\{x\}^{\perp}$. So, a metric is naturally defined on the tangent space to each point of I_n; it is easily verified that this metric is globally differentiable, and therefore I_n is endowed with a Riemannian structure. We shall denote by $\mathbb{H}^n$ the manifold I_n endowed with this structure.

DISC MODEL. Let π be the restriction to $\mathbb{H}^n$ of the stereographic projection with respect to $(0, ..., 0, -1)$ of $\{x \in \mathbb{R}^{n+1} : x_{n+1} > 0\}$ onto $\mathbb{R}^n \times \{0\}$. We omit the last coordinate, so that the range of π is $\mathbb{R}^n$:

$$\pi(x) = \frac{(x_1, ..., x_n)}{1 + x_{n+1}}.$$

It is easily verified that π is a diffeomorphism of $\mathbb{H}^n$ onto the open Euclidean unit ball D^n of $\mathbb{R}^n$. The manifold D^n endowed with the pull-back metric with respect to π^{-1} will be denoted by $\mathbb{D}^n$. This manifold is canonically oriented as a domain of $\mathbb{R}^n$.

HALF-SPACE MODEL. Let us consider the differentiable mapping:

$$i : \mathbb{D}^n \to \mathbb{R}^n \qquad x \mapsto 2\frac{x + e_n}{\|x + e_n\|^2} - e_n$$

where $e_n = (0, ..., 0, 1)$ and $\| . \|$ denotes the Euclidean norm on $\mathbb{R}^n$. (In Sect. A.3 we shall introduce the notion of inversion with respect to a sphere: it is worth remarking early that i is the inversion with respect to the sphere of centre $-e_n$ and radius $\sqrt{2}$.) It is easily checked that i is a diffeomorphism of $\mathbb{D}^n$ onto the open half-space $\Pi^{n,+} = \{x \in \mathbb{R}^n : x_n > 0\}$. We shall denote by $\Pi^{n,+}$ this half-space endowed with the pull-back metric with respect to i^{-1}. $\Pi^{n,+}$ is canonically oriented as a domain of $\mathbb{R}^n$.

PROJECTIVE (OR KLEIN) MODEL. Let p be the restriction to $\mathbb{H}^n$ of the canonical projection of $\mathbb{R}^{n+1}$ onto the real projective n-space $\mathbb{RP}^n$. p is a diffeomorphism onto an open subset of $\mathbb{RP}^n$ (actually, the unit disc in a suitable affine chart of $\mathbb{RP}^n$) which can be endowed with the pull-back metric with respect to p^{-1}. Since we are not going to use this model we do not introduce a specific symbol for this representation of $\mathbb{H}^n$.

Figures 1 and 2 illustrate the geometric construction of the first three models in the 2-dimensional case.

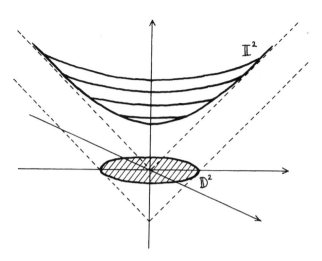

Fig. A.1. Two-dimensional models of hyperbolic space: the hyperboloid and its projection onto the disc

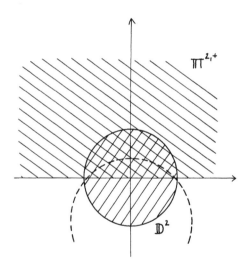

Fig. A.2. Two-dimensional models of hyperbolic space: the disc and its inversion onto the upper half-plane

A.2 Isometries of Hyperbolic Space: Hyperboloid Model

For a Riemannian manifold M we shall denote by $\mathcal{I}(M)$ the set of all isometric diffeomorphisms of M onto itself (briefly: isometries of M). If M is supposed to be oriented, we shall denote by $\mathcal{I}^+(M)$ the set of all isometries of M preserving orientation. $\mathcal{I}(M)$ and $\mathcal{I}^+(M)$ are groups with respect to the operation of composition. In this section we shall determine the groups $\mathcal{I}(\mathbb{I}^n)$ and $\mathcal{I}^+(\mathbb{I}^n)$, while the isometries of $\mathbb{H}^n$ in the other models will be calculated later.

We shall denote the differential of a mapping f in a point x of M by $d_x f$. The scalar product defined on the tangent space $T_x M$ will be denoted by $\langle .|.\rangle_x$ and the quadratic form associated to it by ds_x^2. We recall that the condition that f be an isometry means the following:

$$\langle d_x f(v) | d_x f(w) \rangle_{f(x)} = \langle v | w \rangle_x \quad \forall x \in M, \ v, w \in T_x M.$$

The following result is quite standard, but it will be included for completeness since it is the basis for most of our arguments; we shall use the notion of geodesic and exponential mapping, and the well-known result about existence of normal neighborhoods (see *e.g.* [He]).

Proposition A.2.1. Let M and N be Riemannian manifolds of the same dimension, assume M is connected and let

$$\phi_1 : M \to \phi_1(M) \subseteq N, \quad \phi_2 : M \to \phi_2(M) \subseteq N$$

be local isometries onto their range. If for some $y \in M$ we have $\phi_1(y) = \phi_2(y)$ and $d_y \phi_1 = d_y \phi_2$ then $\phi_1 = \phi_2$.
The conclusion holds in particular if ϕ_1 and ϕ_2 are isometries of M onto N.

Proof. The set

$$S = \left\{ x \in M : \phi_1(x) = \phi_2(x), d_x \phi_1 = d_x \phi_2 \right\}$$

is obviously closed and it contains y, hence we only have to prove that it is open. Let $x \in S$; since M and N have the same dimension the ranges of ϕ_1 and ϕ_2 are open in N; it follows that we can find an open neighborhood V of $\phi_1(x) = \phi_2(x)$ and two open neighborhoods U_1, U_2 of x such that $\phi_i : U_i \to V$ is a surjective isometry for $i = 1, 2$. Let $U \subseteq U_1$ be a normal neighborhood of x and $p : T_x M \supset W \to U$ be the corresponding restriction of the exponential mapping. We set $f = \left(\phi_2 |_{U_2} \right)^{-1} \circ \left(\phi_1 |_{U_1} \right)$; f is an isometry of U_1 onto U_2, $f(x) = x$ and $d_x f = \mathrm{I}$. If γ is a geodesic arc in U starting at x then $f \circ \gamma$ is a geodesic arc starting at x with tangent vector

$$(f \circ \gamma)'(0) = d_x f(\gamma'(0)) = \gamma'(0)$$

and hence $f \circ \gamma = \gamma$, which implies $f \circ p = p$, and finally $f|_U = \mathrm{id}$. It was checked that $\phi_1|_U = \phi_2|_U$, so that $U \subset S$, and the proposition is proved. $\square$

In order to illustrate completely the determination of the isometries of $\mathbb{I}^n$ we start with some elementary facts in linear algebra.

Let V be an n-dimensional real vector space and let $\langle .|.\rangle$ be a non-degenerate bi-linear form on V. It is well-known that there exists a basis $\{v_1, ..., v_n\}$ of V such that

$$\langle v_i | v_j \rangle = \begin{cases} 0 & \text{if } i \neq j \\ +1 & \text{if } i = j \leq p \\ -1 & \text{if } i = j > p. \end{cases}$$

If we set $q = n - p$ the pair (p, q) depends only on $\langle . | . \rangle$ and it is called the signature. If $\mathrm{Gl}(V)$ denotes the linear group on V we set

$$O(V, \langle . | . \rangle) = \left\{ A \in \mathrm{Gl}(V) : \langle Ax | Ay \rangle = \langle x | y \rangle \, \forall x, y \in V \right\}.$$

If $V = W \oplus W'$ and $p : V \to W$ is the associated projection, we shall call reflection the linear mapping $\rho : v \mapsto 2p(v) - v$. It is easily verified that ρ is in $O(V, \langle . | . \rangle)$ if and only if $W' = W^{\perp}$, where $\perp$ denotes orthogonality with respect to $\langle . | . \rangle$. If this is the case we shall say that ρ is the reflection with respect to W, or parallel to $W^{\perp}$. From now on we shall call reflections only those with respect to a hyperplane (i.e. parallel to a vector v such that $\langle v | v \rangle \neq 0$).

Proposition A.2.2. $O(V, \langle . | . \rangle)$ is generated by reflections.

Proof. Let us remark first that $-I$ is generated by the n reflections parallel to the vectors v_i of the basis described above.

We carry out the proof by induction on the dimension n. The first step is obvious. Assume the proposition is true for an integer n, let V have dimension $n + 1$, let $\langle . | . \rangle$ be a non-degenerate bi-linear form on V, let A belong to $O(V, \langle . | . \rangle)$ and choose $v \in V$ such that $\langle v | v \rangle \neq 0$.
We can assume $\langle Av - v | Av - v \rangle \neq 0$: if this is not the case it is easily verified that $\langle -Av - v | -Av - v \rangle \neq 0$, hence we can replace A by $-A$; but by the first remark if $-A$ is a product of reflections then A is too. Let ρ be the reflection parallel to $Av - v$; since

$$v = \frac{1}{2}(Av + v) - \frac{1}{2}(Av - v) \qquad \langle Av + v | Av - v \rangle = 0$$

then $\rho(v) = Av \Rightarrow (\rho \circ A)(v) = v \Rightarrow (\rho \circ A)\big|_{v^{\perp}} \in O(v^{\perp}, \langle . | . \rangle\big|_{v^{\perp} \times v^{\perp}})$. Since every reflection in $v^{\perp}$ extends to a reflection in V the induction hypothesis implies that A is a product of reflections. $\square$

Assume now that $V = \mathbb{R}^{n+1}$ and $\langle . | . \rangle$ is the standard bi-linear form of signature $(n, 1)$, and let I_n be defined as in Sect. A.1. We shall denote by $O(I_n)$ the subgroup of $O(\mathbb{R}^{n+1}, \langle . | . \rangle)$ of those mappings that keep I_n invariant, and by $SO(I_n)$ the intersection of $O(I_n)$ with $Sl(n + 1, \mathbb{R})$. Let us remark that $O(I_n)$ and $SO(I_n)$ are closed subgroups of $\mathrm{Gl}(n + 1, \mathbb{R})$, and hence they are naturally endowed with a Lie group structure.

Proposition A.2.3. $O(I_n)$ is generated by the reflections it contains.

Proof. Every reflection parallel to a vector v with $\langle v | v \rangle \neq 0$ keeps the whole hyperboloid $I_n \cup (-I_n)$ invariant, and it exchanges the two folds if and only if $\langle v | v \rangle < 0$.

Let $A \in O(I_n)$, and write it as a product of reflections: $A = \rho_1 \circ ... \circ \rho_k$. Let ρ_i be parallel to a vector x_i; if $\langle x_i | x_i \rangle < 0$ we can complete x_i to an orthogonal basis $\{x_i, w_1, ..., w_n\}$ of $\mathbb{R}^{n+1}$, with the property that $\langle w_j | w_j \rangle > 0 \, \forall j$. Then ρ_i is given by $-(\sigma_1 \circ ... \circ \sigma_n)$, where σ_j is the reflection parallel to w_j. If we make this substitution for all i's such that $\langle x_i | x_i \rangle < 0$ we obtain that

$$A = \pm(\tau_1 \circ ... \circ \tau_h)$$

where all the τ_k's are reflections and belong to $O(I_n)$. The minus sign is obviously absurd and hence the proposition is proved. $\square$

Theorem A.2.4. $\mathcal{I}(\mathbb{I}^n)$ consists of the restrictions to $\mathbb{I}^n$ of the elements of $O(I_n)$, whence $\mathcal{I}(\mathbb{I}^n) \cong O(I_n)$; in particular $\mathcal{I}(\mathbb{I}^n)$ is generated by reflections. Similarly $\mathcal{I}^+(\mathbb{I}^n) \cong SO(I_n)$.

Proof. Let $f \in \mathcal{I}(\mathbb{I}^n)$ and choose arbitrarily $x \in \mathbb{I}^n$; since $d_x f$ is an isometry of $x^\perp$ onto $f(x)^\perp$ and $\langle x|x \rangle = \langle f(x)|f(x) \rangle = -1$ it is readily checked that the linear mapping

$$A : \mathbb{R}^{n+1} = \mathbb{R}\,x \oplus x^\perp \to \mathbb{R}^{n+1} \qquad \lambda x + v \mapsto \lambda f(x) + d_x f(v)$$

is an element of $O(I_n)$. As the restriction of A to $\mathbb{I}^n$ is obviously an isometry, and $f(x) = Ax$, $d_x f = A\big|_{T_x \mathbb{I}^n}$, then by Proposition A.2.1 f is the restriction of A to $\mathbb{I}^n$. It follows that

$$\mathcal{I}(\mathbb{I}^n) = \big\{ A\big|_{\mathbb{I}^n} : A \in O(I_n) \big\}.$$

Since the linear span of I_n is $\mathbb{R}^{n+1}$, the mapping

$$O(I_n) \ni A \mapsto A\big|_{I_n} \in \mathcal{I}(\mathbb{I}^n)$$

is one-to-one, and hence it is a group isomorphism.

The case of orientation-preserving isometries is a straight-forward consequence of the general one. $\square$

Though we are mainly interested in hyperbolic space we prove an analogue of Theorem A.2.4 for two other very important Riemannian manifolds: the sphere and Euclidean space.

$\mathbb{R}^n$ will be endowed with the standard Euclidean metric, and the unit sphere S^n in $\mathbb{R}^{n+1}$ will be endowed with the restriction of the Euclidean metric to its tangent bundle (the construction is completely analogous to the one we presented in A.1 for $\mathbb{I}^n$: $\langle .|. \rangle_{(n,1)}$ is substituted by the Euclidean metric and -1 by 1). In $\mathbb{R}^n$ reflections with respect to affine hyperplanes are naturally defined, while in S^n we shall consider the restrictions of the reflections of $\mathbb{R}^{n+1}$.

Theorem A.2.5. $\mathcal{I}(S^n) = \big\{ A\big|_{S^n} : A \in O(n+1) \big\}$,
$\mathcal{I}(\mathbb{R}^n) = \big\{ (x \mapsto Ax + b : A \in O(n), b \in \mathbb{R}^n \big\}$.
Both of these groups are generated by reflections.

Proof. The technique is the same as for A.2.4: inclusions $\supseteq$ are obvious, and for the converse it is checked that for each element of the group on the left an element of the group on the right can be found in such a way that the two coincide up to first order.

The last assertion is obvious in the first case, while in the second we only have to remark that the translation of a vector b is the product of the reflections with respect to $b^\perp$ and $b/2 + b^\perp$. $\square$

A.3 Conformal Geometry

In this section we will be concerned with conformal geometry in $\mathbb{R}^n$ and we shall prove an important theorem due to Liouville (see for instance [Ber]). The reason for this long parenthesis is that conformal geometry in $\mathbb{R}^n$ permits a complete calculation of the isometries of $\mathbf{H}^n$ in the disc and half-space model: we shall prove that every isometry with respect to the hyperbolic structure in $\mathbb{D}^n$ and $\mathbb{II}^{n,+}$ is a conformal automorphism with respect to the Euclidean structure naturally defined by the immersion in $\mathbb{R}^n$, and conversely.

Let M and N be Riemannian manifolds: we shall say a diffeomorphism $f : M \to N$ is __conformal__ if there exists a differentiable positive function α on M such that

$$\langle d_x f(v) | d_x f(w) \rangle_{f(x)} = \alpha(x) \langle v | w \rangle_x \quad \forall x \in M, \ v, w \in T_x M$$

(*i.e.* f preserves angles but not necessarily lengths). This definition can be easily generalized to manifolds endowed with a __conformal structure__, *i.e.* manifolds in which the angle between two vectors is defined.

The set of conformal diffeomorphisms of M onto N will be denoted by $\mathrm{Conf}(M, N)$, and by $\mathrm{Conf}(M)$ in case $N = M$; remark that $\mathrm{Conf}(M)$ is a group. As usual, the $+$ superscript will mean that orientation (if any) is preserved.

We introduce now a very important notion for the study of conformal geometry in $\mathbb{R}^n$. If $x_0 \in \mathbb{R}^n$ and $\alpha > 0$ we shall call __inversion__ with respect to the sphere $M(x_0, \alpha)$ of centre x_0 and radius $\sqrt{\alpha}$ the following mapping:

$$i_{x_0, \alpha} : x \mapsto \alpha \cdot \frac{x - x_0}{\|x - x_0\|^2} + x_0.$$

We shall think of $i_{x_0,\alpha}$ both as a mapping of $\mathbb{R}^n \setminus \{x_0\}$ onto itself and as a mapping of $\mathbb{R}^n \cup \{\infty\}$ onto iself, where $\mathbb{R}^n \cup \{\infty\} \cong S^n$ is the one-point compactification of $\mathbb{R}^n$, and $i_{x_0,\alpha}$ exchanges x_0 and ∞. Throughout this section S^n will be endowed with its natural conformal structure; remark that $\mathbb{R}^n = S^n \setminus \{\infty\}$ inherits from S^n its own conformal structure; every open subset of $\mathbb{R}^n$ will be endowed with the conformal structure induced from $\mathbb{R}^n$. Remark that the definition of $i_{x_0,\alpha}$ makes sense also for $\alpha < 0$, and it is easily checked that in this case $i_{x_0,\alpha}$ is the composition of the inversion with respect to $M(x_0, -\alpha)$ with the symmetry centred at x_0. In the following proposition we shall list a few important properties of inversions. We shall say two hyperplanes H_1 and H_2 in $\mathbb{R}^n$ are <u>orthogonal</u> if the lines $H_1^\perp$ and $H_2^\perp$ are orthogonal; consequently we shall say that two intersecting spheres are orthogonal if for any point of their intersection the two tangent hyperplanes are orthogonal in the above sense; that is, if x_0 and x_1 are the centres of the spheres, for each point x of the intersection $\langle x - x_0 | x - x_1 \rangle = 0$.

We are not going to say explicitly if an inversion $i_{x_0,\alpha}$ is considered to be defined on $\mathbb{R}^n \setminus \{x_0\}$ or on S^n, since it will be evident from the context.

Proposition A.3.1. (1) $i_{x_0,\alpha} \circ i_{x_0,\beta}$ is the dilation centred at x_0 of ratio α/β.
(2) $i_{x_0,\alpha}$ is a C^∞ involution (of both $\mathbb{R}^n \setminus \{x_0\}$ and S^n).
(3) $i_{x_0,\alpha}|_{M(x_0,\alpha)} = \mathrm{id}$.
(4) $i_{x_0,\alpha}$ is a conformal mapping.
(5) Given $\alpha, \beta > 0$ and $x_1 \neq x_0$ the following facts are equivalent:
  i) $M(x_1,\beta)$ is $i_{x_0,\alpha}$-invariant;
  ii) $M(x_0,\alpha)$ is $i_{x_1,\beta}$-invariant;
  iii) $\|x_1 - x_0\|^2 = \alpha + \beta$;
  iv) $M(x_1,\beta)$ and $M(x_0,\alpha)$ are orthogonal spheres.
(6) Let $i = i_{x_0,\alpha}$; then
  i) H hyperplane, $H \ni x_0 \Rightarrow i(H) = H$;
  ii) H hyperplane, $H \not\ni x_0 \Rightarrow i(H)$ sphere, $i(H) \ni x_0$;
  iii) M sphere, $M \ni x_0 \Rightarrow i(M)$ hyperplane, $i(M) \not\ni x_0$,
  iv) M sphere, $M \not\ni x_0 \Rightarrow i(M)$ sphere, $i(M) \not\ni x_0$;
  v) i operates bijectively on the set of all open balls and all open half-spaces in $\mathbb{R}^n$.

Proof. First of all we remark that if T is the translation $x \mapsto x + x_0$ we have $i_{x_0,\alpha} = T \circ i_{0,\alpha} \circ T^{-1}$, and $i_{0,\alpha} = \alpha \cdot i_{0,1}$, hence we shall often assume $x_0 = 0$ and $\alpha = 1$.

$$(1) \ (i_{0,\alpha} \circ i_{0,\beta})(x) = \alpha \frac{\beta x / \|x\|^2}{\left\| \beta x / \|x\|^2 \right\|^2} = \alpha\beta^{-1}x.$$

(2) By (1) $i_{x_0,\alpha}$ is an involution; differentiability is evident.
(3) Obvious.
(4) Dilations and translations are conformal, and hence we refer to $i_{0,1}$, which is conformal at $x \neq 0$ since its differential is

$$d_x i_{0,1}(y) = \frac{1}{\|x\|^2} \cdot P_x(y)$$

where P_x is the reflection parallel to x, *i.e.* the reflection with respect to the hyperplane $x^\perp$. Moreover $i_{0,1}$ is the standard chart around ∞, and hence it is by definition conformal at 0.

(5)
  i) $\Rightarrow$ iii). We assume $x_0 = 0$ and $\alpha = 1$. The intersection of $M(x_1,\beta)$ with the line $\mathbb{R}x_1$ consists of the points $(1 \pm \sqrt{\beta}/\|x_1\|)x_1$ and hence $i_{0,1}$ must

exchange them (in fact both the sphere and the line are $i_{0,1}$-invariant, and it is easily checked that it is impossible that both points are fixed). By direct calculation

$$(1 - \sqrt{\beta}/\|x_1\|)x_1 = i_{0,1}((1 + \sqrt{\beta}/\|x_1\|)x_1) = \frac{x_1}{(1 + \sqrt{\beta}/\|x_1\|)\|x_1\|^2}$$

$$\Rightarrow (1 - \sqrt{\beta}/\|x_1\|) \cdot (1 + \sqrt{\beta}/\|x_1\|) \cdot \|x_1\|^2 = 1 \Rightarrow \|x_1\|^2 = 1 + \beta.$$

iii) $\Rightarrow$ i). As above, $x_0 = 0$ and $\alpha = 1$. Let $x \in M(x_1, \beta)$, then

$$\|x_1\|^2 - 1 = \beta = \|x - x_1\|^2 = \|x\|^2 - 2\langle x|x_1\rangle + \|x_1\|^2 \Rightarrow 2\langle x|x_1\rangle = 1 + \|x\|^2$$

and therefore

$$\|i_{0,1}(x) - x_1\|^2 = \left\| \frac{x}{\|x\|^2} - x_1 \right\|^2 = \frac{1}{\|x\|^2} - \frac{2\langle x|x_1\rangle}{\|x\|^2} + 1 + \beta = \beta.$$

ii) $\Leftrightarrow$ iii) is proved in the very same way.

iii) $\Leftrightarrow$ iv). Since $\alpha + \beta < (\sqrt{\alpha} + \sqrt{\beta})^2$, condition iii) implies that the two spheres intersect. Moreover, if x is in the intersection we have

$$\|x_1 - x_0\|^2 = \alpha + \beta \Leftrightarrow \|x_1 - x\|^2 + \|x_0 - x\|^2 = \|x_1 - x_0\|^2 \Leftrightarrow$$
$$\Leftrightarrow -\langle x_1|x\rangle + \langle x|x\rangle - \langle x_0|x\rangle = -\langle x_1|x_0\rangle \Leftrightarrow \langle x_1 - x|x_0 - x\rangle = 0.$$

(6) Since the properties we are considering are invariant under dilations and translations, we take $i = i_{0,1}$.

i) is obvious.

ii). Let $H \doteq h + h^\perp$ with $h \in \mathbb{R}^n \setminus \{0\}$. We set $c = {}^{h}/_{2\|h\|^2}, \gamma = {}^{1}/_{4\|h\|^2}$. For $x \neq 0$ we have

$$i(x) \in M(c, \gamma) \Leftrightarrow \|i(x) - c\|^2 = \gamma \Leftrightarrow \left\| \frac{x}{\|x\|^2} - \frac{h}{2\|h\|^2} \right\| = \frac{1}{4\|h\|^2} \Leftrightarrow$$

$$\Leftrightarrow \frac{1}{\|x\|^2} - \frac{\langle x|h\rangle}{\|x\|^2 \cdot \|h\|^2} = 0 \Leftrightarrow \langle h - x|h\rangle = 0 \Leftrightarrow x \in H.$$

Moreover $i(\infty) = 0 \in M(c, \gamma)$, whence $i(H) = M(c, \gamma)$.

iii). Let $M = M(c, \gamma)$. Since $0 \in M$ we have $\gamma = \|c\|^2$. If we set $h = {}^{c}/_{2\|c\|^2}$ and $H = h + h^\perp$, by ii) we have $i(H) = M$, and then $i(M) = H$.

iv). Let $M = M(c, \gamma)$. Since $0 \notin M$ we have $\|c\|^2 \neq \gamma$. The following holds:

$$i(x) \in M(c, \gamma) \iff \left\| \frac{x}{\|x\|^2} - c \right\|^2 = \gamma \iff$$

$$\iff \frac{1}{\|x\|^2} - \frac{2\langle x | c \rangle}{\|x\|^2} + \|c\|^2 = \gamma \iff$$

$$\iff \|x\|^2 - \frac{2\langle x | c \rangle}{\|c\|^2 - \gamma} + \frac{1}{\|c\|^2 - \gamma} = 0 \iff$$

$$\iff \left\| x - \frac{c}{\|c\|^2 - \gamma} \right\|^2 = \frac{\|c\|^2}{(\|c\|^2 - \gamma)^2} - \frac{1}{\|c\|^2 - \gamma} = \frac{\gamma}{(\|c\|^2 - \gamma)^2} \iff$$

$$\iff x \in M\left(\frac{c}{(\|c\|^2 - \gamma)}, \frac{\gamma}{(\|c\|^2 - \gamma)^2} \right).$$

Therefore $i\big(M(c, \gamma)\big) = i^{-1}\big(M(c, \gamma)\big) = M\left(\frac{c}{(\|c\|^2 - \gamma)}, \frac{\gamma}{(\|c\|^2 - \gamma)^2} \right)$, and by iii) this sphere cannot contain 0.

v). If A is either an open ball or an open half-space we have that ∂A is either a sphere or a hyperplane, and by i)–iv) the same holds for $i(\partial A)$. By (2) $i(A)$ is connected and its boundary is $i(\partial A)$, which implies that it is either a ball or a half-space. $\qquad\square$

Now, for $n \geq 2$ we will deal with the set of all conformal diffeomorphisms between two domains of $\mathbb{R}^n$. The technique is completely different for the case $n = 2$ and the case $n \geq 3$; however, for the particular open sets we are interested in, the result is the same for all integers n.

FIRST CASE: $n = 2$.

We begin by recalling (see [Sp] or [DC]) that a connected oriented Riemannian surface M admits a complex structure (given by isothermal coordinates), and this structure is uniquely determined by the requirement that

$$f : \mathbb{C} \supset U \to M$$

is a holomorphic chart if and only if it preserves orientation and

$$ds^2_{f(z)}\big(d_z f(w)\big) = \alpha(z) \cdot |w|^2 \qquad \forall\, z \in U, w \in \mathbb{C}$$

for some function $\alpha > 0$.

By the following proposition conformal geometry in dimension 2 reduces to a problem in the theory of functions of one complex variable.

Proposition A.3.2. If M and N are connected oriented Riemannian surfaces (naturally endowed with complex structures), the set of all conformal diffeomorphisms of M onto N is the set of all holomorphisms and all anti-holomorphisms of M onto N.

Proof. This fact could be easily deduced from the uniqueness of the complex structure. However, we shall prove it directly: actually, this very argument proves the uniqueness of the complex structure (while existence is much more complicated).

Let $f : M \to N$ be conformal. Since the only both holomorphic and anti-holomorphic functions are the constants, and since holomorphy and anti-holomorphy are closed conditions, by connectedness it suffices to prove that f is locally holomorphic or anti-holomorphic, hence we can assume that M and N are domains of $\mathbb{C}$. The conformality condition is expressed by

$$|df|^2 = \alpha \cdot |dz|^2$$

$$\Rightarrow \left| \frac{\partial f}{\partial z} dz + \frac{\partial f}{\partial \bar{z}} d\bar{z} \right|^2 = \alpha |dz|^2$$

$$\Rightarrow \left(\left| \frac{\partial f}{\partial z} \right|^2 + \left| \frac{\partial f}{\partial \bar{z}} \right|^2 \right) |dz|^2 + 2 \Re \left[\frac{\partial f}{\partial z} \cdot \overline{\left(\frac{\partial f}{\partial \bar{z}} \right)} \cdot (dz)^2 \right] = \alpha |dz|^2.$$

Since $2\Re \left[\partial f / \partial z \cdot \overline{\left(\partial f / \partial \bar{z} \right)} \cdot (dz)^2 \right]$ is not a multiple of $|dz|^2$ it must be zero, whence $\partial f / \partial z \cdot \overline{\left(\partial f / \partial \bar{z} \right)} = 0$. It follows that for each point z_0 we have $\partial f / \partial z (z_0) = 0$ or $\partial f / \partial \bar{z} (z_0) = 0$ (but not both, as $d_{z_0} f \neq 0$). Then

$$M = \left\{ z_0 : \frac{\partial f}{\partial z}(z_0) = 0 \right\} \cup \left\{ z_0 : \frac{\partial f}{\partial \bar{z}}(z_0) = 0 \right\}.$$

Since these two sets are closed and disjoint, one of them must be empty, so that f is holomorphic or anti-holomorphic in M.

The above calculations show as well that if $f : M \to N$ is holomorphic or anti-holomorphic then it is conformal. $\square$

Let us consider the Riemann sphere $S^2 = \mathbb{CP}^1$ naturally identified with the set $\mathbb{C} \cup \{\infty\}$ (where $\infty = 0^{-1}$). We define the two classes of mappings of $\mathbb{CP}^1$ onto itself by

$$\text{homographies}: \qquad z \mapsto \frac{az + b}{cz + d}$$

$$\text{anti} - \text{homographies}: \qquad z \mapsto \frac{a\bar{z} + b}{c\bar{z} + d}$$

where $\begin{pmatrix} a & b \\ c & d \end{pmatrix}$ varies in $\mathrm{Gl}(2, \mathbb{C})$.

The following theorem settles the two-dimensional conformal geometry for the most important domains. We shall identify $\mathbb{R}^2$ with $\mathbb{C}$, in such a way that $\mathbb{R}^2$, D^2 and $\Pi^{2,+}$ are open subsets of $\mathbb{CP}^1$. If F is a set of mappings we denote by $c(F)$ the set $\{(z \mapsto \overline{f(z)}) : f \in F\}$ and by $-F$ the set $\{(z \mapsto -f(z)) : f \in F\}$.

All results we shall need from the theory of one complex variable can be found $e.g.$ in [La] and [Na].

Theorem A.3.3. The group $\mathrm{Conf}^+(S^2)$ consists of all homographies, and the group $\mathrm{Conf}(S^2)$ consists of all homographies and anti-homographies. For $M = \mathbb{R}^2, D^2, \Pi^{2,+}$ we have

$$\mathrm{Conf}^+(M) = \{f|_M : f \in \mathrm{Conf}^+(S^2), f(M) = M\}$$
$$\mathrm{Conf}(M) = \{f|_M : f \in \mathrm{Conf}(S^2), f(M) = M\}.$$

In particular:

$$\mathrm{Conf}^+(\mathbb{C}) = \left\{(z \mapsto az + b) : a, b \in \mathbb{C}, a \neq 0\right\}$$
$$\mathrm{Conf}(\mathbb{C}) = \mathrm{Conf}^+(\mathbb{C}) \bigcup c\left(\mathrm{Conf}^+(\mathbb{C})\right)$$
$$\mathrm{Conf}^+(D^2) = \left\{\left(z \mapsto e^{i\theta} \cdot \frac{z - \alpha}{1 - \bar{\alpha}z}\right) : \theta \in \mathbb{R}, \alpha \in D^2\right\}$$
$$\mathrm{Conf}(D^2) = \mathrm{Conf}^+(D^2) \bigcup c\left(\mathrm{Conf}^+(D^2)\right)$$
$$\mathrm{Conf}^+(\Pi^{2,+}) = \left\{\left(z \mapsto \frac{az + b}{cz + d}\right) : \begin{pmatrix} a & b \\ c & d \end{pmatrix} \in \mathrm{Sl}(2, \mathbb{R})\right\}$$
$$\mathrm{Conf}(\Pi^{2,+}) = \mathrm{Conf}^+(\Pi^{2,+}) \bigcup \left(-c\left(\mathrm{Conf}^+(\Pi^{2,+})\right)\right).$$

Proof. By Proposition A.3.2 we have to determine the set of holomorphisms and anti-holomorphisms of these complex surfaces. We shall refer only to holomorphisms; all the details for the case of anti-holomorphisms can be filled in as an exercise.

We begin with the explicit determination of the holomorphisms in all cases.

If $f : \mathbb{C} \to \mathbb{C}$ is a holomorphism then f cannot have an essential singularity at ∞ (otherwise, by Picard's theorem, it would not be one-to-one); the power series expansion of f at 0

$$f(z) = \sum_{n \geq 0} a_n \cdot z^n$$

coincides with the Laurent expansion of f at ∞, and hence it is finite. It follows that f is a polynomial, and bijectivity immediately implies that $f(z) = az + b$ with $a \neq 0$.

As for $\mathbb{CP}^1$, the set of all homographies is a group of holomorphisms of $\mathbb{CP}^1$. Conversely, since homographies operate transitively, given a holomorphism f we can find a homography ϕ with $(\phi \circ f)(\infty) = \infty$; it follows that $(\phi \circ f)$ is a holomorphism of $\mathbb{C}$, and hence it is a homography, which implies that f is a homography too.

By Schwarz's lemma the group of holomorphisms of D^2 keeping the origin fixed is given by rotations, and the proof works as above since the described set is a group of holomorphisms of D^2 containing rotations and operating transitively.

The determination of the group $\mathrm{Conf}^+(\Pi^{2,+})$ easily follows from that of $\mathrm{Conf}^+(D^2)$ via the Cayley transformation $z \mapsto {(z - i)}/{(z + i)}$, which maps $\Pi^{2,+}$ bi-holomorphically onto D^2.

Now, let $M \in \{\mathbb{C}, D^2, \Pi^{2,+}\}$; we are left to prove that

$$\mathrm{Conf}^+(M) = \{f|_M : f \in \mathrm{Conf}^+(S^2), f(M) = M\}.$$

If f is a homography and $f(M) = M$ the restriction of f to M is obviously a holomorphism of M. As for the converse, it easily follows from the determination of $\mathrm{Conf}^+(M)$ in the three cases that all its elements extend to homographies.   □

As for completeness, we recall the usual representation of the group $\mathrm{Conf}^+(D^2)$ (see [Ve]). After defining

$$ J = \begin{pmatrix} 1 & 0 \\ 0 & -1 \end{pmatrix} \qquad \mathrm{SU}(1,1) = \{A \in \mathrm{Sl}(2, \mathbb{C}) : {}^t\overline{A} J A = J\} $$

it is checked that

$$ \mathrm{SU}(1,1) = \left\{ \begin{pmatrix} \alpha & \beta \\ \overline{\beta} & \overline{\alpha} \end{pmatrix} : \alpha, \beta \in \mathbb{C}, |\alpha|^2 - |\beta|^2 = 1 \right\} $$

$$ \mathrm{Conf}^+(D^2) = \left\{ \left(z \mapsto \frac{az+b}{cz+d} \right) : \begin{pmatrix} a & b \\ c & d \end{pmatrix} \in \mathrm{SU}(1,1) \right\} \cong \mathrm{SU}(1,1) \big/ \{\pm I\}. $$

Proposition A.3.4. If we identify $\mathbb{CP}^1$ with $\mathbb{R}^2 \cup \{\infty\}$ then $\mathrm{Conf}(\mathbb{CP}^1)$ consists of all and only the mappings of the form

$$ x \mapsto \lambda Ai(x) + v $$

where $\lambda > 0$, $A \in O(2)$, i is either the identity or an inversion and $v \in \mathbb{R}^2$.

Proof. Since the conjugation is an element of $O(2)$ we consider an anti-homography

$$ f : z \mapsto \frac{a\overline{z}+b}{c\overline{z}+d} $$

and we show that f can be written as $\lambda Ai + v$. If $c = 0$ this fact is obvious. If $c \neq 0$ we have

$$ \frac{a\overline{z}+b}{c\overline{z}+d} = \frac{a}{c} + \frac{(bc-ad)/c^2}{\overline{z} + d/c}. $$

Let $(bc-ad)/c^2 = \lambda u$, with $\lambda > 0$ and $|u| = 1$ (hence $u \in O(2)$); if we define i to be the inversion with respect to the sphere of centre $-\overline{d/c}$ and radius 1, we have

$$ i(z) = -\overline{d/c} + \frac{z + \overline{d/c}}{\left| z + \overline{d/c} \right|^2} = -\overline{d/c} + \frac{1}{\overline{z} + d/c} $$

and hence

$$ \frac{a\overline{z}+b}{c\overline{z}+d} = \lambda u i(z) + \lambda u \overline{d/c} + a/c. $$

A similar calculation proves that every mapping of the form $\lambda Ai + v$ is a homography or an anti-homography.   □

The following result could be proved as a corollary of A.3.4, but we shall prove it directly from A.3.3. In A.3.9 we shall check that a completely analogous statement holds for $n \geq 3$ too.

Theorem A.3.5. (1) $\mathrm{Conf}(D^2)$ consists of all and only the mappings of the form $x \mapsto Ai(x)$, where $A \in O(2)$ and i is either the identity or an inversion with respect to a sphere orthogonal to ∂D^2.

(2) $\mathrm{Conf}(\Pi^{2,+})$ consists of all and only the mappings of the form

$$x \mapsto \lambda \begin{pmatrix} u & 0 \\ 0 & 1 \end{pmatrix} i(x) + \begin{pmatrix} b \\ 0 \end{pmatrix}$$

where $\lambda > 0$, $u \in O(1) = \{\pm 1\}$, i is either the identity or an inversion with respect to a sphere orthogonal to $\mathbb{R} \times \{0\}$ and $b \in \mathbb{R}$.

Proof. (1) By A.3.1 (4) and (5) every mapping of the form Ai belongs to $\mathrm{Conf}(D^2)$. As for the converse, we remark that the set of all the mappings of the required form is a group: hence, by A.3.3, since the conjugation and all rotations belongs to $O(2)$ we only have to check that for $\alpha \in D^2$ the function $z \mapsto (\bar{z} - \alpha)/(1 - \overline{\alpha}\,\bar{z})$ can be written as Ai. The sphere of centre $1/\alpha$ and squared radius $1/|\alpha|^2 - 1$ is orthogonal to ∂D^2; let i denote the inversion with respect to it; we have

$$i(z) = \frac{1}{\alpha} + \left(\frac{1}{|\alpha|^2} - 1 \right) \frac{1}{\bar{z} - 1/\alpha} =$$

$$= \frac{1}{\alpha} + \frac{1 - |\alpha|^2}{\alpha} \cdot \frac{1}{\overline{\alpha}\bar{z} - 1} = \frac{1}{\alpha} \cdot \frac{\overline{\alpha}\bar{z} - |\alpha|^2}{\overline{\alpha}\bar{z} - 1} = -\frac{\overline{\alpha}}{\alpha} \cdot \frac{\bar{z} - \alpha}{1 - \overline{\alpha}\bar{z}}$$

$$\Rightarrow \frac{\bar{z} - \alpha}{1 - \overline{\alpha}\bar{z}} = -\frac{\alpha}{\overline{\alpha}} \cdot i(z).$$

(2) By A.3.1 (4) and (6)-i) every mapping of the described form belongs to $\mathrm{Conf}(\Pi^{2,+})$. As for the converse, we remark that the set of all the mappings of the required form is a group: then, by A.3.3, as the mapping $z \mapsto -\bar{z}$ is expressed by $\begin{pmatrix} -1 & 0 \\ 0 & 1 \end{pmatrix}$, we only have to check that for $\begin{pmatrix} a & b \\ c & d \end{pmatrix} \in \mathrm{Sl}(2, \mathbb{R})$ the mapping

$$z \mapsto -\frac{a\bar{z} + b}{c\bar{z} + d}$$

can be written in the required form. If $c = 0$ this is obvious. Otherwise we set $\mu = -(bc - ad)/c^2 = 1/c^2 > 0$ and we consider the inversion i with respect to the sphere of centre $-d/c$ and radius 1 (which is orthogonal to $\mathbb{R} \times \{0\}$, since its centre lies on such a line); it is easily checked that

$$-\frac{a\bar{z} + b}{c\bar{z} + d} = \mu i(z) + \mu\frac{d}{c} - \frac{a}{c}$$

and the conclusion follows immediately.   □

Remark A.3.6. Since translations, dilations and elements of $SO(n)$ preserve orientation, while elements of $O(n) \setminus SO(n)$ and inversions with respect to spheres reverse orientation, we have the following:

(1) $Ai \in \text{Conf}(D^2)$ belongs to $\text{Conf}^+(D^2)$ if and only if
$$[A \in SO(2), i = \text{identity}] \text{ or } [A \in O(2) \setminus SO(2), i = \text{inversion}].$$

(2) $\left(\lambda \begin{pmatrix} u & 0 \\ 0 & 1 \end{pmatrix} i + \begin{pmatrix} b \\ 0 \end{pmatrix} \right) \in \text{Conf}(\Pi^{2,+})$ belongs to $\text{Conf}^+(\Pi^{2,+})$ if and only

if $\quad [u = 1, i = \text{identity}] \text{ or } [u = -1, i = \text{inversion}].$

SECOND CASE: $n \geq 3$.

Our aim is to prove an analogue of Theorem A.3.5 for the n-dimensional ball and half-space. We shall use a much more general result due to Liouville, whose long proof we are going to present now. It is worth remarking that an analogue of Liouville's theorem in dimension two is false: we shall point out the steps where the assumption $n \geq 3$ is essential.

Theorem A.3.7 (Liouville). Every conformal diffeomorphism between two domains of $\mathbb{R}^n$ has the form

$$x \mapsto \lambda Ai(x) + b$$

where $\lambda > 0, A \in O(n), i$ is either the identity or an inversion and $b \in \mathbb{R}^n$.

Proof. If U, V are domains in $\mathbb{R}^n$ and $f : U \to V$ is a conformal diffeomorphism, we shall denote by $\mu_f \in C^\infty(U, \mathbb{R}_+)$ the coefficient of dilation of the metric, that is the function satisfying

$$\|d_x f(v)\| = \mu_f(x) \|v\| \qquad \forall x \in U, v \in \mathbb{R}^n.$$

We define ρ_f as $1/\mu_f$.

We shall say f is of type (a) if it is expressed as $\lambda A + b$ with $A \in O(n)$, $b \in \mathbb{R}^n$, and of type (b) if it is expressed as $\lambda Ai + b$, where $A \in O(n), b \in \mathbb{R}^n$ and i is the inversion with respect to a sphere. The theorem can be re-phrased as follows: *every conformal diffeomorphism $f : U \to V$ is either of type (a) or of type (b).*

The proof is a straight-forward corollary of the following partial results:

Step 1.

i) f is of type (a) if and only if ρ_f is constant;

ii) f is of type (b) if and only if there exist $x_0 \in \mathbb{R}^n$ and $\eta \in \mathbb{R} \setminus \{0\}$ such that $\rho_f(x) = \eta \|x - x_0\|^2$.

Step 2. There exist $\eta, \tau \in \mathbb{R}, z \in \mathbb{R}^n$ such that $\rho_f(x) = \eta \|x\|^2 + \langle x | z \rangle + \tau$.

Step 3. If in step 2 it is $\eta \neq 0$ then for some $x_0 \in \mathbb{R}^n$ we have
$$\rho_f(x) = \eta \|x - x_0\|^2.$$

Step 4. In step 2 it cannot occur that $\eta = 0$ and $z \neq 0$.

According to Step 2, we shall say that f is of type I if $\eta = 0$ and $z = 0$, of type II if $\eta = 0$ and $z \neq 0$, of type III if $\eta \neq 0$. By step 1, if f is of type I then it is of type (a), and steps 3 and 4 can be respectively re-phrased as follows:

- if f is of type III then it is of type (b).

- f cannot be of type II.

We turn to the proof of these steps.

Proof of step 1. i) The "only if" part is obvious. As for the "if" part let us remark that

$$\tilde{f} = \rho_f \cdot f : U \to \rho_f \cdot V$$

is an isometry. If $x_0 \in U$ we set

$$\phi : U \to \mathbb{R}^n \qquad x \mapsto d_{x_0}\tilde{f}(x - x_0) + \tilde{f}(x_0);$$

since ϕ is obviously an isometry onto its range, by Proposition A.2.1 ϕ and $\tilde{f}$ coincide, whence f is of type (a).

ii) The "only if" part follows from the fact that the the coefficient of dilation of the metric for an inversion $i_{x_0,\alpha}$ in a point x is given by $\alpha / {\|x - x_0\|^2}$. As for the "if" part we remark that if $\rho_f(x) = \eta \|x - x_0\|^2$ and $i = i_{x_0,1/\eta}$ then $f \circ i$ is an isometry of $i(U)$ onto V with constant dilation coefficient, and the above argument applies.

Proof of step 2. We set $\rho = \rho_f$ and $\mu = \mu_f$. If $x \in U$ and $d_x^2\rho$ denotes the second differential of ρ in x, we shall prove the following facts:

(i) $d_x^2\rho(u, w) = 0$ if $u \perp w$;

(ii) $d_x^2\rho(u, w) = \eta(x)\langle u|w\rangle$ for some $\eta \in C^\infty(U)$;

(iii) η is a constant function;

Then the conclusion follows from the fact that the general solution of the differential equation $d_x^2\rho(u, w) = \eta\langle u|w\rangle$ has the form

$$\rho(x) = \frac{\eta}{2}\|x\|^2 + \langle x|z\rangle + \tau \qquad z \in \mathbb{R}^n, \tau \in \mathbb{R}.$$

(i) Let u, v, w be pairwise orthogonal vectors; we shall often use without mention the fact that they can be taken to be simultaneously non-zero. If we consider the partial derivative in direction w of the identity

$$\langle d_x f(u)|d_x f(v)\rangle = 0$$

we obtain

$$\langle d_x^2 f(u, w)|d_x f(v)\rangle = -\langle d_x f(u)|d_x^2 f(v, w)\rangle$$

If we allow u, v, w to vary (with the condition that they keep pairwise orthogonal), we obtain that the left hand side is symmetric in the pairs (u, w) and (v, w), and skew-symmetric in the pair (u, v), which implies that it is identically zero. Now, let us fix u and w. Since the image under $d_x f$ of the subspace orthogonal to u and w is the subspace orthogonal to $d_x f(u)$ and $d_x f(w)$, we deduce from above that for some real functions α and β depending on u and w

$$d_x^2 f(u,w) = \alpha(x)d_x f(u) + \beta(x)d_x f(w).$$

If we consider the partial derivative in direction w of the identity

$$\|d_x f(u)\|^2 = \mu(x)^2 \|u\|^2$$

we obtain

$$\langle d_x^2 f(u,w) | d_x f(u) \rangle = \mu(x)d_x \mu(w)\|u\|^2$$

therefore

$$\alpha(x) = \frac{d_x \mu(w)}{\mu(x)}$$

and similarly

$$\beta(x) = \frac{d_x \mu(u)}{\mu(x)}.$$

Since $d_x \rho(z) = -\dfrac{d_x \mu(z)}{\mu(x)^2}$ we obtain that

$$\rho(x)d_x^2 f(u,w) + d_x \rho(w)d_x f(u) + d_x \rho(u)d_x f(w) = 0.$$

This identity holds for all $x \in U$ and $u, w \in \mathbb{R}^n$ with the only condition that $\langle u|w \rangle = 0$. If v is orthogonal to u and w and we take the partial derivative in direction v we obtain

$$d_x \rho(v)d_x^2 f(u,w) + \rho(x)d_x^3 f(u,w,v) + d_x^2 \rho(w,v)d_x f(u) +$$
$$+ d_x \rho(w)d_x^2 f(u,v) + d_x^2 \rho(u,v)d_x f(w) + d_x \rho(u)d_x^2 f(w,v) = 0.$$

The second, the fourth and the fifth terms are symmetric in the pair (u,v), and the same holds for the sum of the first and the sixth terms, so that the third term is symmetric in (u,v) too:

$$d_x^2 \rho(w,v)d_x f(u) = d_x^2 \rho(w,u)d_x f(v);$$

but $d_x f(u)$ and $d_x f(v)$ are mutually orthogonal, whence

$$d_x^2 \rho(w,u) = 0 \qquad \forall x \in U \; w, u \in \mathbb{R}^n \; s.t. \; \langle w|u \rangle = 0.$$

(ii) For fixed x $d_x^2 \rho$ is a symmetric bi-linear form on $\mathbb{R}^n$. By (i), if $\{e_1, ..., e_n\}$ is the canonical basis of $\mathbb{R}^n$, we have

$$d_x^2 \rho(e_i, e_j) = k_i \delta_j^i \qquad (k_i \in \mathbb{R}).$$

Moreover for $i \neq j$

$$0 = d_x^2 \rho(e_i + e_j, e_i - e_j) = k_i - k_j \;\Rightarrow\; k_i = k_j$$

and then $d_x^2 \rho$ must be a multiple of the scalar product. The dependence of the multiplying constant on x is obviously differentiable.

(iii) If we consider the partial derivative in an arbitrary direction v of the identity

$$d_x^2 \rho(w, u) = \eta(x)\langle w|u \rangle$$

we obtain
$$d_x^3\rho(w, u, v) = d_x\eta(v)\langle w|u\rangle.$$
Since $d_x^3\rho$ is symmetric we have
$$\langle d_x\eta(v)w - d_x\eta(w)v|u\rangle = 0$$
which implies that $d_x\eta(v)w = d_x\eta(w)v$ and therefore $d_x\eta = 0$, *i.e.* η is a constant.

Proof of step 3. It is readily verified that
$$\eta\|x\|^2 + \langle x|z\rangle + \tau = \eta\|x - x_0\|^2 + \tau'$$
where $x_0 = -{}^z\!/_{2\eta}$, $\tau' = \tau - \eta\|x_0\|^2$. We must check that $\tau' = 0$.

We set $g = f^{-1}$ and we remark that g cannot be of type I (otherwise f would be of type I too). The set
$$\mathcal{F}_1 = \Big\{\{x \in U : \rho_f(x) = \lambda\} : \lambda > 0\Big\}$$
is a family of spheres centred at x_0 intersected with U, while the set
$$\mathcal{F}_2 = \Big\{\{y \in V : \rho_g(y) = \lambda\} : \lambda > 0\Big\}$$
is a family of spheres or hyperplanes intersected with V, according to the fact that g is of type III or II. Moreover, by the obvious relation
$$\rho_g(f(x)) = \rho_f(x)^{-1} \ \ \forall x \in U,$$
f maps $\mathcal{F}_1$ bijectively onto $\mathcal{F}_2$.

Since f is conformal, if γ is an arc in U orthogonal to all the elements of $\mathcal{F}_1$, then $f \circ \gamma$ is an arc in V orthogonal to all the elements of $\mathcal{F}_2$; such an arc can be re-parametrized as $t \mapsto y_0 + tu_2$, where

$$\begin{cases} \text{if } g \text{ is of type III, } y_0 \text{ is s.t. } \rho_g(y) = \eta'\|y - y_0\|^2 + \tau'' \text{ and } u_2 \neq 0 \\ \text{if } g \text{ is of type II, } u_2 \text{ is s.t. } \rho_g(y) = \langle y|u_2\rangle + \tau'' \text{ and } y_0 \in \mathbb{R}^n. \end{cases}$$

Let γ have the form $\gamma(t) = x_0 + tu_1$, $|t - t_0| < \varepsilon$; then we have
$$(f \circ \gamma)(t) = y_0 + \phi(t)u_2,$$
where y_0 and u_2 are as above and ϕ is a diffeomorphism onto an open interval in $\mathbb{R}$. We have

$$|\dot\phi(t)| \cdot \|u_2\| = \|(f \circ \gamma)'(t)\| = \|d_{\gamma(t)}f(u_1)\| = \frac{\|u_1\|}{\rho_f(\gamma(t))}$$

$$\Rightarrow \dot\phi(t) = \pm\frac{\|u_1\|}{\|u_2\| \cdot (\eta t^2\|u_1\|^2 + \tau')}.$$

Now, if we assume by contradiction that $\tau' \neq 0$, we can find $\lambda \in \mathbb{C} \setminus \{0\}$ (real or purely imaginary) and $k \in \mathbb{R} \setminus \{0\}$ such that

$$\dot\phi(t) = \frac{k}{t^2 - \lambda^2}.$$

The image of the interval $(t_0 - \varepsilon, t_0 + \varepsilon)$ under the transformation

$$t \mapsto (t - \lambda)\big/(t + \lambda)$$

is connected and simply connected and it does not contain 0, so that we can find a holomorphic determination log of the logarithm function defined on a neighborhood of it. The function

$$\psi : (t_0 - \varepsilon, t_0 + \varepsilon) \ni t \mapsto \frac{k}{2\lambda} \log \left(\frac{t - \lambda}{t + \lambda} \right) \in \mathbb{C}$$

is well-defined and differentiable, and $\dot\psi = \phi$, which implies that for some $k_1 \in \mathbb{C}$

$$\phi(t) = \frac{k}{2\lambda} \log \left(\frac{t - \lambda}{t + \lambda} \right) + k_1.$$

According to the fact that g is of type III or II, the condition

$$\rho_g(f(\gamma(t))) \cdot \rho_f(\gamma(t)) = 1$$

can be re-written respectively as

$(*)$
$$\left(\eta' \phi(t)^2 \|u_2\|^2 + \tau'' \right) \left(\eta t^2 \|u_1\|^2 + \tau' \right) = 1$$
$$\left(\phi(t) \|u_2\|^2 + \tau''' \right) \left(\eta t^2 \|u_1\|^2 + \tau' \right) = 1.$$

Let us remark that it is known by now that one of these relations is true for $t \in (t_0 - \varepsilon, t_0 + \varepsilon)$. However, by the explicit expression of ϕ, if Ω is an open subset of $\mathbb{C}$ containing $(t_0 - \varepsilon, t_0 + \varepsilon)$ and the image of Ω under the mapping $t \mapsto (t - \lambda)\big/(t + \lambda)$ is connected and simply connected and does not contain 0, the definition of ϕ can be extended holomorphically to Ω, and hence the above relation holds for t in Ω. In particular we can choose Ω in such a way that for some $w \in \mathbb{C} \setminus \{0\}$ and $\delta > 0$ it contains the segment $\{\lambda + sw : 0 < s < \delta\}$. By the choice of λ we have $\eta \lambda^2 \|u_1\|^2 + \tau' = 0$, and then relations $(*)$ can be re-written for $t = \lambda + sw$ as

$$\left(\eta' \phi(\lambda + sw)^2 \|u_2\|^2 + \tau'' \right)(2\lambda + ws)\eta w \|u_1\|^2 \cdot s = 1$$
$$\left(\phi(\lambda + sw) \|u_2\|^2 + \tau''' \right)(2\lambda + ws)\eta w \|u_1\|^2 \cdot s = 1.$$

But now we have that

$$\lim_{s \to 0} \phi(\lambda + sw)s = \lim_{s \to 0} \phi(\lambda + sw)^2 s = 0$$

and hence both the above relations imply the contradiction 0=1.

Proof of step 4. The argument is completely analogous to the one presented for step 3, so we shall work out calculations without comments; we assume by contradiction that $\eta = 0$ and $z \neq 0$.

$$\rho_f(x) = \langle x | u_1 \rangle + \tau. \qquad \gamma(t) = x_0 + t u_1 \implies (f \circ \gamma)(t) = y_0 + \phi(t) u_2.$$

$$\begin{cases} \rho_f(\gamma(t)) = \rho_f(x_0 + t u_1) = t \|u_1\|^2 + \tau', \\ |\dot\phi(t)| \cdot \|u_2\| = \|(f \circ \gamma)'(t)\| = \dfrac{\|u_1\|}{\rho_f(\gamma(t))} \end{cases}$$

$$\Rightarrow \dot{\phi}(t) = \pm \frac{\|u_1\|}{\|u_2\| \cdot (t\|u_1\|^2 + \tau')}$$

$\exists k \in \mathbb{R} \setminus \{0\}$ such that $\dot{\phi}(t) = \frac{k}{t+k'}$ $(k' = \tau'/\|u_1\|^2)$

$$\Rightarrow \phi(t) = k \log(t+k') + k''.$$

$$\begin{cases} (\eta'\phi(t)^2\|u_2\|^2 + \tau'') \cdot (t\|u_1\|^2 + \tau') = 1 & (g \text{ type III}) \\ (\phi(t)\|u_2\|^2 + \tau''') \cdot (t\|u_1\|^2 + \tau') = 1 & (g \text{ type II}) \end{cases}$$

$$\Rightarrow \begin{cases} (\eta'\phi(s-k')^2\|u_2\|^2 + \tau'') \cdot \|u_1\|^2 \cdot s = 1 \\ (\phi(t)\|u_2\|^2 + \tau''') \cdot \|u_1\|^2 \cdot s = 1 \end{cases}$$

$$\lim_{s \to 0} \phi(s-k')s = \lim_{s \to 0} \phi(s-k')^2 s = 0 \Rightarrow 0 = 1. \text{ Absurd.}$$

The proof of Liouville's theorem is now complete. □

It is quite interesting to remark that the assumption that $n \geq 3$ was used only in the proof of Step 2–(i); however, it can be easily checked that the assumption cannot be dropped. For instance, on the unit disc D^2 of $\mathbb{C}$ every one-to-one holomorphic mapping is a conformal diffeomorphism onto its range; and plenty of holomorphic functions are one-to-one on the disc (e.g. if $p \in \mathbb{C}[x]$ and $p'(0) \neq 0$, if $\varepsilon > 0$ is small enough the function $z \mapsto p(\varepsilon z)$ is one-to-one on the disc).

Moreover the Riemann mapping theorem (see [Na]) implies that every simply connected proper domain of $\mathbb{C}$ is conformally equivalent to D^2, while A.3.7 implies that for $n \geq 3$ only open balls and open subspaces are conformally equivalent to D^n. This fact is the first feature of a phenomenon of *rigidity* for the case $n \geq 3$ we shall discuss in Chapt. C.

In the following two results we shall check that in spite of the differences between the two cases $n = 2$ and $n \geq 3$, the determination of the groups $\text{Conf}(S^n)$, $\text{Conf}(D^n)$ and $\text{Conf}(\Pi^{n,+})$ given for $n = 2$ in A.3.4 and A.3.5 can be generalized word-by-word to the case $n \geq 3$.

Corollary A.3.8. $\text{Conf}(S^n)$ consists of all and only the mappings of the form

$$x \mapsto \lambda A i(x) + b$$

where $\lambda > 0$, $A \in O(n)$, i is either the identity or the inversion with respect to a sphere and $b \in \mathbb{R}^n$.

If M and N are domains in S^n then

$$\text{Conf}(M,N) = \{f|_M : f \in \text{Conf}(S^n), f(M) = N\}.$$

Proof. The mappings of the form $\lambda A i + b$ constitute a group of conformal diffeomorphisms operating transitively on S^n and containing the isotropy group of ∞ (that is, $\text{Conf}(\mathbb{R}^n)$), and therefore this group is $\text{Conf}(S^n)$.

The second assertion is a straight-forward consequence of A.3.7. □

Theorem A.3.9. Let $n \geq 2$.

(1) $\mathrm{Conf}(D^n)$ consists of all and only the mappings of the form

$$x \mapsto Ai(x),$$

where $A \in O(n)$ and i is either the identity or an inversion with respect to a sphere orthogonal to ∂D^n.

(2) $\mathrm{Conf}(\Pi^{n,+})$ consists of all and only the mappings of the form

$$x \mapsto \lambda \begin{pmatrix} A & 0 \\ 0 & 1 \end{pmatrix} i(x) + \begin{pmatrix} b \\ 0 \end{pmatrix}$$

where $\lambda > 0$, $A \in O(n-1)$, i is either the identity or an inversion with respect to a sphere orthogonal to $\mathbb{R}^{n-1} \times \{0\}$ and $b \in \mathbb{R}^{n-1}$.

Proof. By A.3.5 we only have to consider the case $n \geq 3$.

(1) By A.3.1 the set of all the mappings of the required form is a group of conformal diffeomorphisms of D^n. As for the converse, let $f \in \mathrm{Conf}(D^n)$. Assume first that $f = \lambda A + b$; since the ball of centre b and radius λ must be D^n, it follows that $b = 0$ and $\lambda = 1$, whence f has the required form. If $f = \lambda Ai + b$, with $i = i_{x_0,r}$, then certainly $x_0 \notin \overline{D^n}$ (otherwise we would have $\infty \in \overline{f(D^n)} = \overline{D^n}$, which is false). Let us consider the inversion j with respect to the sphere centred at x_0 with radius $\sqrt{\|x_0\|^2 - 1}$ (which is orthogonal to ∂D^n); then $f \circ j \in \mathrm{Conf}(D^n)$ and it has the form, $\lambda' A' + b'$, and the conclusion follows from the first part.

(2) By A.3.1 the set of all the mappings of the required form is a group of conformal diffeomorphisms of $\Pi^{n,+}$. As for the converse, let $f \in \mathrm{Conf}(\Pi^{n,+})$. Assume first that $f = \lambda A + b$; since 0 belongs to the boundary of $\Pi^{n,+}$, the same must hold for $f(0) = b$, so the last coordinate of b must be 0. Then $A = \lambda^{-1}(f - b) \in \mathrm{Conf}(\Pi^{n,+})$. If for some $j < n$ the element a_{nj} on the n-th row and j-th column of A does not vanish, the image under A of the j-th element e_j of the canonical basis of $\mathbb{R}^n$ does not belong to the boundary of $\Pi^{n,+}$, while e_j does, and this is absurd. Since the same argument works for $A^{-1} = {}^t A$, A must have the form $\begin{pmatrix} B & 0 \\ 0 & w \end{pmatrix}$, and then we obviously have that $B \in O(n-1)$ and $w = 1$, whence f has the required form. Now, if $f = \lambda Ai + b$, where i is the inversion with respect to a sphere centred at a point x_0, since $f(x_0) = \infty$ belongs to the boundary of $\Pi^{n,+}$, the same must hold for x_0, i.e. $x_0 \in \mathbb{R}^{n-1} \times \{0\}$, and then every sphere centred at x_0 is orthogonal to $\mathbb{R}^{n-1} \times \{0\}$. The conclusion follows from the fact that $f \circ i \in \mathrm{Conf}(\Pi^{n,+})$ has the form $\lambda' A' + b'$. □

Remark A.3.10. Everything we said in A.3.6 about conformal diffeomorphisms preserving or reversing orientation can be repeated word-by-word in the general case, so that in particular:

(1) $Ai \in \mathrm{Conf}(D^n)$ preserves the orientation if and only if

$$\left[A \in \mathrm{SO}(n), i = \mathrm{id}\right] \text{ or } \left[A \notin \mathrm{SO}(n), i \neq id\right].$$

(2) $\lambda \begin{pmatrix} A & 0 \\ 0 & 1 \end{pmatrix} i + \begin{pmatrix} b \\ 0 \end{pmatrix} \in \text{Conf}(\Pi^{n,+})$ preserves the orientation if and only if
$[A \in \text{SO}(n-1), i = \text{id}]$ or $[A \notin \text{SO}(n-1), i \neq \text{id}]$.

We conclude this section with a technical result we shall need in the sequel:

Lemma A.3.11. Let $n \geq 2$. If $\phi \in \text{Conf}(S^n)$ is not the identity and there exists a submanifold N of S^n of codimension 1 such that ϕ is the identity on N, then one (and only one) of the following facts is verified:

(a) N is contained in a sphere, and ϕ is the inversion with respect to this sphere;

(b) N is contained in a hyperplane, and ϕ is the reflection with respect to this hyperplane.

Proof. By A.3.4 and A.3.8 we have the following possibilities:

(a) $\phi = \lambda A + b$;

(b) $\phi = \lambda A i + b$, where i is an inversion.

We consider them separately.

(a) Let $x \in N$, $x \neq \infty$; since $d_x \phi|_{T_x N} = \text{id}$ it must be $\lambda = 1$, hence ϕ is an isometry of $\mathbb{R}^n$. If $\{v_1, ..., v_{n-1}\}$ is an orthonormal basis of $T_x N$, we can complete it with a vector v_n to an orthonormal basis of $\mathbb{R}^n$. If $d_x \phi(v_n) = -v_n$ the reflection with respect to $T_x N$ is an isometry of $\mathbb{R}^n$ which coincides up to first order with ϕ in x, and hence it coincides with ϕ; since ϕ is the identity only on $T_x N$, obviously $N \subseteq T_x N$. The condition $d_x \phi(v_n) = v_n$ would imply that ϕ is the identity, and this is absurd.

(b) Let $i = i_{x_0, \alpha}$. For $x \in N$ and $v \in T_x N$ we have

$$\|v\| = \|d_x \phi(v)\| = \frac{\lambda \alpha}{\|x - x_0\|^2} \cdot \|v\| \;\Rightarrow\; \|x - x_0\|^2 = \lambda \alpha.$$

If we set $\beta = \lambda \alpha$ we have $N \subseteq M(x_0, \beta)$. Moreover $\phi \circ i_{x_0, \beta}$ is of type (a) and it is the identity on N; then it must be necessarily the identity, otherwise N would be contained in a hyperplane too, and this is absurd since the intersection of a sphere and a hyperplane has co-dimension at least 2. $\qquad\square$

A.4 Isometries of Hyperbolic Space: Disc and Half-space Models

In this section the results of the long parenthesis about conformal geometry are used for the determination of the groups $\mathcal{I}(\mathbb{D}^n)$ and $\mathcal{I}(\Pi^{n,+})$.

Theorem A.4.1. $\mathcal{I}(\mathbb{D}^n) = \text{Conf}(D^n)$, $\mathcal{I}^+(\mathbb{D}^n) = \text{Conf}^+(D^n)$. In particular, these groups operate transitively on $\mathbb{D}^n$.

Proof. We start by proving that the restriction of the stereographic projection $p : \Pi^n \to D^n$ used for the definition of $\mathbb{D}^n$ is conformal (Π^n is endowed with the hyperbolic metric and D^n with the Euclidean one). We shall denote by $\langle .|. \rangle_{(n,1)}$ the standard bi-linear symmetric form on $\mathbb{R}^{n+1}$, and by $\langle .|. \rangle$ the usual scalar product in $\mathbb{R}^n$. Since $p(x,t) = {}^x/(1+t)$, we have

$$d_{(x,t)}p(y,s) = \frac{y}{1+t} - \frac{sx}{(1+t)^2};$$

for $(x,t) \in \mathbb{I}^n$, $(y,s),(z,r) \in T_{(x,t)}\mathbb{I}^n$ we have

$$\langle x|x \rangle = t^2 - 1 \qquad \langle x|y \rangle = st \qquad \langle x|z \rangle = rt$$

and hence

$$\langle d_{(x,t)}(y,s)|d_{(x,t)}(z,r) \rangle = \langle {}^y/1+t - {}^{sx}/(1+t)^2 | {}^z/1+t - {}^{rx}/(1+t)^2 \rangle =$$

$$= \frac{\langle y|z \rangle}{(1+t)^2} - \frac{2rst}{(1+t)^3} + \frac{rs(t^2-1)}{(1+t)^4} = \frac{1}{(1+t)^2}(\langle y|z \rangle - rs) =$$

$$= \frac{\langle (y,s)|(z,r) \rangle_{(n,1)}}{(1+t)^2}.$$

We have proved our assertion.

Let $\phi \in \mathcal{I}(\mathbf{D}^n)$; since by definition p is an isometry we have

$$p^{-1} \circ \phi \circ p \in \mathcal{I}(\mathbb{I}^n) \subseteq \mathrm{Conf}(\mathbb{I}^n)$$

and since p is conformal from $\mathbb{I}^n$ to D^n we obtain $\phi \in \mathrm{Conf}(D^n)$. Inclusion $\subseteq$ is proved.

As for the converse, using A.3.9, we shall prove that all the elements of $O(n)$ and all the inversions with respect to spheres orthogonal to ∂D^n belong to $\mathcal{I}(\mathbf{D}^n)$. The first fact is easy: if $A \in O(n)$, then

$$\begin{pmatrix} A & 0 \\ 0 & 1 \end{pmatrix} \in O(I_n) = \mathcal{I}(\mathbb{I}^n) \qquad p \circ \begin{pmatrix} A & 0 \\ 0 & 1 \end{pmatrix} \circ p^{-1} = A.$$

For the second fact we remark that p^{-1} is expressed by

$$p^{-1} : x \mapsto \frac{1}{1 - \|x\|^2} \begin{pmatrix} 2x \\ 1 + \|x\|^2 \end{pmatrix};$$

if $(y,t) \in \mathbb{R}^{n+1}$ and $\langle (y,t)|(y,t) \rangle_{(n,1)} = 1$ the set

$$N = \{x \in D^n : \langle p^{-1}(x)|(y,t) \rangle_{(n,1)} = 0\}$$

is given by the equation

$$2\langle x|y \rangle = (1 + \|x\|^2)\sqrt{\|y\|^2 - 1}$$

i.e., if we set $w = {}^y/\sqrt{\|y\|^2 - 1}$,

$$2\langle x|w \rangle = 1 + \|x\|^2 \quad \Leftrightarrow \quad \|x - w\|^2 = \|w\|^2 - 1$$

and hence N is the intersection of D^n with the sphere S_w of centre w and radius $\sqrt{\|w\|^2 - 1}$, which is orthogonal to ∂D^n.

If $\phi \in \mathcal{I}(\mathbb{I}^n)$ is the reflection parallel to (y, t), then $p \circ \phi \circ p^{-1}$ is a conformal diffeomorphism of D^n different from the identity and such that it is the identity on N; by A.3.5 and A.3.9 it extends to a conformal diffeomorphism of S^n; it follows from A.3.11 that $p \circ \phi \circ p^{-1}$ is the inversion with respect to S_w, and hence the latter belongs to $\mathcal{I}(\mathbb{D}^n)$. For the conclusion we only have to remark that if $w \notin D^n$ it is always possible to find $(y, t) \in \mathbb{R}^{n+1}$ such that $\langle (y, t) | (y, t) \rangle_{(n,1)} = 1$ and $w = y/t$.

The case of orientation-preserving isometries is now straight-forward. $\square$

Theorem A.4.2. $\mathcal{I}(\mathbb{I}^{n,+}) = \mathrm{Conf}(\mathbb{I}^{n,+})$,
$$\mathcal{I}^+(\mathbb{I}^{n,+}) = \mathrm{Conf}^+(\mathbb{I}^{n,+}).$$

Proof. As we remarked in Sect. A.1, the diffeomorphism $i : D^n \to \mathbb{I}^{n,+}$ used for the definition of the hyperbolic structure on $\mathbb{I}^{n,+}$ is the inversion with respect to the sphere of centre $(0, ..., 0, -1)$ and radius $\sqrt{2}$, so it is conformal, and the conclusion follows immediately from A.4.1. $\square$

Using A.4.1 and A.4.2 the hyperbolic metrics on $\mathbb{D}^n$ and $\mathbb{I}^{n,+}$ can be explicitly computed. Since $\mathbb{D}^n$ and $\mathbb{I}^{n,+}$ are open subsets of $\mathbb{R}^n$, their tangent bundles are canonically identified with $\mathbb{D}^n \times \mathbb{R}^n$ and $\mathbb{I}^{n,+} \times \mathbb{R}^n$ respectively.

Theorem A.4.3. For $x \in \mathbb{D}^n$, $(y, t) \in \mathbb{I}^{n,+}$ and $v \in \mathbb{R}^n$ the metrics are explicitly given by
$$ds_x^2(v) = \left(\frac{2}{1 - \|x\|^2} \right)^2 \|v\|^2$$
$$ds_{(y,t)}^2(v) = \frac{\|v\|^2}{t^2}.$$

Proof. The differential of the mapping $p : \mathbb{I}^n \to \mathbb{D}^n$ at the point $(0, ..., 0, 1)$ is half the identity on $\mathbb{R}^n \times \{0\} = T_{(0,1)}\mathbb{I}^n$, and the restriction of $\langle . | . \rangle_{(n,1)}$ to $\mathbb{R}^n \times \{0\}$ is the standard scalar product, so that
$$ds_0^2(v) = 4\|v\|^2.$$

For $x \in \mathbb{D}^n$ the inversion with respect to the sphere orthogonal to $\partial \mathbb{D}^n$ of centre $x/\|x\|^2$ is an isometry of $\mathbb{D}^n$ mapping 0 in x, and its differential in 0 is $(1 - \|x\|^2)$ times an orthogonal operator (compare A.3.1 (4)), and this implies that
$$ds_x^2(v) = \left(\frac{2}{1 - \|x\|^2} \right)^2 \|v\|^2.$$

Similarly, the differential of the inversion $i : \mathbb{D}^n \to \mathbb{I}^{n,+}$ at the point 0 is twice an orthogonal operator, so that
$$ds_{(0,1)}^2(v) = \|v\|^2.$$

Since horizontal translations are isometries we only have to remark that the differential in $(0,1)$ of the dilation of coefficient $t > 0$ is t times the identity, and hence

$$ds^2_{(y,t)}(v) = \frac{\|v\|^2}{t^2}.$$

$\square$

Corollary A.4.4. At any point of $\mathbb{D}^n$ and $\mathbb{I}^{n,+}$ the hyperbolic metric is a positive mutiple of the Euclidean one.

Remark A.4.5. By A.4.4 the notion of conformal diffeomorphism on $\mathbb{D}^n$ and $\mathbb{I}^{n,+}$ is the same if we consider the Euclidean metric and the hyperbolic one. Using A.4.1 and A.4.2, this implies that all conformal diffeomorphisms of $\mathbb{D}^n$ and $\mathbb{I}^{n,+}$ with respect to the hyperbolic metric are isometries. This fact does not depend on the concrete representation of $\mathbb{H}^n$ (only the hyperbolic metric is used), so that we have the following proposition: *a sufficient condition for a diffeomorphism of $\mathbb{H}^n$ to preserve lengths is that it preserves angles* (the converse being true for all Riemannian manifolds). We can interpret heuristically this fact in the following way: *a unit of measure is intrinsically defined on $\mathbb{H}^n$, and it cannot be changed.* We shall prove other facts explaining this assertion.

A.5 Geodesics, Hyperbolic Subspaces and Miscellaneous Facts

We start with the hyperboloid model.

Proposition A.5.1. If $x \in \mathbb{I}^n$, $y \in T_x\mathbb{I}^n$, $\langle y|y \rangle_{(n,1)} = 1$ the geodesic starting at x with velocity y is given by

$$\mathbb{R} \ni t \mapsto \cosh(t) \cdot x + \sinh(t) \cdot y.$$

In particular as a set it is given by the intersection of $\mathbb{I}^n$ with the linear subspace of $\mathbb{R}^{n+1}$ generated by x an y.

Proof. Let W be the plane generated by x and y, and let ω be the maximal geodesic starting at x with velocity y; we shall confuse ω with its support. Let $\phi \in O(I_n)$ be defined by $\phi|_W = \mathrm{id}$ and $\phi|_{W^\perp} = -\mathrm{id}$. Since $\phi(x) = x$ and $d_x\phi(y) = y$, ω is ϕ-invariant, and therefore $\omega \subseteq W \cap \mathbb{I}^n$. Moreover it is easily checked that the mapping

$$\mathbb{R} \ni t \mapsto \cosh(t) \cdot x + \sinh(t) \cdot y$$

gives a parametrization of $W \cap \mathbb{I}^n$ with velocity of length identically 1, and therefore it coincides with ω. $\square$

By the above result every geodesic in $\mathbb{H}^n$ is defined on the whole real line, and therefore the Hopf-Rinow theorem (see [He]) yields the following:

Corollary A.5.2. $\mathbb{H}^n$ is a complete Riemannian manifold.

An easy argument proves the following further consequence of A.5.1:

Corollary A.5.3. There exists one and only one geodesic line passing through any two different points of $\mathbb{H}^n$.

As in A.2.5 we record an analogy between $\mathbb{H}^n$ and S^n. We omit the proof since it works just like the one presented above for A.5.1.

Proposition A.5.4. If $x \in S^n$, $y \in T_x S^n$, $\langle y|y \rangle = 1$ the geodesic starting at x with velocity y is given by

$$\mathbb{R} \ni t \to \cos(t) \cdot x + \sin(t) \cdot y.$$

In particular as a set it is given by the intersection of S^n with the linear subspace of $\mathbb{R}^{n+1}$ generated by x and y.

We shall say a subset N of $\mathbb{H}^n$ is a <u>hyperbolic subspace</u> if it contains the entire geodesic passing through any two of its points. Remark that points and entire geodesics are hyperbolic subspaces. Proposition A.5.1 imply the following:

Corollary A.5.5. $N \subset \mathbb{I}^n$ is a hyperbolic subspace if and only if it is the intersection of $\mathbb{I}^n$ with a linear subspace of $\mathbb{R}^{n+1}$. In particular hyperbolic subspaces are submanifolds of $\mathbb{I}^n$, and hence their dimension is well-defined.

We consider now the other models of $\mathbb{H}^n$. Before stating the result we introduce some terminology.

– We shall say an affine subspace Y of $\mathbb{R}^n$ is <u>vertical</u> if it has the form $Y' + \mathbb{R}\, e_n$, where Y' is an affine subspace of $\mathbb{R}^{n-1} \times \{0\}$ and $e_n = (0, ...0, 1)$.

– From now on we shall allow a <u>sphere</u> in $\mathbb{R}^n$ to have dimension lower than $n - 1$; this is obtained simply by considering the intersection of an $(n-1)$-dimensional sphere with an affine subspace passing through its centre. However, when speaking of inversion with respect to a sphere, we shall always mean that the sphere has maximal dimension.

– Let M_1 and M_2 be spheres or affine subspaces (or parts of) in $\mathbb{R}^n$, and let m_1, m_2 be their respective dimensions. We shall say that M_1 and M_2 are <u>orthogonal</u> if for each $x \in M_1 \cap M_2$ the linear space $W = T_x M_1 \cap T_x M_2$ has dimension $\max\{0, m_1 + m_2 - n\}$ and the orthogonal complements of W in $T_x M_1$ and $T_x M_2$ are orthogonal to each other.

Proposition A.5.6. (1) $N \subset \mathbb{D}^n$ is a hyperbolic subspace if and only if it is the intersection of $\mathbb{D}^n$ with a linear subspace of $\mathbb{R}^n$ or with a sphere orthogonal to $\partial \mathbb{D}^n$. In particular geodesics are obtained by parametrization of diameters of $\mathbb{D}^n$ and circles orthogonal to $\partial \mathbb{D}^n$.

(2) $N \subset \mathbb{I}^{n,+}$ is a hyperbolic subspace if and only if it is the intersection of $\mathbb{I}^{n,+}$ with an affine vertical subspace or with a sphere orthogonal to $\mathbb{R}^{n-1} \times \{0\}$. In particular geodesics are obtained by parametrization of vertical lines and circles orthogonal to $\mathbb{R}^{n-1} \times \{0\}$.

Proof. It is easily established that the image under the stereographic projection $p : \mathbf{I}\!\mathbf{I}^n \to \mathbf{D}^n$ of a linear subspace W passing through $(0, ..., 0, 1)$ is the intersection of $\mathbf{D}^n$ with a linear subspace of $\mathbb{R}^n$. Hence the hyperbolic subspaces of $\mathbf{D}^n$ passing through 0 are the intersections of $\mathbf{D}^n$ with the linear subspaces of $\mathbb{R}^n$. If $x \in \mathbf{D}^n$ the inversion i with respect to the sphere of centre ${}^x/_{\|x\|^2}$ and squared radius ${}^1/_{\|x\|^2} - 1$ is an isometry of $\mathbf{D}^n$ and it maps 0 to x; this implies that it maps the set of all hyperbolic subspaces through 0 onto the set of all hyperbolic subspaces through x. Now, if Y is a linear subspace of dimension p of $\mathbb{R}^n$, we have two possibilities:

(i) $x \in Y$; in this case $i(Y) = Y$.

(ii) $x \notin Y$. If we consider the subspace X generated by Y and x, X is i-invariant and hence $i(Y) \subset X$. Since Y is a hyperplane in X, Proposition A.3.1 (6) implies that $i(Y)$ is a sphere in X, hence $i(Y)$ is a p-dimensional sphere in $\mathbb{R}^n$. Since i is conformal and Y is orthogonal to $\partial\mathbf{D}^n$, $i(Y)$ is orthogonal to $\partial\mathbf{D}^n$ too.

It is easily verified with the same method that i maps linear subspaces and spheres orthogonal to $\partial\mathbf{D}^n$ passing through x onto linear subspaces, and hence (1) is proved.

(2) The case of the half-space is a direct consequence of the previous one. In fact the mapping $i : \mathbf{D}^n \to \mathbf{I}\!\mathbf{I}^{n,+}$ used for the definition is an inversion, and hence by the above argument it maps the set of spheres and affine subspaces onto itself; moreover it preserves orthogonality and the conclusion follows at once. $\square$

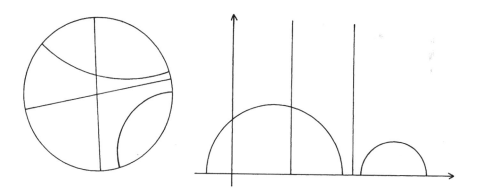

Fig. A.3. Geodesics in the two-dimensional disc and half-space

Corollary A.5.7. A p-dimensional hyperbolic subspace in $\mathbf{H}^n$ is isometrically diffeomorphic to $\mathbb{H}^p$.

Proof. We consider the disc model and assume that the hyperbolic subspace contains 0. By A.5.6 it is a p-dimensional disc, and by A.4.3 it inherits the same metric as that of $\mathbf{D}^p$. $\square$

Proposition A.5.6 allows us to compute explicitly the hyperbolic distance in $\mathbf{D}^n$ and $\mathbf{II}^{n,+}$. We shall denote by th the hyperbolic tangent function and by ath its inverse function.

Corollary A.5.8. (1) If $x, y \in \mathbf{D}^n$ we have

$$d(x,y) = 2\mathrm{ath}\left(\frac{\|x-y\|}{\left(1 - 2\langle x|y\rangle + \|x\|^2 \cdot \|y\|^2\right)^{1/2}}\right).$$

(2) If $(x,t),(y,s) \in \mathbf{II}^{n,+}$ we have

$$d((x,t),(y,s)) = 2\mathrm{ath}\left(\frac{\|x-y\|^2 + (t-s)^2}{\|x-y\|^2 + (t+s)^2}\right)^{1/2}.$$

Proof. (1) Let $v \in \mathbb{R}^n$, $\|v\| = 1$. A parametrization of the diameter determined by v is given by

$$\gamma : \mathbb{R} \ni t \to \mathrm{th}(t/2) \cdot v$$

and it is a straight-forward computation that

$$ds^2_{\gamma(t)}(\dot\gamma(t)) = 1.$$

It follows that for $x \in \mathbf{D}^n$ we have

$$d(0, x) = 2\mathrm{ath}\|x\|.$$

Now, let $x, y \in \mathbf{D}^n$. The inversion $i \in \mathcal{I}(\mathbf{D}^n)$ with respect to the sphere centred at $x/\|x\|^2$ with squared radius $1/\|x\|^2 - 1$ can be written explicitly as

$$i(z) = (1 - \|x\|^2)\frac{z\|x\|^2 - x}{\|z\|x\|^2 - x\|^2} + \frac{x}{\|x\|^2}.$$

Since $i(x) = 0$ we have

$$d(x, y) = d(0, i(y)) = 2\mathrm{ath}\|i(y)\|$$

and an easy calculation completes the proof.

(2) We only have to consider the inversion mapping $\mathbf{D}^n$ isometrically onto $\mathbf{II}^{n,+}$: explicit computations will be omitted. $\square$

Remark A.5.9. It is well known (see [Si] or [Ve]) that a differential metric on D^2 with respect to which holomorphisms are isometries must be a multiple of the Poincaré one $\omega(z) = 4/|z|^2$: by A.3.2 and A.4.1 this is the case for the hyperbolic metric. Proposition A.4.3 proves that in fact the hyperbolic metric and the Poincaré one coincide, and hence hyperbolic two-space is nothing but the Poincaré disc. The above calculation of the hyperbolic distance is consistent with this fact: it is well-known that the Poincaré distance is given by

$$d(w, z) = 2 \text{ath} \left| \frac{w - z}{1 - w\,\overline{z}} \right|$$

and our formula is a natural generalization of this one.

We shall discuss now the notion of <u>boundary of hyperbolic space</u>. The symbol $\partial \mathbb{D}^n$ was already used to denote the boundary of D^n in $\mathbb{R}^n$, $i.e.$ the sphere S^{n-1}, and similarly we can define $\partial \mathbb{H}^{n,+}$ as the boundary of $\mathbb{H}^{n,+}$ in S^n, $i.e.$ $(\mathbb{R}^{n-1} \times \{0\}) \cup \{\infty\}$. However a concrete representation of hyperbolic space is essential for these definitions. We shall prove now that the notion of boundary is intrinsically defined.

Consider the set $\mathcal{S}$ of all geodesic closed half-lines in $\mathbb{H}^n$ parametrized by arc length on $[0, \infty)$, and define an equivalence relation R on $\mathcal{S}$ in the following way:

$$\gamma_1 R \gamma_2 \iff \sup_{t \geq 0} d(\gamma_1(t), \gamma_2(t)) < \infty.$$

Set $\partial \mathbb{H}^n = \mathcal{S}/R$ and $\overline{\mathbb{H}^n} = \mathbb{H}^n \cup \partial \mathbb{H}^n$. We define a topology on $\overline{\mathbb{H}^n}$ in such a way that $\mathbb{H}^n$ is open and inherits its own topology, and a neighborhood of $p \in \partial \mathbb{H}^n$ is obtained in the following way: choose γ in the class of p, and let x be its starting point, let V be a neighborhood of $\dot{\gamma}(0)$ in the unit sphere of $T_x \mathbb{H}^n$ and let $r > 0$; then we set

$$U(\gamma, V, r) = \left\{ \gamma_1(t) : \gamma_1 \in \mathcal{S}, \gamma_1(0) = x, \dot{\gamma}_1(0) \in V, t > r \right\} \bigcup$$
$$\bigcup \left\{ \langle \gamma_1 \rangle_R : \gamma_1 \in \mathcal{S}, \gamma_1(0) = x, \dot{\gamma}_1(0) \in V \right\}.$$

(We omit the proof that when γ, V and r vary, $\{U(\gamma, V, r)\}$ satisfies the axioms of a fundamental system of neighborhoods of p.)

Proposition A.5.10. $\partial \mathbb{H}^n$ is homeomorphic to S^{n-1} and $\overline{\mathbb{H}^n}$ is homeomorphic to $\overline{D^n}$. Moreover if we consider the disc model $\mathbb{D}^n$ of hyperbolic space, $\overline{\mathbb{D}^n}$ is canonically identified with the closure of $\mathbb{D}^n$ as a subset of $\mathbb{R}^n$.

Proof. We shall prove the second fact, which implies the first. Given a geodesic half-line, since it is an arc of diameter or circle, it determines a unique point on ∂D^n. Moreover, two geodesic half-lines are in relation R if and only if they determine the same point on ∂D^n. Then $\overline{\mathbb{D}^n}$ is canonically identified with $\overline{D^n}$, and it is straight-forward that this identification is a homeomorphism with respect to the topology defined above. $\square$

Remark A.5.11. Since the inversion mapping $\mathbb{D}^n$ onto $\mathbb{H}^{n,+}$ maps $\partial \mathbb{D}^n$ onto $(\mathbb{R}^{n-1} \times \{0\}) \cup \{\infty\}$, this set is the natural boundary of $\mathbb{H}^{n,+}$.

Remark also that ∂D^n and $(\mathbb{R}^{n-1} \times \{0\}) \cup \{\infty\}$ are two models for the sphere S^{n-1}, and hence in the disc and half-space model we can endow the boundary of $\mathbb{H}^n$ with the conformal structure of S^{n-1}.

We shall refer to the points of $\partial \mathbb{H}^n$ as the <u>points at infinity</u> of $\mathbb{H}^n$.

If p is a point at infinity in $\mathbb{H}^n$ we shall say a geodesic γ <u>passes through</u> p if p is the equivalence class of $\gamma|_{[0, \infty)}$ or $\gamma|_{(-\infty, 0]}$; equivalently, we shall say that p is an <u>endpoint</u> of γ. It is readily verified that all geodesics have exactly two endpoints, and moreover given $p, q \in \partial \mathbf{H}^n$, $p \neq q$, there exists one and only one geodesic having endpoints p and q.

Remark that a hyperbolic subspace N can be completed to a closed submanifold of $\overline{\mathbf{H}^n}$ by adding to it all the endpoints of the geodesics it contains; these points will be called the points at infinity of N.

We shall say two geodesics in $\mathbf{H}^n$ are:

– incident, if they have a common point in $\mathbf{H}^n$,

– asymptotically parallel, if they have one common endpoint,

– ultra-parallel, if they have no intersection in $\overline{\mathbf{H}^n}$.

Remark that ultra-parallel geodesics exist even in $\mathbb{H}^2$, and this is a deep difference between the hyperbolic plane and the Euclidean one. We shall discuss now the essential properties concerning the mutual position of two hyperbolic subspaces.

Proposition A.5.12. Let $N, M \subset \mathbf{H}^n$ be hyperbolic subspaces;
(a) if N and M meet in $\partial \mathbf{H}^n$ and not in $\mathbf{H}^n$ then they have exactly one common point at infinity, and there exists no geodesic line orthogonal to both N and M;
(b) if N and M do not meet in the whole $\overline{\mathbf{H}^n}$ then there exists a geodesic γ orthogonal to both N and M, and the distance between N and M equals the length of the arc on γ lying through N and M.

Proof. (a) If N and M have two common points p and q at infinity, then both N and M contain the geodesic line having endpoints p and q, and hence they have non-trivial intersection in $\mathbb{H}^n$, which is absurd. In the half-space model we can assume the common point of N and M is ∞, *i.e.* N and M are affine vertical subspaces; $N = N_1 \times \mathbb{R}_+$, $M = M_1 \times \mathbb{R}_+$. A geodesic orthogonal to N is a circle centred at a point of N_1; but $N_1 \cap M_1 = \emptyset$ and hence no geodesic can be orthogonal to both N and M.

(b) It is easily checked that N and M have positive distance (otherwise they would have a common point somewhere: recall that $\overline{\mathbf{H}^n}$ is compact). Let δ be their distance, and let $\{a_n\}$ and $\{b_n\}$ be sequences in N and M respectively such that $d(a_n, b_n) \to \delta$. We can assume that these sequences converge in $\overline{\mathbf{H}^n}$; if one of the limits is a point p at infinity, the condition on the distance implies that the other limit is p too, and this is absurd since p would be a point at infinity of both N and M. It follows that $a_n \to a \in \mathbf{H}^n$ and $b_n \to b \in \mathbf{H}^n$, the convergence being with respect to the hyperbolic distance; it follows in particular that $d(a, b) = \delta$. Let γ denote the geodesic line passing through a and b; let us remark at once that the arc of γ lying through N and M is nothing but the arc from a to b, and hence it has length δ. Assume for instance that γ is not orthogonal to N in a. Choose the model $\mathbb{D}^n$. Since the hyperbolic distance is locally approximated by a multiple of the Euclidean distance (A.5.8), then we can find a point a' on γ (near a) such that its distance from N is strictly less than its distance from a (moreover we can choose a' on the side of M); this implies that the distance from N to M is strictly less than δ, and this is absurd. □

Proposition A.5.13. (1) All isometries of $\mathbf{H}^n$ extend to homeomorphisms of $\overline{\mathbf{H}^n}$, and hence they have some fixed point in $\overline{\mathbf{H}^n}$.

(2) $\mathcal{I}(\mathbf{H}^n)$ and $\mathcal{I}^+(\mathbf{H}^n)$ operate transitively on $\partial\mathbf{H}^n$ and on the set

$$\{(x,v) : x \in \mathbf{H}^n, v \in T_x\mathbf{H}^n, ds_x^2(v) = 1\}$$

where the action is defined by $f(x,v) = (f(x), d_x f(v))$.

(3) an element of $\mathcal{I}(\mathbf{H}^n)$ is uniquely determined by its trace on $\partial\mathbf{H}^n$;

(4) if M is either the disc model or the half-space model the restriction to the boundary is an isomorphism of $\mathcal{I}(M)$ onto $\mathrm{Conf}(\partial M)$.

Proof. (1) It is enough to consider the model $\mathbf{D}^n$, where this fact follows at once from the explicit determination of the isometries. The second assertion is an immediate consequence of A.5.10 and Brouwer's fixed point theorem (see [Mi2] or [Greenb1]).

(2) We consider the disc model again, where the first fact is obvious. As for the second, we only need to remark that $\mathcal{I}^+(\mathbf{D}^n)$ operates transitively on $\mathbf{D}^n$ and $\mathrm{SO}(n)$ operates transitively on S^{n-1}.

(3) Once again we consider the disc model, where this fact is obvious.

(4) By A.4.1 and A.4.2 we have to check that for $N \in \{D^n, \mathrm{II}^{n,+}\}$ the restriction to the boundary is an isomorphism of $\mathrm{Conf}(N)$ onto $\mathrm{Conf}(\partial N)$, and this is a straight-forward corollary of the explicit determination of these groups (A.3.4, A.3.5, A.3.8 and A.3.9). □

We shall give now a classification of the isometries of $\mathbf{H}^n$ with respect to their fixed points.

Proposition A.5.14. If $\phi \in \mathcal{I}(\mathbf{H}^n)$ the following mutually excluding possibilities are given:

(1) ϕ has some fixed point in $\mathbf{H}^n$;

(2) ϕ has no fixed points in $\mathbf{H}^n$, and exactly one fixed point at infinity;

(3) ϕ has no fixed points in $\mathbf{H}^n$, and exactly two fixed points at infinity.

Proof. We only have to check that if ϕ has no fixed point in $\mathbf{H}^n$ then it has at most two fixed points at infinity. In the half-space model, let us assume that ϕ has no fixed point in $\mathrm{II}^{n,+}$, and $0, \infty$ are fixed. Then ϕ can be written as

$$\phi : (y,t) \mapsto \lambda(Ay, t).$$

Since $\phi(0,1) \neq (0,1)$ we have $\lambda \neq 1$, and this implies that ϕ fixes only 0 and ∞. □

According to the above proposition we shall say $\phi \in \mathcal{I}(\mathbf{H}^n)$ is:

– of <u>elliptic</u> type if (1) occurs;

– of <u>parabolic</u> type if (2) occurs;

– of <u>hyperbolic</u> type if (3) occurs.

Remark A.5.15. If ϕ is an isometry of hyperbolic type then there exists one and only one ϕ-invariant geodesic line, whose endpoints are the fixed points at infinity for ϕ.

Now we shall specialize this classification to the case of dimensions 2 and 3 for orientation-preserving isometries.

For $n = 2$ we shall give a geometric and an algebraic classification of the isometries. Since in $\mathbb{H}^2$ the notions of length, angle, and (geodesic) line are defined, the concepts of bisecting line of an angle and axis of a segment are naturally defined too. We shall denote by $[a, b]$ the closed segment with endpoints a and b, and by (a, b, c) the angle between $[b, a]$ and $[b, c]$.

Proposition A.5.16. Let $\phi \in \mathcal{I}^+(\mathbb{H}^2) \setminus \{\mathrm{id}\}$, let x be a non-fixed point of ϕ, and let l_1 be the bisecting line of the angle $(x, \phi(x), \phi^2(x))$ and l_2 the axis of the segment $[\phi(x), \phi^2(x)]$. Then the following holds:

(1) if l_1 and l_2 are incident, ϕ is elliptic;

(2) if l_1 and l_2 are asymptotically parallel, ϕ is parabolic,

(3) if l_1 and l_2 are ultra-parallel, ϕ is hyperbolic.

Proof. Let us remark first that if ϕ is elliptic then it has only one fixed point, otherwise it would be the reflection with respect to a geodesic line which is not in $\mathcal{I}^+(\mathbb{H}^2)$. Moreover the relative position of l_1 and l_2 is invariant under the action of $\mathcal{I}(\mathbb{H}^2)$, so we can choose the fixed point(s) of ϕ in a suitable way. We carry out the proof by pictures by considering the three possible cases.

(1) ϕ elliptic; we choose $0 \in \mathbb{D}^2$ as fixed point and we obtain the situation of Fig. A.4.

(2) ϕ parabolic; we choose $\infty \in \mathbb{H}^{2,+}$ as fixed point and we obtain the situation of Fig. A.5.

(3) ϕ hyperbolic; we choose $0, \infty \in \mathbb{H}^{2,+}$ as fixed points and we obtain the situation of Fig. A.6.

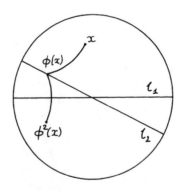

Fig. A.4. Geometric classification of isometries in dimension two: elliptic case

□

According to A.4.2 and A.3.3 every orientation-preserving isometry of $\mathbb{H}^{2,+}$ is represented by a 2×2 real matrix with determinant 1, and it is

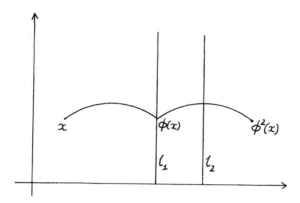

Fig. A.5. Geometric classification of isometries in dimension two: parabolic case

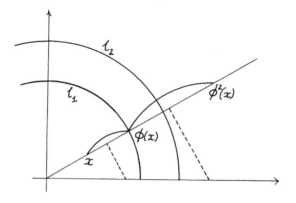

Fig. A.6. Geometric classification of isometries in dimension two: hyperbolic case

easily checked that two matrices A and B represent the same isometry if and only if $A = \pm B$. We shall denote by tr the trace of a matrix.

Proposition A.5.17. Let $\phi \in \mathcal{I}^+(\mathbb{H}^{2,+}) \setminus \{\mathrm{id}\}$ be represented by a matrix $A \in \mathrm{Sl}(2, \mathbb{R})$; then

(1) if $|tr(A)| < 2$, ϕ is elliptic;
(2) if $|tr(A)| = 2$, ϕ is parabolic;
(3) if $|tr(A)| > 2$, ϕ is hyperbolic.

Proof. Let $A = \begin{pmatrix} a & b \\ c & d \end{pmatrix}$. We recall that $\phi(z) = (az + b)/(cz + d)$. If $c = 0$ then ∞ is a fixed point, $\phi(z) = a^2 z + ab$ and $tr(A) = a + 1/a$. If $a = \pm 1$ then $b \neq 0$ (otherwise $\phi = \mathrm{id}$, which is absurd). It follows that

$$tr(A) = \pm 2 \Leftrightarrow a = \pm 1 \Leftrightarrow \phi \text{ parabolic}$$
$$tr(A) > 2 \Leftrightarrow a \neq \pm 1 \Leftrightarrow \phi \text{ hyperbolic.}$$

If $c \neq 0$ we consider the equation

$$\phi(z) = z \;\Leftrightarrow\; cz^2 + (d - a)z - b = 0$$

having discriminant $\Delta = (d - a)^2 + 4bc = tr(A)^2 - 4$, and the above cases are easily discussed.   □

Now we turn to the three-dimensional case. By A.4.2 and A.5.13, the restriction of an orientation-preserving isometry of $\mathbb{H}^3$ to the boundary is a conformal diffeomorphism of S^2, and conversely every element of $\mathrm{Conf}^+(S^2)$ can be extended in a unique way to an element of $\mathcal{I}^+(\mathbb{H}^3)$. Moreover by A.3.3

$$\mathrm{Conf}^+(S^2) \cong \mathrm{Sl}(2, \mathbb{C}) \big/ \{\pm I\}$$

and hence

$$\mathcal{I}^+(\mathbb{H}^3) \cong \mathrm{Sl}(2, \mathbb{C}) \big/ \{\pm I\}.$$

Proposition A.5.18. Let $\phi \in \mathcal{I}^+(\mathbb{H}^3) \setminus \{\mathrm{id}\}$ be represented by a matrix $A \in \mathrm{Sl}(2, \mathbb{C})$. Then
(1) if $tr(A) \in \mathbb{R}$, $|tr(A)| < 2$, ϕ is elliptic;
(2) if $tr(A) = \pm 2$, ϕ is parabolic;
(3) if $tr(A) \notin \mathbb{R}$ or $tr(A) \in \mathbb{R}$, $|tr(A)| > 2$, ϕ is hyperbolic.

Proof. Let $A = \begin{pmatrix} a & b \\ c & d \end{pmatrix}$, $ad - bc = 1$; since the equation

$$\frac{az + b}{cz + d} = z$$

has some solution in $\mathbb{C} \cup \{\infty\}$, then A has a fixed point in $\mathbb{C} \cup \{\infty\}$. Moreover we have that:
– $\mathrm{Sl}(2, \mathbb{C})$ operates transitively on $\mathbb{C} \cup \{\infty\}$;
– $tr(B^{-1}AB) = tr(A) \;\forall B \in \mathrm{Sl}(2, \mathbb{C})$;
– ϕ and $\psi^{-1}\phi\psi$ are of the same type $\forall \psi \in \mathcal{I}(\mathbb{H}^3)$.
Therefore we can assume A fixes ∞, *i.e.* $c = 0$, $d = 1/a$, $tr(A) = a + 1/a$, $A(z) = a^2 z + ab$. The isometry ϕ of $\mathbb{H}^{3,+}$ (characterized by the fact that it extends A) is then given by

$$\phi(z, t) = (a^2 z + ab, |a|^2 t).$$

As in A.5.17, if $a = \pm 1$ then $b \neq 0$, and therefore we have:
– ϕ elliptic $\;\Leftrightarrow\; |a| = 1, a \neq \pm 1$,
– ϕ parabolic $\;\Leftrightarrow\; a = \pm 1$,
– ϕ hyperbolic $\;\Leftrightarrow\; |a| \neq 1$,
and the conclusion follows easily.   □

We introduce now a new geometric notion: the horosphere. Given $p \in \partial \mathbb{H}^n$ we shall say a closed hypersurface N in $\mathbb{H}^n$ is a <u>horosphere</u> centred at p if N is orthogonal to all geodesic lines with endpoint p.

Proposition A.5.19. Given $p \in \partial \mathbb{H}^n$, $\mathbb{H}^n$ is the disjoint union of the horospheres centred at p. These horospheres inherit from $\mathbb{H}^n$ the Riemannian structure of $\mathbb{R}^{n-1}$.

Proof. The notions we are considering are invariant under isometries, and hence we shall assume in the half-space model that $p = \infty$. Since geodesics with endpoint ∞ are given by vertical lines $\{x_0\} \times \mathbb{R}_+$, horospheres centred at ∞ are horizontal hyperplanes $\mathbb{R}^{n-1} \times \{t_0\}$ (and conversely such a hyperplane is a horosphere centred at ∞). The first assertion is proved. As for the second, using A.4.3, we only have to remark that a positive multiple of the standard Riemannian metric of $\mathbb{R}^{n-1}$ is equivalent to it. $\square$

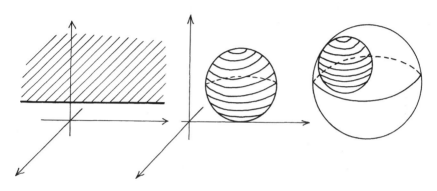

Fig. A.7. Horospheres in the three-dimensional disc and half-space models

Remark A.5.20. A horosphere in $\mathbb{H}^n$ inherits the Riemannian structure of $\mathbb{R}^{n-1}$, but not the metric space structure (that is, the distance of $\mathbb{H}^n$ restricted to a horosphere is not Euclidean). The reason is that if we integrate the metric (in order to obtain the distance) before considering the restriction to the horosphere, we use geodesics which are not contained in the horosphere, and hence if we perform the two operations in the opposite order we obtain a different result.

According to the above characterization a horosphere centred at $p \in \partial \mathbb{H}^n$ divides $\mathbb{H}^n$ into two connected regions homeomorphic to n-balls: we shall call the one meeting $\partial \mathbb{H}^n$ in p a <u>horoball</u> centred at p.

The next result provides an alternative proof of the fact that $\mathbb{H}^n$ is complete (we have already proved this in A.5.2).

Proposition A.5.21. A hyperbolic ball in $\mathbb{D}^n$ (or $\mathbb{II}^{n,+}$) is a Euclidean ball with different centre and radius, whose closure is compact in $\mathbb{D}^n$ (or $\mathbb{II}^{n,+}$).

Proof. By A.5.8 a ball of radius r centred at 0 in $\mathbb{D}^n$ is a Euclidean ball of centre 0 and radius $\operatorname{th}(r/2) < 1$ and hence its closure is compact in $\mathbb{D}^n$ (recall that the hyperbolic distance induces the standard topology on D^n). Since inversions with respect to spheres orthogonal to $\partial \mathbb{D}^n$ and elements of $O(n)$ are homeomorphisms of $\overline{\mathbb{D}^n}$ and they map balls into balls, the proposition holds for every ball in $\mathbb{D}^n$. Moreover the mapping used for the definition of $\mathbb{II}^{n,+}$ is an inversion too, and the proposition is proved. $\square$

We give now a description of the "banana" neighborhoods of geodesics in $\mathbb{H}^n$. Given a geodesic γ in $\mathbb{H}^n$ and $\varepsilon > 0$ we consider the closed tubular ε-neighborhood of γ:

$$N_\varepsilon(\gamma) = \big\{ x \in \mathbb{H}^n : d(x,\gamma) \leq \varepsilon \big\};$$

if in $\mathbb{II}^{n,+}$ we have $\gamma = \{0\} \times \mathbb{R}_+$ then $N_\varepsilon(\gamma)$ is invariant under dilations (as γ is) and it is easily checked (by A.5.8) that $N_\varepsilon(\gamma)$ is the infinite cone based on a horizontal closed $(n-1)$-disc. Now, an immediate argument based on the properties of inversions proves that the general shape of $N_\varepsilon(\gamma)$ in $\mathbb{D}^n$ and $\mathbb{II}^{n,+}$ is the one described in Fig. A.8.

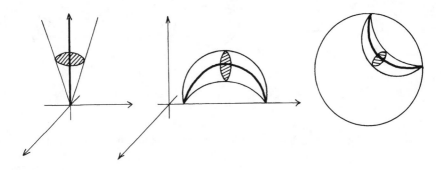

Fig. A.8. Neighborhoods of geodesics in the three-dimensional disc and half-space

Remark A.5.22. As a conclusion of the section we point out some peculiarities of the the projective model for $\mathbb{H}^n$ we introduced at the beginning. The first feature is the following: if we canonically identify it with the unit disc D^n we have that the conformal structure D^n inherits from the hyperbolic structure is not equivalent to the usual one. Moreover it is not difficult to check that the geodesic subspaces are given in this model by the intersections of D^n with the affine subspaces of $\mathbb{R}^n$; we deduce from this for instance that a geodesic polyhedron in this model is a Euclidean geodesic polyhedron. This fact is sometimes useful as some arguments applying to Euclidean geodesic polyhedra generalize to the hyperbolic case (provided they do not involve the notion of measure of an angle).

A.6 Curvature of Hyperbolic Space

We shall prove that in every point of $\mathbf{H}^n$ the sectional curvature of $\mathbf{H}^n$ with respect to any section is -1. Before proving this in detail we give a result on the strict convexity of the distance function, which expresses qualitatively the fact that $\mathbf{H}^n$ is negatively curved. If $x, y \in \mathbf{H}^n$ we shall denote by $(x + y)/2$ the middle point of the geodesic arc joining x and y.

Proposition A.6.1. Let γ_1, γ_2 be closed geodesic arcs in $\mathbf{H}^n$ having in common at most one endpoint and such that they are not arcs of the same maximal geodesic; let $x, x' \in \gamma_1$ and $y, y' \in \gamma_2$ with $x \neq x'$ and $y \neq y'$ and set $p = (x + x')/2$, $q = (y + y')/2$; then

$$2d(p, q) < d(x, y) + d(x', y').$$

Proof. Since $p \in \gamma_1$ and $q \in \gamma_2$ we cannot have $p = q$, otherwise this point would be an endpoint of γ_1, whence $x = x' = p$, which is absurd. Let δ be the maximal geodesic passing through p and q; if both x and y belong to δ then γ_1 and γ_2 are arcs of δ, which is absurd. Let us assume that $x \notin \delta$ (whence $x' \notin \delta$). Let σ be the symmetry with respect to p, i.e. the isometry of $\mathbf{H}^n$ characterized by the relation $p = (w + \sigma(w))/2 \ \forall w \in \mathbf{H}^n$, and similarly let ρ be the symmetry with respect to q. Set $\tau = \rho \circ \sigma$ and $z = \tau(x) = \rho(x')$, $z' = \tau(x') = \rho(x)$, $r = \tau(p) = \rho(p)$; then we have

$$2d(p, q) = d(p, r)$$
$$d(y', z') = d(\rho(y), \rho(x)) = d(y, x)$$
$$d(x', z') \leq d(x', y') + d(y', z').$$

So it is enough to prove that $d(p, r) < d(x', z')$, i.e.

$$d(p, \tau(p)) < d(x', \tau(x')).$$

Let us assume that in the half-space model $\delta = \{0\} \times \mathbb{R}_+$; since τ is the product of two symmetries with respect to different points of δ, it is easily calculated that τ is a dilation of coefficient $\lambda \neq 1$. Moreover p has the form $(0, t_1)$ and x' has the form (a, t_2) with $a \in \mathbb{R}^{n-1} \setminus \{0\}$; by A.5.8 it is easily verified that

$$d\big((0, t_1), (0, \lambda t_1)\big) < d\big((a, t_2), (\lambda a, \lambda t_2)\big). \qquad \square$$

The situation considered in the next result was not included in A.6.1 for technical reasons and in order to emphasize it with a specific statement. We shall say three points in $\mathbf{H}^n$ are non-aligned if each geodesic in $\mathbf{H}^n$ contains at most two of them.

Corollary A.6.2. Let x, x', y be non-aligned points of $\mathbf{H}^n$ and define p as $(x + x')/2$; then

$$2d(p, y) < d(x, y) + d(x', y).$$

Proof. Set $y' = q = y$. Since x does not belong to the geodesic passing through p and q, the above proof works in this case too. □

Remark A.6.3. With the same symbols as in E.6.1 we have in $\mathbb{R}^n$ the same inequality

$$2d(p,q) \le d(x,y) + d(x',y')$$

but we do have non-trivial cases when equality holds; moreover in S^n (endowed with the metric it inherits from $\mathbb{R}^{n+1}$) we have cases when the opposite inequality

$$2d(p,q) > d(x,y) + d(x',y')$$

holds. These facts express qualitatively the fact that $\mathbb{H}^n$, $\mathbb{R}^n$ and S^n have different curvature.

We are now going to prove the assertion about the sectional curvatures of $\mathbb{H}^n$. We recall that the <u>curvature</u> at a point of an oriented Riemannian surface can be defined in the following equivalent ways (see for instance [Boo], [Ga-Hu-La], [Ko-No] or [Sp]):

– via the definition of the Levi-Civita connection and the Riemann tensor associated to it; in particular, if Ω is a domain in $\mathbb{R}^2$, $\alpha : \Omega \to \mathbb{R}_+$ is a C^∞ function and a differential metric on Ω is defined by $ds_x^2(v) = \alpha(x)^2 \cdot \|v\|^2$, this procedure yields the following expression of the curvature at a point x:

$$k(x) = -\frac{1}{\alpha(x)^2} \cdot (\Delta \log \alpha)(x)$$

where Δ denotes the Laplace operator.

– via the definition of parallel transport and of a function ϕ associating to each pre-compact domain D with smooth boundary the number

$$\phi(D) = \sphericalangle (v, P_{\partial D}(v)),$$

where $P_{\partial D}(v)$ denotes the parallel transport along ∂D in the positive direction of a vector v tangent to a point of ∂D, and $\sphericalangle (v, w)$ denotes the measure with sign of the angle between v and w: it is shown that for a suitable function k on the surface

$$\phi(D) = \int_D k(x) dm(x)$$

where $dm(x)$ denotes the element of area at x.

The sectional curvature in a point x of a Riemannian manifold M with respect to a 2-subspace $V \subset T_x M$ (called a section at x) is defined as the curvature at x of the oriented Riemannian surface obtained as the image under the exponential mapping of a suitably small neighborhood of 0 in V.

Lemma A.6.4. The sectional curvature of $\mathbb{H}^n$ in a point x with respect to a section $V \subset T_x \mathbb{H}^n$ is independent of V, x and n.

Proof. Independence of V and x follows at once from the isometry-invariance of the curvature and from the fact that $\mathcal{I}(\mathbb{H}^n)$ operates transitively on the pairs

x, V. As for the last assertion we only need to recall that by A.5.7 everything reduces to the case $n = 2$: in fact the image under the exponential mapping of a section at any point of $\mathbf{H}^n$ is a hyperbolic 2-subspace. ☐

Lemma A.6.5. Let T be a geodesic triangle in $\mathbb{H}^2$ with inner angles α, β, γ, let $m(T)$ denote its measure and let $\phi(T)$ be defined as above; then
    (a) $m(T) = \pi - (\alpha + \beta + \gamma)$;
    (b) $\phi(T) = \alpha + \beta + \gamma - \pi$.

Proof. (a) Let us consider the half-space model. We start by computing the area of a geodesic triangle Δ having a vertex at ∞ and inner angles $\alpha, \beta, 0$.

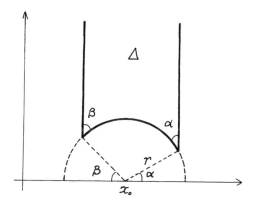

Fig. A.9. Computation of the area of a geodesic triangle in hyperbolic two-space; the case of a vertex at infinity

We have that $m(\Delta) = \int_\Delta \frac{dx\,dy}{y^2}$. Since

$$\Delta = \left\{ (x_0 + r\cos\vartheta, y) : \alpha \leq \vartheta \leq \pi - \beta,\ y \geq r\sin\vartheta \right\}$$

we have that

$$m(\Delta) = \int\limits_{\alpha}^{\pi-\beta} r\sin\vartheta\, d\vartheta \int\limits_{r\sin\vartheta}^{\infty} \frac{dy}{y^2} = \int\limits_{\alpha}^{\pi-\beta} d\vartheta = \pi - \alpha - \beta.$$

    Now, for the general case, we remark that the area of T can be expressed as the algebraic sum of the areas of three geodesic triangles having a vertex at ∞, and the conclusion follows quite easily by considering the different possibilities. For instance in the situation of Fig. A.10 we consider the triangles:
Δ_1 with vertices x, y, ∞ and inner angles $\alpha_1, \beta_1, 0$;
Δ_2 with vertices x, z, ∞ and inner angles $\alpha_2, \gamma_1, 0$;
Δ_3 with vertices y, z, ∞ and inner angles $\beta_2, \gamma_2, 0$;
as suggested in Fig. A.11.

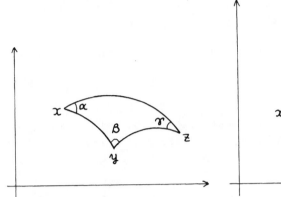

Fig. A.10. Computation of the area of a geodesic triangle in hyperbolic two-space; general case

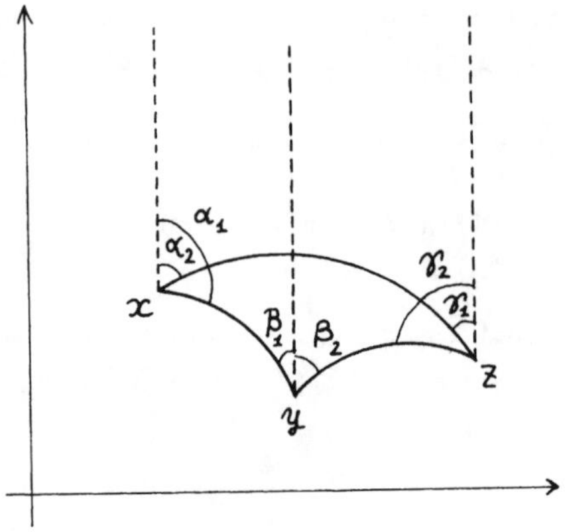

Fig. A.11. Computation of the area of a geodesic triangle in hyperbolic two-space; how to divide a general triangle

We have the obvious relations

$$\alpha_1 = \alpha + \alpha_2 \qquad \beta = \beta_1 + \beta_2 \qquad \gamma_2 = \gamma + \gamma_1$$

and hence

$$m(T) = m(\Delta_1) + m(\Delta_3) - m(\Delta_2) =$$
$$= \pi - \alpha_1 - \beta_1 + \pi - \beta_2 - \gamma_2 - (\pi - \alpha_2 - \gamma_1) =$$
$$= \pi - \alpha - \beta - \gamma.$$

(b) In $\mathrm{III}^{2,+}$ we can assume that one of the sides of T is vertical; the proof is contained in Fig. A.12. $\qquad\qquad\square$

Remark A.6.6. The above result holds also if T is assumed to be a geodesic triangle in $\overline{\mathrm{IH}}^2$, *i.e.* the vertices are allowed to be points at infinity. (The proof works just as above.) In particular, the area of any geodesic triangle having all vertices at infinity is π. We shall use this fact while discussing the rigidity theorem (Sect. C.2).

Theorem A.6.7. All sectional curvatures of $\mathbf{H}^n$ are -1.

Proof. We present two proofs, according to the two possible definitions of the sectional curvature given above. In both cases, recalling A.6.3, we consider only $n = 2$.

   $-\mathbf{D}^2$ is the Riemannian surface associated to the function $\alpha : D^2 \to \mathbb{R}_+$, $\alpha(x) = {2}/{(1 - \|x\|^2)}$, and it easily checked that

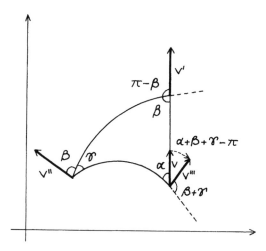

Fig. A.12. Calculation of the parallel transport along the boundary of a geodesic triangle in hyperbolic two-space

$$-\frac{1}{\alpha(0)^2} \cdot (\Delta \log \alpha)(0) = -1.$$

– For $x \in \mathbb{H}^2$ we can consider a sequence of geodesic triangles $\{T_n\}$ such that $x \in T_n$ and

$$\lim_{n \to \infty} diam(T_n) = 0.$$

Then, by A.6.5,

$$k(x) = \lim_{n \to \infty} \phi(T_n)/m(T_n) = -1.$$

$\square$

Remark A.6.8. Lemma A.6.5 (a) provides another evidence of the fact that a unit of measure is intrinsically defined in $\mathbb{H}^n$ (compare A.4.5).

The following determination of the curvature of S^n and $\mathbb{R}^n$ is easily obtained:

Theorem A.6.9. All sectional curvatures of S^n are 1 and all sectional curvatures of $\mathbb{R}^n$ are 0.

We conclude this chapter with the construction of a surface in $\mathbb{R}^3$ having constant curvature -1 (with respect to the Riemannian metric induced from $\mathbb{R}^3$). Let us consider the Euclidean plane $\mathbb{R}^2 = \mathbb{R}_x \times \mathbb{R}_y$; we shall call tractrix a curve α in the open first quadrant such that the distance between $\alpha(t)$ and the intersection of the y-axis with the tangent line to α at $\alpha(t)$ is identically 1. It is quite easily verified that the tractrix exists and is unique up to change of parameter, and in particular it is the graphic of the function

$$y(x) = \int_x^1 \sqrt{1/_{t^2} - 1}\, dt \qquad x \in (0,1).$$

(The integral can be explicitly computed, but it does not say very much.)

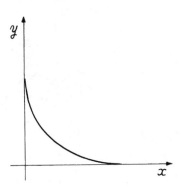

Fig. A.13. The tractrix

In $\mathbb{R}^3$ we shall call <u>pseudo-sphere</u> the surface generated by the rotation of the tractrix around the y-axis.

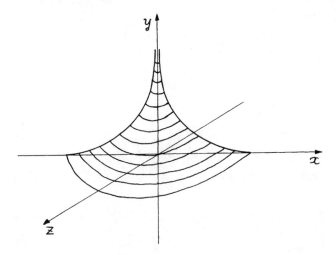

Fig. A.14. The pseudo-sphere

The pseudo-sphere in a neighborhood of a point $(x_0, y(x_0), 0)$ is the graphic of the function

$$(x, z) \mapsto \sqrt{y(x)^2 - z^2}$$

and a straight-forward application of the general formulae for the curvature of a graphic yields:

Proposition A.6.10. The curvature of the pseudo-sphere is identically -1.

Of course the pseudo-sphere is not a complete Riemannian manifold; the following result implies that this fact cannot be avoided (see [DC]):

Proposition A.6.11. No complete surface in $\mathbb{R}^3$ can have strictly negative curvature everywhere.

Chapter B.
Hyperbolic Manifolds and
the Compact Two-dimensional Case

In this chapter we are going to introduce the notion of hyperbolic manifold (*i.e.* a manifold modeled on hyperbolic space) via the introduction of a much more general class of manifolds. We shall prove the first essential properties of such manifolds (namely, the fact that if a hyperbolic manifold is complete then it can be obtained as a quotient of hyperbolic space). Afterwards we shall consider the special case of compact surfaces and we shall give a complete classification of the hyperbolic structures on a surface of fixed genus (that is we shall give a parametrization of the so-called Teichmüller space).

B.1 Hyperbolic, Elliptic and Flat Manifolds

This section is essentially based on [Th1, ch. 3].

Let X be a connected, simply connected, oriented n-dimensional manifold (as in Chapt. A, we shall consider only $n \geq 2$), and let G be a group of diffeomorphisms of X onto itself; we shall say a differentiable n-manifold M is endowed with an (X, G)-structure if we are given an open covering $\{U_i\}$ of M and a set of differentiable open mappings $\{\varphi_i\}$ (with $\varphi_i : U_i \to X$) such that

1) $\varphi_i : U_i \to \varphi_i(U_i)$ is a diffeomorphism;

2) if $U_i \cap U_j \neq \phi$ then the restriction of $\varphi_i \circ \varphi_j^{-1}$ to each connected component of $\varphi_j(U_i \cap U_j)$ is the restriction of an element of G.

$\{(U_i, \varphi_i)\}$ will be called an <u>atlas</u> defining the (X, G)-structure; of course such an atlas is contained in a unique maximal one enjoying the same properties.

Remark B.1.1. If M is an (X, G)-manifold and M is oriented then M is also an (X, G^+)-manifold, where G^+ consists of all the elements of G which preserve orientation.

According to this fact we shall always assume that if M is an oriented (X, G)-manifold then $G = G^+$.

Remark B.1.2. If X is a Riemannian manifold and G is a subgroup of $\mathcal{I}(X)$, each (X, G)-manifold is naturally endowed with the Riemannian structure defined by the requirement that all the φ_i's above are isometries.

From now on we shall assume that X is a Riemannian manifold and that $G = \mathcal{I}(X)$ (or $G = \mathcal{I}^+(X)$, in case we are considering an oriented (X, G)-manifold). The extension of some of the results to a slightly different (and more general) situation is sketched at the end of the section, after Example B.1.16.

We will say an n-dimensional (X, G)-manifold M is hyperbolic if $X = \mathbf{H}^n$, elliptic if $X = S^n$ and flat if $X = \mathbb{R}^n$.

We are now going to characterize all complete connected (X, G)-manifolds for arbitrary X (we recall that we are supposing X to be an n-dimensional, connected, simply connected, oriented Riemannian manifold).

Proposition B.1.3. Let M be a connected and simply connected (X, G)-manifold, let $\varphi_0 : U_0 \to X$ be an isometry defined on the connected open subset U_0 of M; then there exists one and only one local isometry $D : M \to X$ extending φ_0.

Proof. Uniqueness follows at once from A.2.1.

As for existence, fix a point p_0 in U_0; if p is an arbitrary point of M, we can find a differentiable path $\gamma : [0,1] \to M$ with $\gamma(0) = p_0$, $\gamma(1) = p$; we also require that $\gamma^{-1}(U_0)$ is an interval. If $L = \gamma([0,1])$, we can find a subset $\{(U_i, \varphi_i)\}_{i=1,\ldots,m}$ of the maximal atlas defining the (X, G)-structure such that:

1) $\gamma^{-1}(U_i) = I_i$ is an interval for $i = 1, \ldots, m$ (we already supposed this for $i = 0$);

2) $L \subset \cup_{i=0}^m U_i$;

3) $I_i \cap I_j = \phi$ if $|i - j| \geq 2$;

4) $U_i \cap U_j$ is connected for all $i, j = 0, \ldots, m$.

Now, let $g_i \in G$ be the isometry of X extending $\varphi_i \circ \varphi_{i-1}^{-1}$; we define D on L in the following way:

$$D|_{U_0 \cap L} = \varphi_0|_L$$
$$D|_{U_1 \cap L} = g_1^{-1} \circ (\varphi_1|_L) \quad (\ldots)$$
$$D|_{U_m \cap L} = g_1^{-1} \circ \ldots \circ g_m^{-1} \circ (\varphi_m|_L).$$

(The definition obiously matches on the intersections.) This method allows us in particular to define $D(p)$; we are now going to prove that:

(A) $D(p)$ does not depend on $U_1, \ldots, U_m$;

(B) $D(p)$ does not depend on γ.

After this the proof will be over, since D obviously extends φ_0, and, with the above symbols, in the neighborhood of the point p it can be written as $g_1^{-1} \circ \ldots \circ g_m^{-1} \circ \varphi_m$, which implies that it is a local isometry.

(A) It suffices to consider the case of two coverings $\{U_0, U_1\}$ and $\{U_0, \tilde{U}_1\}$ of L (the proof is then completed by induction); we assume also by simplicity that $U_1 \cap \tilde{U}_1$ is connected. Since $U_0 \cap U_1 \cap \tilde{U}_1 \neq \phi$, if $g \in G$ is the isometry extending $\varphi_1 \circ \tilde{\varphi}_1^{-1}$, we have $g = g_1 \circ \tilde{g}_1^{-1}$; then we have:

$$D\big|_{U_1 \cap \tilde{U}_1 \cap L} = g_1^{-1} \circ \varphi_1\big|_{U_1 \cap \tilde{U}_1 \cap L}$$

$$\tilde{D}\big|_{U_1 \cap \tilde{U}_1 \cap L} = \tilde{g}_1^{-1} \circ \tilde{\varphi}_1\big|_{U_1 \cap \tilde{U}_1 \cap L} =$$

$$= g_1^{-1} \circ g \circ \tilde{\varphi}_1\big|_{U_1 \cap \tilde{U}_1 \cap L} = g_1^{-1} \circ \varphi_1\big|_{U_1 \cap \tilde{U}_1 \cap L}$$

$$D\big|_{U_0 \cap L} = \tilde{D}\big|_{U_0 \cap L} = \phi_0\big|_{U_0 \cap L} \, .$$

It follows that $D = \tilde{D}$ on L, and hence in particular in p.

(B) Let γ_0 and γ_1 be two differentiable paths joining p_0 to p, and let H be a differentiable homotopy between γ_0 and γ_1. We can find two subdivisions of $[0,1]$

$$0 = t_0 < t_1 < \ldots < t_\nu = 1 \qquad\qquad 0 = s_0 < s_1 < \ldots < s_\mu = 1$$

in such a way that $\forall i, j$ the set $H([t_{i-1}, t_i] \times [s_{j-1}, s_j])$ is contained in some coordinate set of the maximal atlas defining the (X, G)-structure. Hence we can think that γ_0 is obtained from γ_1 by a sequence of modifications as shown in Fig. B.1.

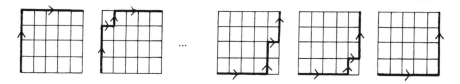

Fig. B.1. Step by step modification γ_1 to γ_0

Now, it is easily checked by the choice of the t_i's and the s_j's that the definition of $D(p)$ remains unaltered at all steps (the argument is similar to the one presented for (A)). The proof is complete. □

The mapping D described in the above proposition will be called the <u>developing function</u> of M with respect to φ_0.

Remark B.1.4. If D and D' are developing functions of M then there exists $g \in G$ such that $D = g \circ D'$.

Proof. Let U be a small connected open susbset of M such that $D\big|_U$ and $D'\big|_U$ are both isometries; then $D\big|_U \circ (D'\big|_U)^{-1}$ is the restriction of an element g of G. Since $g \circ D'$ is a developing function of M with respect to $D\big|_U$, it coincides with D. □

Theorem B.1.5. Every complete, connected, simply connected (X, G)-manifold is isometrically diffeomorphic to X. (In particular, if such a manifold exists then X is complete.)

Proof. Let D be any developing function of M. We first check that it suffices to prove that D has the following property:

$$(*) \quad \begin{cases} \forall x \in D(M), \tilde{x} \in D^{-1}(x), \gamma \in C^1([0,1], X) \text{ with } \gamma(0) = x \\ \exists! \tilde{\gamma} \in C^1([0,1], M) \text{ with } \tilde{\gamma}(0) = \tilde{x}, \ D \circ \tilde{\gamma} = \gamma. \end{cases}$$

In fact, if $(*)$ holds, C^1 paths in X starting in $D(M)$ can be lifted (via D) in a unique way to C^1 paths in M, and hence also C^1 homotopies between loops starting in $D(M)$ can be lifted in a unique way to C^1 homotopies. It follows that if $D(x) = D(x')$ and γ is a C^1 path in M joining x to x' (so that $D \circ \gamma$ is a loop), we can find a C^1 homotopy H between $D \circ \gamma$ and the constant loop $D(x)$; since H can be lifted to M, the constant loop $D(x)$ is lifted to the constant loop x and the loop $D \circ \gamma$ is lifted to γ, we obtain that γ is a loop, i.e. $x = x'$. We have proved that $(*)$ implies that D is one-to-one; obviously $(*)$ implies also that D is onto, and then D is an isometric diffeomorphism between M and X.

We are left to prove $(*)$. For fixed $x, \tilde{x}$ and γ we start by remarking that if for $0 < t \le 1$ there exists $\tilde{\gamma}^{(t)} \in C^1([0,t], M)$ such that $\tilde{\gamma}^{(t)}(0) = \tilde{x}$ and $D \circ \tilde{\gamma}^{(t)} = \gamma|_{[0,t]}$, then $\tilde{\gamma}^{(t)}$ is unique (recall that D is a local isometry); in particular if $\tilde{\gamma}^{(t_1)}$ and $\tilde{\gamma}^{(t_2)}$ are defined for $t_1 < t_2$ then $\tilde{\gamma}^{(t_2)}|_{[0,t_1]} = \tilde{\gamma}^{(t_1)}$. Then we set

$$t_0 = \sup \left\{ t \in [0,1] : \exists \tilde{\gamma} \in C^1([0,t], M) \ s.t. \ \tilde{\gamma}(0) = \tilde{x}, D \circ \tilde{\gamma} = \gamma|_{[0,t]} \right\}.$$

Since D is a local isometry then necessarily $t_0 > 0$. The proof of $(*)$ reduces to check that $t_0 = 1$.

We claim now that there exists $\tilde{\gamma}$ defined on $[0, t_0]$ with the required properties (i.e. the sup is a maximum); once this fact is verified the proof will be over, as D is an isometry in the neighborhood of $\tilde{\gamma}(t_0)$, and then necessarily $t_0 = 1$.

By the above remark $\tilde{\gamma}(t)$ is well-defined for all $t < t_0$; we claim that if $\{t_n\}$ is an increasing sequence converging to t_0 then $\tilde{\gamma}(t_n)$ is a Cauchy sequence in M.

If by contradiction $\tilde{\gamma}(t_n)$ were not a Cauchy sequence then we could find $\varepsilon > 0$ such that $\forall n_0$ there exist $n, m \ge n_0$ with $d(\tilde{\gamma}(t_n), \tilde{\gamma}(t_m)) \ge \varepsilon$. In particular $\tilde{\gamma}$ would have infinite length; we are now going to prove that this is absurd. We can find a sequence $\{I_j\}$ of half-open intervals such that $[0, t_0)$ is the disjoint union of the I_j's and

$$\tilde{\gamma}|_{I_j} = \left(D|_{U_j} \right)^{-1} \circ \gamma|_{I_j}$$

where U_j is an open set on which D is an isometry. Then we have

$$L(\tilde{\gamma}) = \sum_j \int_{I_j} \|\dot{\tilde{\gamma}}(s)\| \, ds = \sum_j \int_{I_j} \|\dot{\gamma}(s)\| \, ds = L(\gamma|_{[0,t_0)}) \le L(\gamma) < \infty.$$

Our claim is proved, and hence we can set $\tilde{\gamma}(t_0) = \lim \tilde{\gamma}(t_n)$. Since $\tilde{\gamma}$ is a lifting of γ on the whole $[0, t_0]$, it is differentiable in t_0 too, which implies that the sup is indeed a maximum, and the proof is over.   □

From now on we shall also assume that X is complete.

We recall now a few facts about groups operating on a topological space. We shall be definitely not exhaustive (even for the case of hyperbolic space): we confine ourselves to the facts we are going to need explicitly.

Let T be a locally compact Hausdorff topological space; a group Γ of homeomorphisms of T is said to operate:

<u>freely</u> if $\gamma \in \Gamma$, $x \in T$ and $\gamma(x) = x$ implies $\gamma = \mathrm{id}$;

<u>properly</u> <u>discontinuously</u> if for any pair H, K of compact subsets of T the set $\Gamma(K, H) = \big\{ \gamma \in \Gamma : \gamma(K) \cap H \neq \phi \big\}$ is finite.

We are now going to relate these notions to the notion of covering, in particular for our three geometric models $\mathbf{H}^n$, S^n and $\mathbb{R}^n$. We recall that by A.2.4 and A.2.5 we have

$$\mathcal{I}(\mathbb{I}^n) \cong O(I_n)$$
$$\mathcal{I}(S^n) \cong O(n+1)$$
$$\mathcal{I}(\mathbb{R}^n) \cong O(n) \amalg \mathbb{R}^n$$

(we denoted by $\amalg$ the semi-direct product of groups). It follows that these three groups can be naturally endowed with a topology (inducing a topological group structure).

Proposition B.1.6. Let T be a connected Hausdorff locally compact topological space, and let Γ be a group of homeomorphisms of T. The following conditions are pairwise equivalent:

(1) Γ operates freely and properly discontinuously on T;

(2) T/Γ is a Hausdorff space and given $x \in T$ there exists a neighborhood U of x such that $\gamma(U) \cap U = \phi$ whenever $\gamma \in \Gamma \setminus \{\mathrm{id}\}$;

(3) T/Γ is a Hausdorff space and the quotient projection $\pi : T \to T/\Gamma$ is a covering mapping.

Moreover if $T \in \{\mathbb{I}^n, S^n, \mathbb{R}^n\}$ and Γ is a subgroup of $\mathcal{I}(T)$ the following condition is equivalent to the previous ones:

(4) Γ operates freely on T and it is a discrete subset of $\mathcal{I}(T)$ (with respect to the above natural topology).

Proof. We shall always denote by π the projection of T onto T/Γ.

(1) $\Rightarrow$ (2). We begin with the second fact. For $x \in T$ let V be a neighborhood of x such that $\overline{V}$ is compact and suppose

$$\Gamma(V, V) = \big\{\mathrm{id}, \gamma_1, ..., \gamma_k\big\}.$$

Let W be another neighborhood of x with compact closure and such that $\overline{W}$ does not contain $\gamma_i^{-1}(x)$ $\forall i$. Then if we set

$$U = W \setminus \bigcup_{i=1}^{k} \gamma_i(\overline{W})$$

everything works.

The first fact is similar: if $x, y \in T$ are such that $\pi(x) \neq \pi(y)$, or equivalently $y \notin \Gamma(x)$, we take two neighborhoods $U \ni x$ and $V \ni y$ having compact closure. Finiteness of $\Gamma(U, V)$ easily implies that we can find other neighborhoods $U' \subset U$ and $V' \subset V$ such that $\Gamma(U') \cap V' = \phi$, and hence $\pi(U')$ and $\pi(V')$ are disjoint neighborhoods of $\pi(x)$ and $\pi(y)$.

Equivalence of (2) and (3) is quite evident, so we are left to prove the implication (2) $\Rightarrow$ (1). Of course the action is free. As for proper discontinuity it follows from the very definition of compactness that we can confine ourselves to the case where H and K are compact sets having open neighborhoods $U \supset H$ and $V \supset K$ such that

$$U \cap \gamma(U) = V \cap \gamma(V) = \phi \qquad \forall \gamma \neq \text{id}.$$

The set $C = \{\gamma(H) \cap K : \gamma \in \Gamma(H, K)\}$ consists of pairwise disjoint non-empty closed subsets of K. Remark that

$$\pi(\Gamma(H) \cap K) = \pi(H) \cap \pi(K)$$

so that $\pi(\Gamma(H) \cap K)$ is compact; since T/Γ is Hausdorff then $\pi(\Gamma(H) \cap K)$ is closed, so that $\pi^{-1}(\pi(\Gamma(H) \cap K))$ is closed and hence

$$\pi^{-1}(\pi(\Gamma(H) \cap K)) \cap K = \Gamma(H) \cap K$$

is closed too. It follows that the union of all the elements of C is closed and then compactness of K implies that C is finite, so that $\Gamma(H, K)$ is finite too.

We shall prove the implications (4) $\Rightarrow$ (2) and (1) $\Rightarrow$ (4) for the hyperbolic space, and leave the other cases as exercises.

(4) $\Rightarrow$ (2). We start with the second fact. Assume by contradiction that there exists $x \in \mathbb{H}^n$ and a sequence $\{\gamma_n\} \subset \Gamma \setminus \{\text{id}\}$ such that $\gamma_n(x) \to x$. Let $r > 0$ and consider the closed balls $B_r(x)$ and $B_{2r}(x)$ in $\mathbb{H}^n$. For $y \in B_r(x)$ we have

$$d(\gamma_n(y), x) \leq d(\gamma_n(y), \gamma_n(x)) + d(\gamma_n(x), x) =$$
$$= d(y, x) + d(\gamma_n(x), x) \leq r + d(\gamma_n(x), x) \to r$$

and hence if n is big enough we have

$$\gamma_n(\overline{B_r(x)}) \subset \overline{B_{2r}(x)}$$

Since $\overline{B_r(x)}$ spans $\mathbb{R}^{n+1}$ and $\overline{B_{2r}(x)}$ is closed and bounded in $\mathbb{R}^{n+1}$, the subset of $Gl(n+1)$

$$\{A \in Gl(n+1) : \det A = \pm 1, \ A(\overline{B_r(x)}) \subset \overline{B_{2r}(x)}\}$$

is compact (condition $A(\overline{B_r(x)}) \subset \overline{B_{2r}(x)}$ defines a compact subset of $\mathbb{R}^{(n+1)^2}$ and condition $\det A = \pm 1$ implies that the set is bounded away from the hyperplane of singular matrices). By construction this subset of $Gl(n+1)$ contains all the γ_n's for sufficiently large n. Since $\mathcal{I}(\mathbb{H}^n)$ is closed in $Gl(n+1)$ we have that Γ is discrete in $Gl(n+1)$ too, so that the sequence $\{\gamma_n\}$ contains finitely many different isometries. This implies that if n is big enough we have $\gamma_n(x) = x$, and this is absurd.

We are left to prove that T/Γ is Hausdorff. Let $x, y \in T$, $y \notin \Gamma(x)$. Since we just proved that $\Gamma(x)$ is discrete we have $d(\Gamma(x), y) \geq 2\varepsilon > 0$. We can take ε small enough that

$$B_\varepsilon(x) \cap \gamma(B_\varepsilon(x)) = B_\varepsilon(y) \cap \gamma(B_\varepsilon(y)) = \phi \qquad \forall \gamma \neq \mathrm{id}.$$

Since Γ is a group of isometries we have

$$B_\varepsilon(y) \cap \gamma(B_\varepsilon(x)) = \phi \qquad \forall \gamma \in \Gamma$$

and the conclusion follows easily.

(1) $\Rightarrow$ (4). Consider a compact set $K \subset \mathbb{I}^n$ such that the linear span of K is $\mathbb{R}^{n+1}$, and take another compact set $H \subset \mathbb{I}^n$ such that $K \subset \overset{\circ}{H}$. Then

$$\left\{ A \in \mathrm{Gl}(n+1) : A(K) \subset H \right\}$$

is a neighborhood of the identity in $\mathrm{Gl}(n+1)$, and hence

$$\left\{ \gamma \in \Gamma : \gamma(K) \subset H \right\}$$

is a neighborhood of the identity in Γ, but it is contained in $\Gamma(K, H)$ and then it is finite, so that Γ is discrete. $\qquad\square$

In the sequel when saying a group of isometries of $\mathbb{H}^n$, S^n or $\mathbb{R}^n$ is discrete we shall think of the above topologies; however B.1.6 implies that the property of a group of isometries Γ to be discrete and operate freely can be re-phrased without mentioning any topology. Remark as well that it may be proved that the above topologies coincide with the topology of uniform-local convergence, so that in particular the topology on $\mathcal{I}(\mathbb{H}^n)$ is intrinsically defined regardless of the model for $\mathbb{H}^n$.

Theorem B.1.7. If M is a complete connected (X, G)-manifold then $\Pi_1(M)$ can be identified with a subgroup of G acting freely and properly discontinuously on X, and M is isometrically diffeomorphic to the quotient Riemannian manifold $X/\Pi_1(M)$. (The converse being obvious: if $\Gamma < G$ acts freely and properly discontinuously then the quotient Riemannian manifold X/Γ is a complete connected (X, G)-manifold and its fundamental group can be identified with Γ.)

Proof. The universal covering $\tilde{M}$ of M is naturally endowed with an (X, G)-structure; moreover by the Hopf-Rinow theorem $\tilde{M}$ is complete too, and hence by B.1.5 it can be identified with X and it follows from B.1.6 that $\Pi_1(M)$ operates freely and properly discontinuously on $\tilde{M}$, and hence everything works. $\qquad\square$

B.1.7 implies in particular that every complete hyperbolic, elliptic or flat manifold M is a quotient of $\mathbb{H}^n$, S^n or $\mathbb{R}^n$ under the action of $\Pi_1(M)$ which is identified with a discrete group of isometries acting freely.

The proof of the following interesting result can be found in [Hic] and [Ga-Hu-La]:

Theorem B.1.8. A complete, connected, simply connected Riemannian n-manifold whose sectional curvatures are constantly -1 (respectively: 1, 0) is isometrically diffeomorphic to $\mathbb{H}^n$ (respectively: S^n, $\mathbb{R}^n$).

This result together with the above remarks imply the following:

Theorem B.1.9. If M is a complete connected Riemannian n-manifold whose sectional curvatures are constantly -1 (respectively: 1, 0) then M is the quotient of $\mathbb{H}^n$ (respectively: S^n, $\mathbb{R}^n$) under the action of a discrete group of isometries acting freely. In particular, M is a hyperbolic (respectively: elliptic, flat) manifold.

Now we go back to non-necessarily complete structures; we shall prove that under a suitable hypothesis on X (verified by our geometric models) all (X, G)-structures on a manifold M determine (up to conjugation) a group homomorphism of $\Pi_1(M)$ into G; of course the range of this homomorphism will not be in general a group acting freely and properly discontinuously on X. We will denote by $B_r(x)$ the open ball of radius r and centre x in X. We will say X has the isometries-extension property (IEP) if the following holds: there exists $r_0 > 0$ such that for $0 < r < r_0$ given $x, y \in X$ and an isometry φ of $B_r(x)$ onto $B_r(y)$, there exists $\phi \in \mathcal{I}(X)$ extending φ.

Lemma B.1.10. $\mathbb{H}^n$, S^n and $\mathbb{R}^n$ have the IEP.

Proof. This fact holds in any case with $r_0 = \infty$, and the proof follows from A.3.9 and the explicit determination of the isometries. We confine ourselves to $\mathbb{H}^n$ and leave the other cases as exercises.

Let $x, y \in \mathbb{H}^n$, $r > 0$ and $\varphi : B_r(x) \to B_r(y)$ be an isometry. Then certainly $\varphi(x) = y$. By composition on the left with an element of $\mathcal{I}(\mathbb{H}^n)$ mapping y to x we can assume that $\varphi(x) = x$. In the disc model we take $x = 0$; then we have that $B_r(0)$ is a Euclidean ball $B_\rho^{(e)}(0)$ of a suitable radius ρ centred at 0. Moreover $\varphi \in \mathrm{Conf}(B_\rho^{(e)}(0))$ and $\varphi(0) = 0$; it follows immediately from A.3.9 that φ is the restriction of an element of $O(n)$ and the proof is complete. $\square$

Now, let M be a connected (X, G)-manifold.

Lemma B.1.11. Assume X has the IEP. If $\varphi \in \mathcal{I}(M)$, $x \in M$ and $(U, \alpha), (V, \beta)$ are elements of the atlas defining the (X, G)-structure with U, V connected and $x \in U$, $\varphi(U) \subseteq V$, then $\beta \circ \varphi \circ \alpha^{-1}$ is the restriction of an element of $\mathcal{I}(X)$.

Proof. We can assume that $V = \varphi(U)$ and $\alpha(U), \beta(V)$ are open balls in X. The proof is then a direct application of the definition of the IEP. $\square$

As above, we denote by $\tilde{M}$ the universal covering of M (naturally endowed with an (X, G)-structure), and by D a fixed developing function of $\tilde{M}$. We remark that an atlas defining the (X, G)-structure on $\tilde{M}$ is given by

$$\{(U, D|_U) : U \subset \tilde{M} \text{ open } s.t. \ D|_U \text{ is one-to-one}\}.$$

Lemma B.1.12. Assume X has the IEP. If $\varphi \in \mathcal{I}(\tilde{M})$, $x, y \in \tilde{M}$ and we have $D(x) = D(y)$, then $D(\varphi(x)) = D(\varphi(y))$.

Proof. Let $\gamma : [0, 1] \to \tilde{M}$ be a differentiable path with $\gamma(0) = x$ and $\gamma(1) = y$. According to B.1.11 for each $t \in [0, 1]$ we can find $\phi_t \in \mathcal{I}(X)$ such that if U_t is a small neighborhood of $\gamma(t)$ then

$$D \circ \varphi \circ \left(D\big|_{U_t} \right)^{-1} = \phi_t$$

The function $t \mapsto \phi_t$ is obviously locally constant, and hence it is globally constant, which implies that

$$D(\varphi(x)) = \phi_0(D(x)) = \phi_1(D(y)) = D(\varphi(y)). \qquad \square$$

Proposition B.1.13. Assume X has the IEP. There exists one and only one group homomorphism

$$D_* : \mathcal{I}(\tilde{M}) \to \mathcal{I}(X)$$

such that $D \circ \varphi = D_*(\varphi) \circ D \ \forall \varphi \in \mathcal{I}(\tilde{M})$.

Proof. If $\varphi \in \mathcal{I}(\tilde{M})$ and $x \in D(\tilde{M})$, according to B.1.12 the point

$$\varphi_*(x) = D(\varphi(x')) \qquad (\text{for } x' \in D^{-1}(x))$$

is well-defined (*i.e.* it does not depend on the choice of x'). B.1.11 implies that the mapping $x \mapsto \varphi_*(x)$ is locally the restriction of an element of $\mathcal{I}(X)$. Since $D(\tilde{M})$ is connected there exists one and only one element $D_*(\varphi)$ of $\mathcal{I}(X)$ such that

$$D_*(\varphi)(x) = \varphi_*(x) \quad \forall x \in D(\tilde{M}).$$

Existence and uniqueness of D_* are proved, and it is immediately checked that it must be a group homomorphism. $\qquad \square$

Theorem B.1.14. Assume X has the IEP. If M is a connected manifold, each (X, G)-structure on M determines the conjugacy class of a group homomorphism

$$\Psi : \Pi_1(M) \to G.$$

Proof. Assume M has a fixed (X, G)-structure and define $\tilde{M}$, D and D_* as above. A conjugacy class of group homomorphism

$$\pi^* : \Pi_1(M) \to \mathcal{I}(\tilde{M})$$

is naturally associated to the covering. Since two different developing functions of $\tilde{M}$ are obtained from each other by composition with an element of G, it follows that D_* is well-defined up conjugation in $\mathcal{I}(X)$. This implies that $\Psi = D_* \circ \pi^*$ is a well-defined conjugacy class of homomorphism of $\Pi_1(M)$ into $\mathcal{I}(X)$.

If M is oriented then D preserves orientation and hence it is easily checked that Ψ is the conjugacy class of a group homomorphism $\Pi_1(M) \to \mathcal{I}^+(X)$, so that the theorem is proved. $\qquad \square$

The conjugacy class of homomorphism $\Pi_1(M) \to G$ associated to an (X, G)-structure will be called the <u>holonomy</u> of the structure.

Remark B.1.15. It follows from A.3.2 and A.4.1 that the group $\mathcal{I}^+(\mathbb{D}^2)$ coincides with $\mathrm{Aut}(D^2)$, where Aut denotes the group of all holomorphic automorphisms; moreover it may be checked quite easily that if we consider on $S^2 = \mathbb{CP}^1$ and $\mathbb{R}^2 = \mathbb{C}$ their standard metrics, we have again canonical identifications $\mathcal{I}^+(S^2) = \mathrm{Aut}(\mathbb{CP}^1)$, $\mathcal{I}^+(\mathbb{R}^2) = \mathrm{Aut}(\mathbb{C})$. It follows that every complete oriented hyperbolic, elliptic or flat surface can be canonically endowed with a complex structure. Moreover the classical uniformization theorem states that *every connected and simply connected complex surface is holomorphically equivalent to D^2, $\mathbb{CP}^1$ or $\mathbb{C}$*. It follows that every oriented complex surface can be endowed with a complete hyperbolic, elliptic or flat structure. Hence (recalling B.1.7 too) we have that the set of all oriented surfaces which can be endowed with a complete Riemannian structure of constant curvature -1, 1 or 0 is canonically identified with the set of all complex surfaces; moreover it may be shown that the equivalence relation in the first set defined as "existence of an orientation-preserving isometry" corresponds in the second set to the equivalence relation "existence of an holomorphism".

Example B.1.16. $D^2 \setminus \{0\}$ can be endowed with a non-complete hyperbolic structure given by the inclusion in $\mathbb{D}^2$ and with a complete hyperbolic structure associated to the covering function

$$\pi : \Pi^{2,+} \to D^2 \setminus \{0\} \qquad z \mapsto \mathrm{e}^{2\pi i z},$$

which allows to identify $D^2 \setminus \{0\}$ with $\Pi^{2,+}/\mathbb{Z}$ where $\mathbb{Z}$ operates on $\Pi^{2,+}$ by $n(z) = z + n$.

It is worth remarking that these two structures induce equivalent complex structures, while they are certainly not equivalent metrics; this fact does not contradict B.1.15, as one of the structures considered is not complete.

Finally, we observe that if we take as universal covering of $D^2 \setminus \{0\}$ the half-plane $\Pi^{2,+}$ (with the same projection π as above), the pull-back hyperbolic metric with respect to π of the non-complete hyperbolic metric on $D^2 \setminus \{0\}$ is certainly different from the standard hyperbolic metric of $\Pi^{2,+}$. Moreover it is easily verified that a developing function of $\Pi^{2,+}$ with respect to this non-complete structure is obtained simply by considering the composition of π with the inclusion of $D^2 \setminus \{0\}$ into $\mathbb{D}^2$.

We are now going to sketch a generalization of the situation considered in the most part of this section: we are not going to assume any more that X is a Riemannian manifold, but we require X to be an analytic (oriented, connected and simply connected) manifold and G to be a group of analytic diffeomorphisms of X.

If M and N are manifolds endowed with an (X, G)-structure we will denote by $\mathcal{J}(M, N)$ the group of all diffeomorphisms of M onto N which are locally expressed as restrictions of elements of G (*i.e.* $f \in \mathcal{J}(M, N)$ if and only if given $x \in M$ we can find $g \in G$ and two connected charts (U, φ) and (V, ψ) in the fixed atlases of M and N respectively, such that $x \in U$, $f(U) \subset V$ and $\psi \circ f \circ \varphi^{-1}$ is the restriction of g). Existence of <u>developing</u> functions is deduced with the very same argument presented in B.1.3, (where we only need to replace A.2.1 by the analytic continuation principle):

Proposition B.1.17. Let M be a connected, simply connected (X, G)-manifold, let $U \subset M$ be open and connected and let $\varphi \in \mathcal{J}(U, X)$; then there exists a unique mapping $D : M \to X$ extending φ and such that $\forall x \in M$ there exists $V \ni x$ with $D|_V \in \mathcal{J}(V, X)$. Given two such mappings D and D' (obtained from different φ's) there exists $g \in G$ with $D' = g \circ D$.

Let M be a connected (X, G)-manifold, let $\tilde{M}$ be the universal covering of M and let D be a developing function of $\tilde{M}$. We will say M is a <u>complete</u> (X, G)-manifold if D is a covering mapping (of course this notion is independent of D).

Remark B.1.18. According to B.1.5, in case X is Riemannian and $G < \mathcal{I}(X)$ this notion of completeness of M matches the usual one with respect to the Riemannian structure naturally defined.

Theorem B.1.6 can be generalized word-by-word in the present situation, and the definition of the <u>holonomy</u> of a non-complete structure can be repeated too (remark that we do not need to assume any extension property since it was implicit in the definition of $\mathcal{J}$).

An interesting example of (X, G)-manifolds in a non-Riemannian situation is represented by <u>affine</u> manifolds: we take $\mathbb{R}^n$ as X and the set of all affine automorphisms of $\mathbb{R}^n$ as G. We shall meet again this notion in Chapt. F while dealing with flat fiber bundles.

B.2 Topology of Compact Oriented Surfaces

In the following sections we shall be concerned with the hyperbolic, elliptic and flat structures supported by a fixed connected, compact, oriented differentiable surface. In this section we recall a few well-known facts about the elementary topology of such surfaces. We are not going to give any proof, since the results we will state are quite standard; they may be found for instance in [Mass1] and [Hir].

Throughout this section we shall use the word *surface* to denote a connected, compact and oriented 2-manifold.

Theorem B.2.1. Every surface is diffeomorphic either to the sphere S^2 or to the g-<u>holes</u> <u>torus</u> T_g, for some integer $g \geq 1$.

We can associate to each surface an integer number, called the <u>genus</u> of the surface, which is the number of its holes, *i.e.*:

$$\begin{cases} g(S^2) = 0 \\ g(T_g) = g. \end{cases}$$

Fig. B.2. Tori with $1, 2$ and g holes

(This definition could be given intrinsically for each surface, and it turns out that the genus is the only topological (and differential) invariant of the surface: we are not interested now in this theory.)

Theorem B.2.2. Every surface can be triangulated, *i.e.* it is homeomorphic to a polyhedron (of maximal dimension 2, of course).

We recall that if we are given a surface M and a triangulation of M containing V vertices, S sides and F faces, the <u>Euler-Poincaré characteristic</u> of the surface M (which is assumed to be well-defined) can be calculated in the following way:

$$\chi(M) = V - S + F.$$

Our definition of the genus of a surface leads quite immediately to the following:

Proposition B.2.3. $\chi(M) = 2(1 - g(M))$.

Proposition B.2.4. If $g \geq 1$ the surface of genus g can be realized as the quotient of a polygon of $4g$ sides, where:
– all the vertices are identified with each other;
– the sides are identified in pairs.

Proposition B.2.5. If M is a surface and $g(M) \geq 2$ then there exists a finite set $\{\alpha_i\}_{i=1,\dots,h}$ of pairwise disjoint loops on M such that

$$M \setminus \left(\bigcup \{\alpha_i\}_{i=1,\dots,h} \right)$$

consists of k connected components, each of which has closure in M homeomorphic to a <u>pant</u> (*i.e.* a closed disc with two little open discs removed). The integers h and k are uniquely determined by this condition, and they are given by

$$h = 3(g(M) - 1) \qquad k = 2(g(M) - 1) = -\chi(M).$$

Moreover the loops can be chosen to be smooth, and in this case the closures of the connected components described above are diffeomorphic to pants (as manifolds with boundary).

We will call the closed covering of M described in the above proposition a <u>pant decomposition</u> of M.

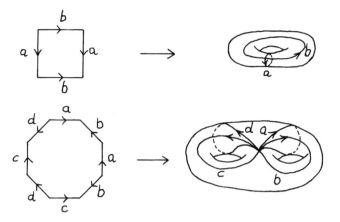

Fig. B.3. Geometric meaning of B.2.4 in the case of genus 1 and 2

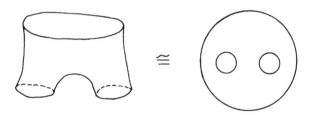

Fig. B.4. A pant

Fig. B.5. A pant decomposition of the surface of genus g

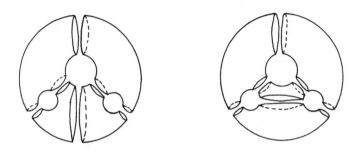

Fig. B.6. Two pant decompositions of the surface of genus 3

B.3 Hyperbolic, Elliptic and Flat Surfaces

In this section we shall describe the constant curvature Riemannian structures with which a (connected, compact, oriented) surface can be endowed. According to B.2.1 we shall refer only to S^2 and the T_g's, since diffeomorphic surfaces can be considered as equal.

Proposition B.3.1. (1) S^2 is the only surface supporting an elliptic structure.

(2) T_1 supports a flat structure.

(3) For $g \geq 2$, T_g supports a hyperbolic structure.

Proof. We start by remarking that the structures we are considering must necessarily be complete (since the surfaces are compact).

(1) Of course S^2 supports its own elliptic structure. Conversely, an elliptic surface is the quotient of S^2 by a group $\Gamma < \mathcal{I}^+(S^2) = \mathrm{SO}(3)$; since every element of $\mathrm{SO}(3)$ has some fixed point on S^2, Γ must be trivial, and hence the surface must coincide with S^2.

(2) T_1 can be obtained as the quotient of $\mathbb{R}^2$ by the action of the discrete and free subgroup of $\mathcal{I}^+(\mathbb{R}^2)$ generated by

$$(x,y) \mapsto (x+1,y) \quad \text{and} \quad (x,y) \mapsto (x,y+1).$$

This implies that T_1 supports a flat structure.

We shall give now a slightly different proof of this fact, since a similar method will allow us to prove (3). Let us represent T_1 as the quotient of $Q = [-1,1] \times [-1,1] \subset \mathbb{R}^2$, with the natural identifications on the boundary. We define explicitly four mappings from D^2 to Q; they are geometrically described in Fig. B.7.

$$\hat{\phi}_1(x,y) = (x,y)$$

$$\hat{\phi}_2(x,y) = \begin{cases} (x, y+1) & \text{if } y \leq 0 \\ (x, y-1) & \text{if } y > 0 \end{cases}$$

$$\hat{\phi}_3(x,y) = \begin{cases} (x+1, y) & \text{if } x \leq 0 \\ (x-1, y) & \text{if } x > 0 \end{cases}$$

$$\hat{\phi}_3(x,y) = \begin{cases} (x+1, y+1) & \text{if } x \leq 0, y \leq 0 \\ (x+1, y-1) & \text{if } x \leq 0, y > 0 \\ (x-1, y+1) & \text{if } x > 0, y \leq 0 \\ (x-1, y-1) & \text{if } x > 0, y > 0. \end{cases}$$

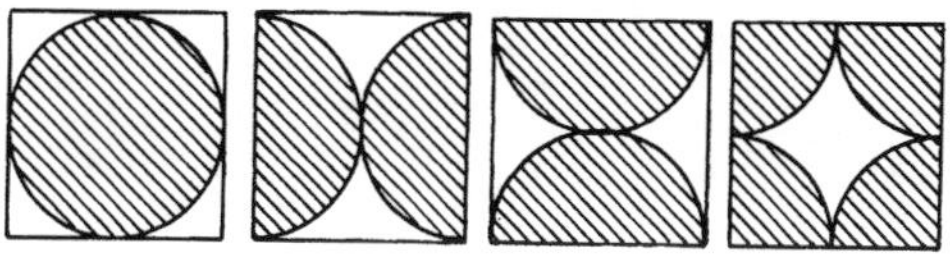

Fig. B.7. Four charts defining a flat structure on the surface of genus 1

If we consider the mappings $\phi_i : D^2 \to T_1$ obtained as quotient of the $\hat{\phi}_i$'s, it is quite easily verified that they define a differentiable atlas on T_1 and moreover the changes of chart are isometries with respect to the flat metric on D^2, *i.e.* these mappings endow T_1 with a flat structure. It is quite interesting to remark that this method works because the sum of the inner angles of Q is 2π: a similar method cannot work for $g \geq 2$. For instance for $g = 2$ we can consider a regular octagon in $\mathbb{R}^2$, and define a flat structure on the punctured T_2 via the charts depicted in Fig. B.8.

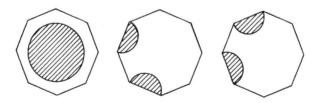

Fig. B.8. Charts defining a flat structure on the punctured surface of genus 2

However, since the sum of the inner angles of the polygon exceeds 2π, there is no way to define the last chart, *i.e.* to globalize the flat structure.

(3) According to the above discussion we only need to find in $\mathbb{H}^2$ a regular polygon with $4g$ sides (*i.e.* a polygon with $4g$ geodesic sides all having the same length) in which the sum of the inner angles is 2π. In the disc model $\mathbb{D}^2$ viewed as a subset of $\mathbb{C}$ we consider for $0 < r < 1$ the geodesic polygon P_r having vertices

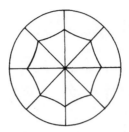

Fig. B.9. A regular octagon centred at the origin

$$r \cdot \exp\left(\frac{2\pi ik}{4g}\right) \qquad k = 0, 1, ..., 4g - 1.$$

If S_r denotes the sum of the inner angles of P_r it is easily verified that:

(a) $\lim_{r \to 1} S_r = 0$;

(b) $\lim_{r \to 0} S_r = (4g - 2)\pi$ (in fact at 0 the hyperbolic metric is a multiple of the Euclidean one);

(c) $r \mapsto S_r$ is a continuous function.

Since $(4g-2)\pi > 2\pi$, for suitable r we have $S_r = 2\pi$. Figure B.10 illustrates the charts used for the definition of the hyperbolic structure on T_2.

Fig. B.10. Charts defining a hyperbolic structure on the surface of genus 2

□

We are now going to prove that if a surface supports a hyperbolic, elliptic or flat structure, then it supports only structures of the same type. We shall need a special version of the following Gauss-Bonnet formula (see [DC] for a general proof):

Theorem B.3.2. If M is a surface of Riemannian curvature k and dS denotes the element of area on M then

$$\int_M k\, dS = 2\pi \chi(M).$$

We prove B.3.2 in the special cases we need.

Proposition B.3.3. If M is a hyperbolic surface and $\mathcal{A}(M)$ denotes the area of M then

$$\mathcal{A}(M) = -2\pi\chi(M).$$

Proof. Let us consider a triangulation of M consisting of F geodesic triangles $\Delta_1, ..., \Delta_F$ (compactness and B.2.2 imply quite easily that such a triangulation exists); let V and S be the numbers of vertices and sides appearing in the triangulation; we remark first that $2S = 3F$. Let $\alpha_i^{(j)}$, $i = 1, 2, 3$ denote the measure of the inner angles of Δ_j; by A.6.4 we have

$$-\mathcal{A}(\Delta_j) = \sum_{i=1}^{3} \alpha_i^{(j)} - \pi$$

and therefore

$$-\mathcal{A}(M) = -\sum_{j=1}^{F} \mathcal{A}(\Delta_j) = \sum_{j=1}^{F} \left(\sum_{i=1}^{3} \alpha_i^{(j)} - \pi \right) =$$

$$= 2\pi V - \pi F = 2\pi(V - S + F) = 2\pi\chi(M). \qquad \square$$

Proposition B.3.4. If M is a flat surface then $\chi(M) = 0$.

Proof. As above we can consider a triangulation of M consisting of F geodesic triangles $\Delta_1, ..., \Delta_F$. Since Δ_j is isometric to a triangle in $\mathbb{R}^2$ the sum of its inner angles is π. It follows that the sum of all the inner angles of the Δ_j's is πF; since this sum can be calculated also as $2\pi V$, we obtain that $2V = F$. Hence

$$\chi(M) = V - S + F = \frac{1}{2} \cdot F - \frac{3}{2} \cdot F + F = 0.$$

$$\square$$

Theorem B.3.5. Let M be a compact, connected, oriented surface; then:
(1) M supports elliptic structures if and only if $g(M) = 0$;
(2) M supports flat structures if and only if $g(M) = 1$;
(3) M supports hyperbolic structures if and only if $g(M) \geq 2$.

Proof. The "if" parts in (1), (2) and (3) and the "only if" part in (1) were already established in B.3.1. The "only if" part in (3) follows from B.3.3, since $\mathcal{A}(M)$ must be positive. The "only if" part of (2) follows from B.3.4. $\square$

B.4 Teichmüller Space

In this section we shall consider, for a fixed integer number $g \geq 2$, all the hyperbolic structures supported by a surface M of genus g, with a suitable equivalence relation. A connected, compact, oriented surface M_0 of genus g will be considered as fixed (for instance, by a fixed immersion in $\mathbb{R}^3$). The parametrization we shall give of the hyperbolic structures on M_0 is the Fenchel-Nielsen one, and it is based on the existence of a pant decomposition of M_0; this is not the only possible method (others are due to Fricke-Klein and Teichmüller but we shall not say anything about them). Our treatment is not far from the one presented in [Fa-La-Po] (see also [Th1, ch. 5]). We address the reader interested in the other parametrizations of the Teichmüller space to [Ab].

We define Teichmüller space τ_g in the following equivalent ways:

(A) We consider the set $\mathcal{H}$ of all hyperbolic metrics on M_0 (*i.e.* the metrics of constant Riemannian curvature -1) and we define on $\mathcal{H}$ the following equivalence relation: $h_1 R h_2$ if there exists an isometric diffeomorphism $\phi : (M_0, h_1) \to (M_0, h_2)$ which is isotopic to the identity (in particular, it must be orientation-preserving). Then τ_g is defined as the quotient set $\mathcal{H}/R$.

(We recall that two diffeomorphisms ϕ_0 and ϕ_1 of M onto itself are called <u>isotopic</u> if there exists a diffeomorphism

$$\Phi : M \times [0,1] \to M \times [0,1]$$

such that $\Phi(x, t) = (\phi_t(x), t) \ \forall x, t$, and the definition of ϕ_0 and ϕ_1 matches the original one.)

(B) We consider the set $\mathcal{H}'$ of all the triples (M, h, f), where M is a connected, compact, oriented surface of genus g, h is a hyperbolic metric on M and f is an orientation-preserving diffeomorphism of M onto M_0. An equivalence relation R' is defined on $\mathcal{H}'$ in the following natural way: $(M_1, h_1, f_1) R'(M_2, h_2, f_2)$ if there exists an isometric diffeomorphism

$$\phi : (M_1, h_1) \to (M_2, h_2)$$

such that $f_2 \circ \phi \circ f_1^{-1}$ is isotopic to the identity on M_0. Then τ_g is the quotient set $\mathcal{H}'/R'$. (Equivalence with definition (A) is evident.)

(C) We consider on the group $\text{Diff}^+(M_0)$ the topology of C^∞ convergence and we denote by $\text{Diff}^0(M_0)$ the connected component of the identity with respect to this topology (it is a subgroup of $\text{Diff}^+(M_0)$). If $\mathcal{H}$ is as in (A) the group $\text{Diff}^+(M_0)$ operates on $\mathcal{H}$ via the pull-forward of the metrics: for $f \in \text{Diff}^+(M_0)$ and $h \in \mathcal{H}$

$$f_*(h)_x(v, w) = h_{f^{-1}(x)}\big(d(f^{-1})_x(v), d(f^{-1})_x(w)\big).$$

Then τ_g is given by the quotient space $\mathcal{H}/\text{Diff}^0(M_0)$ (the reason is that a diffeomorphism of M_0 is isotopic to the identity if and only if it belongs to $\text{Diff}^0(M_0)$).

Remark B.4.1. Maybe the most natural equivalence relation to consider on $\mathcal{H}$ is the existence of an orientation-preserving isometry (without the requirement that such an isometry is isotopic to the identity); in this case the space to be studied would be $\mathcal{H}/\mathrm{Diff}^+(M_0)$. This task is much more difficult and in any case it requires a preliminary study of τ_g. At the end of the present section we shall outline a few results about this space.

Remark B.4.2. Construction (C) of τ_g allows to endow it in a natural way with a topology; in fact $\mathcal{H}$ is a subset of the set of all sections of the fiber bundle $T(M_0)^* \otimes T(M_0)^*$, and hence it can be endowed with the C^∞ topology too; $\mathrm{Diff}^+(M_0)$ operates continuously as a group of homeomorphisms of $\mathcal{H}$, and hence the quotient set is naturally endowed with a topology. Since we will not care about the topology, we should have defined τ_g as Teichmüller *set* instead of Teichmüller *space*. We shall prove in the following that τ_g can be identified with $\mathbb{R}_+^{3(g-1)} \times \mathbb{R}^{3(g-1)}$; our identification will be only a bijective mapping, but it could be proved with some more effort that it is a homeomorphism too.

For the study of τ_g we shall need a few preliminaries; the first results do not require the hyperbolic manifold in question to be a surface.

We begin with a very general fact. Let X be a connected manifold (much less is necessary, but we confine ourselves to this case). We shall say two loops $\alpha, \beta : [0,1] \to X$ are <u>free-homotopic</u> if there exists a continuous mapping $F : [0,1]^2 \to X$ such that

$$F(t,0) = \alpha(t) \quad F(t,1) = \beta(t) \quad F(0,s) = F(1,s) \quad \forall\, t, s$$

It is immediately checked that free-homotopy is an equivalence relation: we will denote by $\tilde{\Pi}_1(X)$ the quotient set (*i.e.* the set of all loops up to free-homotopy). We recall that if G is a group $\mathrm{Int}(G)$ denotes the group of all automorphisms of G of the form $g \mapsto h^{-1} \cdot g \cdot h$, for suitable $h \in G$.

Proposition B.4.3. If $\pi : \tilde{X} \to X$ is a universal covering, π induces a natural bijection

$$\Phi : \left.\Pi_1(X)\right/\mathrm{Int}(\Pi_1(X)) \to \tilde{\Pi}_1(X).$$

Proof. Let us remark first that if α is a loop and β is a path in X such that $\beta(1) = \alpha(0)$, then $\beta^{-1} \cdot \alpha \cdot \beta$ is free-homotopic to α. (The proof is easily deduced from Fig. B.11.)

π induces an action of $\Pi_1(X)$ as a group of diffeomorphisms of $\tilde{X}$. Given $T \in \Pi_1(X)$, choose $\tilde{x} \in \tilde{X}$, let $\tilde{\alpha}$ be a path joining $\tilde{x}$ to $T(\tilde{x})$ and set $\alpha = \pi \circ \tilde{\alpha}$. We will denote by $\langle T \rangle$ the class of T up to $\mathrm{Int}(\Pi_1(X))$ and by $\langle \alpha \rangle$ the class of α up to free-homotopy. We claim that the mapping

$$\Phi : \langle T \rangle \mapsto \langle \alpha \rangle$$

is well-defined.

Independence of $\langle \alpha \rangle$ on $\tilde{\alpha}$ follows at once from simple connectedness of $\tilde{X}$, while independence on $\tilde{x}$ follows from simple connectedness and the first remark we made ($\beta^{-1} \cdot \alpha \cdot \beta$ free-homotopic to α). In the situation of Fig. B.12 we have

$$\langle \pi \circ \tilde{\alpha}' \rangle = \langle \pi \circ (\tilde{\beta}^{-1} \cdot \alpha \cdot T(\tilde{\beta})) \rangle = \langle \pi \circ \tilde{\beta} \rangle^{-1} \cdot \langle \pi \circ \tilde{\alpha} \rangle \cdot \langle \pi \circ \tilde{\beta} \rangle = \langle \pi \circ \tilde{\alpha} \rangle$$

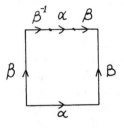

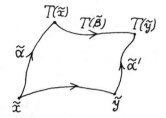

Fig. B.11. $\beta^{-1} \cdot \alpha \cdot \beta$ is free-homotopic to α

Fig. B.12. Independence on $\tilde{\alpha}$ of the definition of Φ

If we consider $S^{-1} \circ T \circ S$ instead of T we can take the path $S^{-1}(\tilde{\alpha})$ starting at $S^{-1}(\tilde{x})$, so that in X we obtain the same loop.

Our claim is proved.

Now, given a loop α in X we take $\tilde{x} \in \pi^{-1}(\alpha(0))$ and we lift α to $\tilde{\alpha}$ starting at $\tilde{x}$; then there exists a unique $T \in \Pi_1(X)$ such that $\tilde{\alpha}(1) = T(\tilde{x})$. We claim that the mapping

$$\Psi : \langle \alpha \rangle \mapsto \langle T \rangle$$

is well-defined. If we start at another point $S(\tilde{x}) \in \pi^{-1}(\alpha(0))$ we obtain $S^{-1} \circ T \circ S$ instead of T, *i.e.* an element of $\langle T \rangle$. If we take another loop γ in $\langle \alpha \rangle$ then we can lift the whole free-homotopy between α and γ as suggested by Fig. B.13, so that $\langle T \rangle$ remains the same.

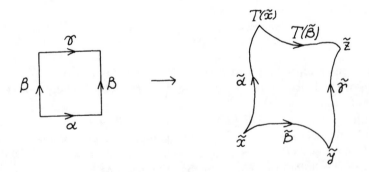

Fig. B.13. Independence on α of the definition of Ψ

It follows that Φ and Ψ are both well-defined, and it is easily checked that they are the inverse of each other. $\qquad\square$

From now on we shall refer to loops in X as subsets of X which may be parametrized on a circle: this is quite natural (especially when dealing with free-homotopy) as the starting point of a loop is not very relevant (remark however that the choice of a basepoint allows one to define a group structure on $\Pi_1(X)$, while in general $\tilde{\Pi}_1(X)$ cannot be endowed with a natural product). Of course loops must be given a fixed orientation.

Now we consider a connected hyperbolic n-manifold M and we fix a universal covering $\pi : \mathbf{H}^n \to M$.

Lemma B.4.4. If M is compact then the non-trivial elements of $\Pi_1(M)$ operate on $\mathbb{H}^n$ as isometries of hyperbolic type.

Proof. Let $T \in \Pi_1(M)$ be viewed as an isometry of $\mathbb{H}^n$, $T \neq \mathrm{id}$. Since $\Pi_1(M)$ operates freely, T is certainly not of elliptic type. Let us assume by contradiction that T is of parabolic type. Then in $\mathbf{H}^n$ we can find a sequence of points $\{x_n\}$ such that $d(x_n, T(x_n))$ becomes arbitrarily small as $n \to \infty$ (assume in the half-space model that T fixes ∞: then the sequence $(0, n)$ works). It follows that in M we can find arbitrarily short non-trivial loops. It is readily checked that this is a contradiction, since compactness implies that M is covered by a finite number of open sets isometric to balls in $\mathbb{H}^n$, and hence if a loop is short enough it is contained in one of these open sets, which implies that it is trivial. $\qquad\square$

Lemma B.4.5. If M is compact each non-trivial free-homotopy class in M contains a unique geodesic loop.

Proof. Let α be a non-trivial loop, and let $T \in \Phi^{-1}(\langle \alpha \rangle)$ (we recall that T is such that $T(\tilde{\alpha}(0)) = \tilde{\alpha}(1)$ for some lifting $\tilde{\alpha}$ of α). We have checked above that T is an isometry of hyperbolic type, and hence there exists one and only one T-invariant geodesic line $\tilde{\gamma}$. The projection of $\tilde{\gamma}$ in M is a geodesic loop representing the same class as α.

Conversely, let γ_1 be a geodesic loop in M representing the same class of α, and let $\tilde{\gamma}_1$ denote a lifting of γ_1 (extended to the whole real line); if $T' \in \Phi^{-1}(\langle \alpha \rangle)$ is such that $T'(\tilde{\gamma}_1(0)) = \tilde{\gamma}_1(1)$, then the fact that $\tilde{\gamma}_1$ is a geodesic line implies that it is T'-invariant. Moreover $T' = S^{-1} \circ T \circ S$ for suitable S, whence $\tilde{\gamma}_1 = S^{-1}(\tilde{\gamma})$, $\gamma_1 = \gamma$. $\qquad\square$

Though we are not going to need it, we remark that an analogue of the above lemma holds for trivial loops too (provided one allows geodesics to be defined on intervals of length zero, *i.e.* to be points). In fact if γ is a trivial geodesic loop in M, γ can be lifted to a geodesic path in $\mathbb{H}^n$; since γ is trivial $\tilde{\gamma}$ is a loop too, which implies that it consists of a single point, and hence γ consists of a single point too. It follows that the only geodesic loops in M representing the trivial free-homotopy class are the one-points loops.

We recall that two curves $\alpha_0, \alpha_1 : [0, 1] \to M$ are said to be isotopic if there exists a diffeomorphism

$$\Phi : M \times [0, 1] \to M \times [0, 1] \text{ such that } \Phi(x, t) = (\phi_t(x), t)\ \forall x, t \quad \text{and}$$

$$\phi_0 = \mathrm{id}, \quad \phi_1(\alpha_0(s)) = \alpha_1(s)\ \forall s \in [0, 1].$$

In the following we will need this result which may be found in [Ep1]:

Proposition B.4.6. Two free-homotopic simple loops on a connected surface are isotopic too (the converse being obvious).

Proposition B.4.7. If M is a (connected, oriented) compact hyperbolic surface and $\alpha_1, ..., \alpha_\nu$ are pairwise non-intersecting and non-isotopic non-trivial simple loops in M then we can find pairwise non-intersecting and non-isotopic simple geodesic loops $\gamma_1, ..., \gamma_\nu$ such that γ_i is isotopic to α_i for $i = 1, ..., \nu$. Moreover the γ_i's are uniquely determined by these conditions.

Proof. Since isotopy implies free-homotopy the only possible choice of γ_i is that of the geodesic loop in the free-homotopy class of α_i, so that uniqueness is proved.

Conversely we check that this choice of the γ_i's works, *i.e.*

(i) γ_i is isotopic to α_i;

(ii) if $i \neq j$, γ_i is not isotopic to γ_j;

(iii) if $i \neq j$, γ_i does not intersect γ_j;

(iv) γ_i is a simple loop.

Let us prove these facts.

(i) This is a direct corollary of B.4.6.

(ii) If γ_i were isotopic to γ_j, then α_i would be isotopic to α_j, and this is absurd.

(iii) Let us omit for a moment the index of the curves, and let us go back to the construction of γ (B.4.5). We started with a lifting $\tilde{\alpha}$ of α, considered the isometry $T \in \Pi_1(M)$ such that $T(\tilde{\alpha}(0)) = \tilde{\alpha}(1)$ and defined $\tilde{\gamma}$ as the only T-invariant geodesic line. We state now that if $\tilde{\alpha}$ is extended in the natural way to the whole real line, then $\tilde{\alpha}$ and $\tilde{\gamma}$ have the same points at infinity (the points at infinity of $\tilde{\alpha}$ being defined as those points $p \in \partial \mathbf{H}^n$ such that there exists a sequence $\{t_n\}$ of real numbers with $\tilde{\alpha}(t_n) \to p$ in the topology of $\overline{\mathbf{H}^n}$). In fact if $\tilde{\alpha}_0$ and $\tilde{\gamma}_0$ are the restrictions of $\tilde{\alpha}$ and $\tilde{\gamma}$ to $[0,1]$, we have (identifying the curves with their supports):

$$\sup_{x \in \tilde{\alpha}_0} d(x, \tilde{\gamma}_0) < \infty \qquad \sup_{x \in \tilde{\gamma}_0} d(x, \tilde{\alpha}_0) < \infty$$

$$\tilde{\alpha} = \bigcup_{n \in \mathbb{Z}} T_\alpha^n(\tilde{\alpha}_0) \qquad \tilde{\gamma} = \bigcup_{n \in \mathbb{Z}} T_\alpha^n(\tilde{\gamma}_0)$$

$$\Rightarrow \sup_{x \in \tilde{\alpha}} d(x, \tilde{\gamma}) < \infty \qquad \sup_{x \in \tilde{\gamma}} d(x, \tilde{\alpha}) < \infty$$

and this implies quite easily our assertion.

Let us remark also that for any lifting $\tilde{\gamma}$ of γ there exists some lifting $\tilde{\alpha}$ of α having the same points at infinity as $\tilde{\gamma}$ (this fact is easily deduced from what we just proved with the same construction as in B.4.3 and B.4.5: remark that α and γ are being kept fixed).

Now, let us assume that γ_i and γ_j intersect somewhere; then we can consider two liftings $\tilde{\gamma}_i$ and $\tilde{\gamma}_j$ intersecting too (we start the lifting at the common point). If $\tilde{\alpha}_i$ and $\tilde{\alpha}_j$ are liftings of α_i and α_j having the same points at infinity as $\tilde{\gamma}_i$ and $\tilde{\gamma}_j$ respectively, then $\tilde{\alpha}_i$ and $\tilde{\alpha}_j$ intersect, and hence α_i and α_j intersect, which is absurd, as it easily follows from Fig. B.14.

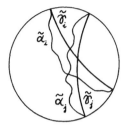

Fig. B.14. The geodesics representing the homotopy classes of two disjoint loops are disjoint

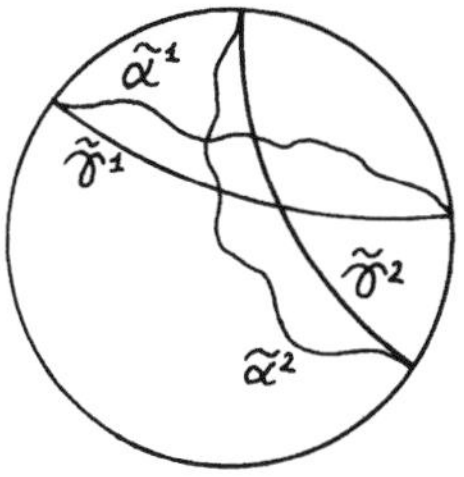

Fig. B.15. The geodesic representing the homotopy class of a simple loop is simple

(iv) We omit the index of the curves. If γ is not simple, then we can find a point $x_0 \in M$ such that two branches of γ pass through x_0 with distinct tangent vectors. If $\tilde{x}_0 \in \pi^{-1}(x_0)$ we can consider two different liftings $\tilde{\gamma}^1$ and $\tilde{\gamma}^2$ of γ passing through $\tilde{x}_0$ (corresponding to the different tangent vectors to γ at x_0). If $\tilde{\alpha}^1$ and $\tilde{\alpha}^2$ are liftings of α having the same points at infinity as $\tilde{\gamma}^1$ and $\tilde{\gamma}^2$ respectively, then, as it easily follows from Fig. B.15, $\tilde{\alpha}^1$ and $\tilde{\alpha}^2$ intersect, which implies that α is not a simple loop, and this is absurd. $\square$

We recall that a connected, oriented compact surface M_0 of genus $g \geq 2$ was fixed at the beginning of the section.

In the following we shall consider as fixed up to isotopy a pant decomposition of M_0, i.e. we shall fix the isotopy class of $3(g-1)$ loops $\alpha_1, ..., \alpha_{3(g-1)}$ giving a pant decomposition of M_0 (in particular an orientation is fixed on these loops). The interior of each pant will be given the orientation it inherits from M_0: remark that the orientation of an edge of a pant inherited from the orientation of the α_i's is not necessarily the orientation it is given as boundary of the oriented pant (indeed, this is true for exactly half of the edges).

In our investigation of τ_g we shall use definition (A).

Remark B.4.8. The loops giving the pant decomposition of the surface M_0 satisfy the hypotheses of B.4.7. Hence, if M_0 is endowed with a hyperbolic structure we can perform the pant decomposition by geodesic loops.

If $h \in \mathcal{H}$ we shall denote by $\gamma_1^{(h)}, ..., \gamma_{3(g-1)}^{(h)}$ the h-geodesic loops as described in B.4.7, corresponding to the fixed loops $\alpha_1, ..., \alpha_{3(g-1)}$. If $L^{(h)}(\beta)$ denotes the length with respect to h of a (piecewise differentiable) curve β, we define the following function:

$$L : \mathcal{H} \to \mathbb{R}_+^{3(g-1)} \qquad h \mapsto \left(L^{(h)}(\gamma_1^{(h)}), ..., L^{(h)}(\gamma_{3(g-1)}^{(h)})\right).$$

Proposition B.4.9. If $h_1 R h_2$ then $L(h_1) = L(h_2)$, so that L is well-defined on the set τ_g.

Proof. By definition there exists an isometry $\phi : (M_0, h_1) \to (M_0, h_2)$ which is isotopic to the identity. The simple loops

$$\phi\big(\gamma_1^{(h_1)}\big), ..., \phi\big(\gamma_{3(g-1)}^{(h_1)}\big)$$

are geodesics with respect to h_2; moreover they are pairwise disjoint and non-isotopic, and $\phi\big(\gamma_i^{(h_1)}\big)$ is isotopic to α_i. This implies that $\gamma_i^{(h_2)} = \phi\big(\gamma_i^{(h_1)}\big)$. Since

$$L^{(h_2)}\big(\phi\big(\gamma_i^{(h_1)}\big)\big) = L^{(h_1)}\big(\gamma_i^{(h_1)}\big)$$

we have that $L(h_1) = L(h_2)$ and the proof is over.   □

According to the above construction, for any fixed hyperbolic structure h on M_0 we can perform the fixed pant decomposition of M_0 via h-geodesic loops: each pant is then endowed with a hyperbolic structure with respect to which the edges are geodesics (the definition of a <u>hyperbolic structure with geodesic boundary</u> on a manifold with boundary is straight-forward: the manifold is required to be locally isometric to an open set of a closed half hyperbolic space, *i.e.* the closure of one of the domains $\mathbb{H}^n$ is divided into by a hyperbolic hyperplane). We are now going to study these structures.

We fix an oriented pant P_0 (with edges $\partial_1, \partial_2, \partial_3$) and we consider the set $\mathcal{P}$ of all the hyperbolic structures on P_0 with respect to which the edges are geodesics. An equivalence relation S is defined on $\mathcal{P}$ in the following way: $k_1 S k_2$ if there exists an isometry $\phi : (P_0, k_1) \to (P_0, k_2)$ which is isotopic to the identity via an isotopy Φ enjoying $\Phi(\partial_i \times \{t\}) = \partial_i \times \{t\}$ for $i = 1, 2, 3$ and all $t \in [0, 1]$.

Remark B.4.10. Given $k \in \mathcal{P}$ we can consider two different copies of P_0 (both endowed with the structure k) and glue them along the edges having the same index (in an isometric way, *i.e.* starting at two arbitrary points and then going on with the arc length as parameter). Since the edges are geodesics it is easily checked that this method produces a hyperbolic structure on the surface of genus 2.

We can observe as well that the only necessary condition for this method to work is that the edges having the same index have the same length: this fact will be extensively used in the sequel.

Lemma B.4.11. For $k \in \mathcal{P}$ and $1 \leq i < j \leq 3$ there exists one and only one k-geodesic arc $C_{i,j}^{(k)}$ joining ∂_i to ∂_j and being orthogonal to both of them. Moreover, if $\{i, j\} \neq \{i', j'\}$ then $C_{i,j}^{(k)} \cap C_{i',j'}^{(k)} = \emptyset$. In particular the endpoints of the $C_{i,j}^{(k)}$'s divide each of the ∂_l's into two non-trivial arcs.

Proof. Let α be a simple path joining ∂_i to ∂j as represented in Fig. B.17. If we double the pant P_0 as described in the above remark (the glueing on the edges being given by the identity mapping) we obtain a surface Q of genus 2 endowed with a hyperbolic structure: moreover α produces a non-trivial

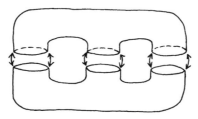

Fig. B.16. A hyperbolic structure on a pant induces a hyperbolic structure on the surface of genus two

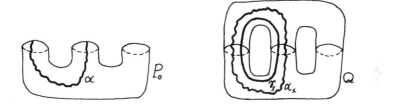

Fig. B.17. Construction of the geodesic which joins orthogonally two components of the boundary of a pant

simple loop α_1 on Q, and hence we can consider the only geodesic γ_1 in the isotopy class of α_1.

If γ is the restriction of γ_1 to P_0, γ is certainly a k-geodesic arc joining ∂_i to ∂_j. Moreover the hyperbolic structure on Q is symmetric with respect to the identified edges, $i.e.$ if we interchange the two copies of P_0 the structure is the same: since γ_1 is unique it must be symmetric with respect to the ∂_l's too. This implies that it is orthogonal to both ∂_i and ∂_j.

Uniqueness follows by the same construction: given $C_{ij}^{(k)}$ with the prescribed conditions, the loop it produces in Q is geodesic and it is in the same isotopy class of γ_1, which implies that it coincides with γ_1.

The fact that $C_{i,j}^{(k)} \cap C_{i',j'}^{(k)} = \phi$ follows from the same construction again: the doubles of $C_{i,j}^{(k)}$ and $C_{i',j'}^{(k)}$ are the geodesic free-homotopy representatives of two non-intersecting non-isotopic non-trivial simple loops, and hence by B.4.7 they do not intesect.

The last assertion is straight-forward: the endpoints of the $C_{i,j}^{(k)}$'s determine on each of the ∂_l's two distinct points.   □

According to the above results, given $k \in \mathcal{P}$ the pant P_0 can be cut (via geodesic lines) into two hexagons as shown in Fig. B.18.

Remark that each hexagon is endowed with a hyperbolic structure with respect to which the edges are geodesics and all the inner angles are right.

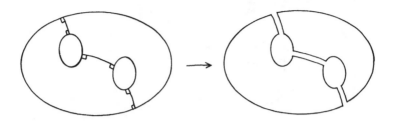

Fig. B.18. How to cut a pant

We are now going to study such hyperbolic hexagons. We fix a hexagon E_0 with a fixed orientation and a preferred vertex: we denote by a_1, b_1, a_2, b_2, a_3 and b_3 the edges of E_0 starting from the preferred vertex and following the orientation of the boundary of E_0.

Fig. B.19. The fixed hexagon E_0

We introduce the set $\mathcal{E}$ of all the hyperbolic structures on E_0 with respect to which all the edges are geodesics and all the inner angles are right. The natural equivalence relation T to consider on $\mathcal{E}$ is the existence of an isometry isotopic to the identity, where the isotopy (at any time) keeps the vertices fixed and maps each edge onto itself. As for definition (B) of τ_g one could define $\mathcal{E}/T$ using all the hyperbolic hexagons (diffeomorphic to E_0) with the corresponding equivalence relation. We shall implicitly assume this definition (the same argument can be repeated word for word for $\mathcal{P}/S$, *i.e.* for the hyperbolic structures on a pant).

Lemma B.4.12. If $e \in \mathcal{E}$, (E_0, e) can be isometrically embedded in $\mathbb{H}^2$.

Proof. Consider two copies of E_0 (both endowed with the structure e), and glue them (isometrically) along the a_i's, in order to obtain a pant with a hyperbolic structure with geodesic edges; then glue two of such pants along the edges, and obtain a hyperbolic structure on a surface of genus 2. Such a surface is a quotient of $\mathbb{H}^2$, and (E_0, e) is isometrically embedded in it. Since E_0 is a simply connected space it can be globally lifted to $\mathbb{H}^2$, and the proof is over. $\square$

We consider now the mapping:

$$A : \mathcal{E} \to \mathbb{R}^3_+ \qquad e \mapsto \left(L^{(e)}(a_1), L^{(e)}(a_2), L^{(e)}(a_3) \right).$$

Proposition B.4.13. If $e_1 T e_2$ then $A(e_1) = A(e_2)$, *i.e.* A induces a well-defined mapping $\hat{A}$ on the quotient set. Moreover

$$\hat{A} : \mathcal{E}\!/\!_T \to \mathbb{R}^3_+$$

is a bijective mapping.

Proof. If $e_1 T e_2$ then there exists an isometry $\phi : (E_0, e_1) \to (E_0, e_2)$ such that $\phi(a_i) = a_i$, and hence $A(e_1) = A(e_2)$.

$\hat{A}$ *is onto:* let us fix three positive numbers l_1, l_2, l_3 and consider two orthogonal geodesic lines γ_1 and γ_2 in $\mathbf{D}^2$ passing through a point x_0. Let x_2 be a point on γ_2 at (hyperbolic) distance l_1 from x_0, and let γ_3 be the geodesic line orthogonal to γ_2 passing through x_2. For $x \in \gamma_1$ we consider the geodesic line β_x orthogonal to γ_1 passing through x; as shown in Fig. B.20, we can find x_1 in such a way that β_{x_1} is asymptotically parallel to γ_3. We denote by $w(l_1)$ the hyperbolic distance of x_1 from x_0 (it is immediately verified that this number does not depend on the choice of the initial point x_0, of the geodesics γ_1, γ_2 and of x_2). In $\mathbf{D}^2$ we consider an arbitrary geodesic line and (for $\lambda > 0$) we fix on it a segment of length $w(l_1) + \lambda + w(l_3)$. We can perform the above construction for l_1 and l_3 starting from the endpoints of this segment, as shown in Fig. B.21.

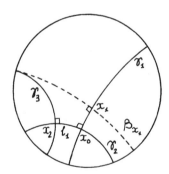

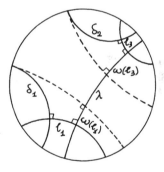

Fig. B.20. Construction of the hyperbolic hexagon having edges of given length: first step (definition of w)

Fig. B.21. Construction of the hyperbolic hexagon having edges of given length: second step and conclusion

According to the definition of w, δ_1 and δ_2 have positive distance. Moreover this distance is a continuous function of λ, which we shall denote by μ. Since

$$\lim_{\lambda \to 0} \mu(\lambda) = 0 \qquad \lim_{\lambda \to \infty} \mu(\lambda) = \infty$$

for a suitable λ we have $dist(\delta_1, \delta_2) = l_2$; according to A.5.12 there exists a distance-realizing geodesic arc from δ_1 to δ_2 orthogonal to both of them. This

completes the construction of a hyperbolic hexagon in which the lengths of
three alternate sides are l_1, l_2 and l_3. This hexagon can be identified with E_0,
and it induces on E_0 a structure e such that $A(e) = (l_1, l_2, l_3)$.

$\hat{A}$ *is one-to-one:* with the above notations, it is very easily checked that
μ is a strictly increasing function, so that λ can be chosen in a unique way.
It follows that the triple (l_1, l_2, l_3) determines uniquely the length of b_3 too
(and similarly it determines uniquely the length of b_1 and b_2). Now, recalling
B.4.12, it suffices to remark that two geodesic hexagons in $\mathbf{D}^2$ with right inner
angles and edges of the same length can be mapped one onto the other by an
isometry of $\mathbf{D}^2$, so that they are equivalent as elements of $\mathcal{E}$.   □

Remark B.4.14. Of course the mapping $e \mapsto (L^{(e)}(b_i))_{i=1,2,3}$ has the same
property as A, *i.e.* it can be used to parametrize the hyperbolic structures on
the hexagon.

Now the hyperbolic structures on an hexagon are parametrized we can go
back to the pant.

Proposition B.4.15. The mapping

$$B : \mathcal{P} \to \mathbb{R}^3_+ \qquad k \mapsto \left(L^{(k)}(\partial_1), L^{(k)}(\partial_2), L^{(k)}(\partial_3) \right)$$

induces well-defined mapping $\hat{B}$ on the quotient set. Moreover

$$\hat{B} : \mathcal{P}/S \to \mathbb{R}^3_+$$

is a bijective mapping.

Proof. Of course if $k_1 S k_2$ then $B(k_1) = B(k_2)$ so that $\hat{B}$ is well-defined.
Bijectivity is deduced from the following facts:

(1) If $k \in \mathcal{P}$ we can associate to k two elements of $\mathcal{E}$: we described above
the way k determines the subdivision of P_0 into two hexagons with right inner
angles and geodesic edges; moreover we can choose as preferred vertex (in
both the hexagons) the one corresponding to the second endpoint of $C^{(k)}_{1,3}$,
and we can give the positive orientation to the hexagon lying on the right of
$C^{(k)}_{1,3}$, and the negative one to the other.

(2) As is evident from the construction we have

$$L(a_1) = L(a'_1) = L(C^{(k)}_{1,3})$$
$$L(a_2) = L(a'_2) = L(C^{(k)}_{1,2})$$
$$L(a_3) = L(a'_3) = L(C^{(k)}_{2,3})$$

and hence by B.4.13 the two hexagons are equivalent to each other. It follows
in particular that $L(b_i) = L(b'_i)$, and hence we have $L(b_i) = L(b'_i) = L(\partial_i)/2$.

(3) By B.4.14 the parameters $L(b_i)$ for $i = 1, 2, 3$ determine uniquely the
hyperbolic structure on the hexagons, and hence on the pant. (Injectivity is
proved.)

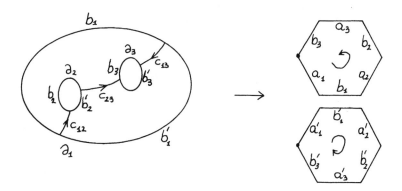

Fig. B.22. How to associate to a hyperbolic pant two (equivalent) hyperbolic hexagons

(4) Given $l_1, l_2, l_3 > 0$ we can find a hyperbolic hexagon E (with geodesic edges, right angles, a fixed orientation and a preferred vertex) such that for $l(b_i) = {}^{l_i}\!/_2$. Then we only have to glue two copies of this hexagon along the a_i's, starting at the preferred vertex and following the orientation: the outcome is a hyperbolic structure on P_0 such that $L(\partial_i) = l_i$. (Surjectivity is proved.) □

We recall that a function $L : \tau_g \to \mathbb{R}_+^{3(g-1)}$ was defined in B.4.9.

Proposition B.4.16. L is onto.

Proof. Choose arbitrarily $l_i > 0$ for $i = 1, ..., 3(g-1)$.

For $1 \le i_1, i_2, i_3 \le 3(g-1)$, according to B.4.15, we can find a hyperbolic structure on a pant with edges ∂_{i_j} $(j = 1, 2, 3)$ in such a way that $L(\partial_{i_j}) = l_{i_j}$. In particular we can endow each pant appearing in the decomposition of M_0 performed via the α_i's (fixed at the beginning), with a hyperbolic structure in such a way that each α_i have length l_i in both the pants it is an edge of. Now we only have to glue these structures together: as we already remarked in B.4.10, this is possible since we only need to identify geodesic edges having the same length, and we can do this simply by picking two points arbitrarily and then go on with the arc length as parameter towards the positive direction of the loops. Of course the resulting hyperbolic structure h on M_0 satisfies the condition $L(h) = (l_1, , ..., l_{3(g-1)})$. □

The study of τ_g is now almost complete; we anticipate heuristically the meaning of the following results.

The mapping L is onto, so we only have to study its fibers; since the length of the edges determines uniquely a hyperbolic structure on a pant, we must see how many different hyperbolic structures can be obtained by glueing $2(g-1)$ hyperbolic pants. As we saw above, the only way to glue together two edges is to choose a point and then proceed: so the only thing we can do is to twist of a certain angle one of the edges before glueing it to the other; and we can do *this for each of the $3(g-1)$ pairs we must glue, so that we have to consider $3(g-1)$ real parameters. With this method we certainly obtain all the possible structures corresponding to a fixed value of L, and it turns out that these structures are pairwise non-equivalent, so that τ_g can be parametrized by*

$$\mathbb{R}_+^{3(g-1)} \times \mathbb{R}^{3(g-1)}.$$

We turn to a more precise discussion of the above ideas. We shall denote by $\sigma : \mathbb{R}_+^{3(g-1)} \to \tau_g$ an arbitrary mapping such that $L \circ \sigma = \text{id}$ (as we will see, σ could be defined in an intrinsic way, but we only need now to know that it exists, which follows from the axiom of choice).

Theorem B.4.17. There exists an action Θ of $\mathbb{R}^{3(g-1)}$ on τ_g such that the mapping

$$\Psi : \mathbb{R}_+^{3(g-1)} \times \mathbb{R}^{3(g-1)} \to \tau_g \qquad (l, \vartheta) \mapsto \Theta_\vartheta(\sigma(l))$$

is bijective.

Proof. We shall define Θ in such a way that for each fixed l the mapping

$$\mathbb{R}^{3(g-1)} \ni \vartheta \mapsto \Theta_\vartheta(\sigma(l)) \in L^{-1}(l)$$

is a bijection (of course this is enough).

For $\vartheta \in \mathbb{R}^{3(g-1)}$ we shall first define (improperly) Θ_ϑ as a mapping of $\mathcal{H}$ onto itself: afterwards we shall check that the Θ_ϑ is well-defined on the quotient set τ_g. Let $h \in \mathcal{H}$ be a fixed hyperbolic metric on M_0, and let $\{\gamma_i\}$ be the h-geodesic loops giving the pant decomposition of M_0 (recall B.4.8). The metric $\Theta_\vartheta(h)$ will differ from h only in a small neighborhood of the γ_i's. Let us fix i. For $x \in \gamma_i$ we denote by δ_x the geodesic line orthogonal to γ_i starting at x in such a way that the pair $(\dot{\gamma}_i(x), \dot{\delta}_x(x))$ is a positive one; we shall assume that both γ_i and δ_x are parametrized by arc length. Since γ_i is compact we can find $\varepsilon > 0$ such that the mapping

$$\gamma_i \times [0, 3\varepsilon] \ni (x, t) \mapsto \delta_x(t)$$

is a diffeomorphism onto its range $\mathcal{C}_i$ (we have constructed a collar based on γ_i, i.e. a half tubular neighborhood of γ_i).

We define now a diffeomorphism ϕ_i (depending on ϑ_i) of $\mathcal{C}_i$ onto itself; of course it suffices to define ϕ_i on $\gamma_i \times [0, 3\varepsilon]$ instead of $\mathcal{C}_i$. Consider the universal covering $\mathbb{R}$ of γ_i, chosen in such a way that the projection is an orientation-preserving local isometry; remark that γ_i is obtained from $\mathbb{R}$ under the action of the translation $x \mapsto x + a$, where a is the length of γ_i. Let us consider a diffeomorphism $\tilde{\phi}_i$ of the universal covering $\mathbb{R} \times [0, 3\varepsilon]$ of $\gamma_i \times [0, 3\varepsilon]$ onto itself with the following properties:
(1) $\tilde{\phi}_i(t, s) = (t + \vartheta_i, s)$ for $s \leq \varepsilon$;
(2) $\tilde{\phi}_i(t, s) = (t, s)$ for $s \geq 2\varepsilon$;
(3) $\tilde{\phi}_i(t_1 + t_2, s) = \phi_i(t_1, s) + (t_1, 0)$ for all t_1, t_2, s.

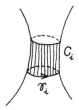

Fig. B.23. A collar based on a geodesic loop

Fig. B.24. Action of $\tilde{\phi}_i$

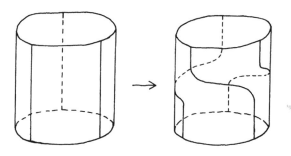

Fig. B.25. Action of ϕ_i

(Of course such a $\tilde{\phi}_i$ exists: Fig. B.24 suggests how to construct it.)

According to (3), $\tilde{\phi}_i$ induces a diffeomorphism ϕ_i of $\gamma_i \times [0, 3\varepsilon]$ onto itself.

If we consider the metric h on $\mathcal{C}_i$ it easily follows from the definition that ϕ_i is an isometry in the neighborhood of γ_i and is the identity in the neighborhood of the other component of the boundary of the collar; hence if ϕ_i^* denotes the pull-forward, the metric

$$h_i' = \begin{cases} h & \text{outside } \mathcal{C}_i \\ \phi_i^*(h) & \text{on } \mathcal{C}_i \end{cases}$$

is well-defined. Of course we can take the $3(g-1)$ collars to be pairwise non-intersecting: then the above procedure applied to all the collars produces a new hyperbolic metric on M_0 which we shall call $\Theta_\vartheta(h)$. Remark that $\Theta_\vartheta(h)$

is well-defined only in τ_g (and not in $\mathcal{H}$) since different choices of the ε's and the ϕ_i's produce different (but obviously equivalent) metrics.

Conversely it is easily checked that if h_1 and h_2 are equivalent metrics then the same holds for $\Theta_\vartheta(h_1)$ and $\Theta_\vartheta(h_2)$: in fact if $\phi : (M_0, h_1) \to (M_0, h_2)$ is an isometry isotopic to the identity and if $\{C_i\}$ are collars for h_1, then $\{\phi(C_i)\}$ are collars for h_2; if ϕ_i is the function used to perturb h_1 on C_i then

$$\phi|_{C_i} \circ \phi_i \circ \left(\phi|_{C_i}\right)^{-1}$$

can be used to perturb h_2 on $\phi(C_i)$. This construction of $\Theta_\vartheta(h_1)$ and $\Theta_\vartheta(h_2)$ satisfies

$$\Theta_\vartheta(h_2) = \phi_*(\Theta_\vartheta(h_1)),$$

hence ϕ itself provides the equivalence between the two metrics.

The definition of Θ is now complete. Let us remark that the γ_i's of the above construction are geodesics for the new metric too, and their length does not change, so that

$$L(\Theta_\vartheta(h)) = L(h) \qquad \forall \vartheta.$$

Hence Θ operates on each fiber of L.

Surjectivity of Ψ is now easily established: in fact, given any element of τ_g, a hyperbolic structure is determined on each pant of the fixed decomposition. The hyperbolic structure on M_0 is then obtained by glueing the structures on the pants along the edges, and the action Θ provides all the possible twists the edges can be given before being glued up.

Injectivity of Ψ is a much more difficult task. Since Θ keeps the fibers of L invariant, it suffices to prove that for $h \in \tau_g$ and $\vartheta \neq \vartheta'$, the metrics $\Theta_\vartheta(h)$ and $\Theta_{\vartheta'}(h)$ are not equivalent. If β is a non-trivial loop in M_0 we shall denote by $\tilde{\Lambda}^h(\beta)$ the length with respect to h of the only h-geodesic loop in the free-homotopy class of β. Let us assume for the moment the following:

Lemma B.4.18. $\forall i = 1, ..., 3(g-1)$ the loop β_i described in Fig. B.26 satisfies the following: $\forall h \in \tau_g$ the function

$$\mathbb{R} \ni \vartheta_i \mapsto \tilde{\Lambda}^{\Theta_{0...\vartheta_i...0}(h)}(\beta_i)$$

is strictly convex and has a minimum.

We can complete the proof of B.4.17: assume by contradiction that $h \in \tau_g$, $\vartheta \neq \vartheta'$ and $\Theta_\vartheta(h) = \Theta_{\vartheta'}(h)$. Then we can find i such that $\vartheta_i \neq \vartheta_i'$. If $\hat{\vartheta}$ and $\hat{\vartheta}'$ represent the vectors ϑ and ϑ' where the i-th element is replaced by 0, we set

$$h_1 = \Theta_{\hat{\vartheta}}(h) \qquad h_2 = \Theta_{\hat{\vartheta}'}(h)$$

so that

$$\Theta_{0...\vartheta_i...0}(h_1) = \Theta_{0...\vartheta_i'...}(h_2)$$

Hence for every integer n

$$\tilde{\Lambda}^{\Theta_{0...n+\vartheta_i...0}(h_1)}(\beta_i) = \tilde{\Lambda}^{\Theta_{0...n+\vartheta_i'...0}(h_2)}(\beta_i)$$

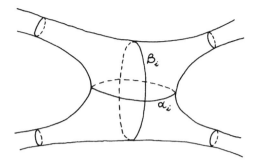

Fig. B.26. The fixed loop whose length defines a strictly convex function under the action of the twist

Since (by definition of Θ) h_1 and h_2 coincide with h in the neighborhood of α_i, in the above formula we can substitute h_1 and h_2 by h. Hence:

$$\tilde{\Lambda}^{\Theta_0 \dots n + \vartheta_i \dots 0(h)}(\beta_i) = \tilde{\Lambda}^{\Theta_0 \dots n + \vartheta'_i \dots 0(h)}(\beta_i).$$

Since a strictly convex function with minimum is strictly increasing on a positive half-line, B.4.18 implies that for n is sufficiently large this formula cannot hold.   □

For the proof of B.4.18 we shall need the following technical result:

Lemma B.4.19. If M is a connected oriented compact hyperbolic surface and γ is a non-trivial simple geodesic loop in M, two different liftings of γ to $\mathbb{IH}^2$ cannot meet in the whole $\overline{\mathbb{IH}^2}$.

Proof. Let $\tilde{\gamma}_1$ and $\tilde{\gamma}_2$ be such liftings.

Certainly $\tilde{\gamma}_1$ and $\tilde{\gamma}_2$ do not meet in $\mathbb{IH}^2$, otherwise γ would not be simple.

In $\mathbb{III}^{2,+}$ assume that $\tilde{\gamma}_1$ and $\tilde{\gamma}_2$ have ∞ as common endpoint. Then the associated isometries T_1 and $T_2 \in \mathrm{II}_1(M)$ have ∞ as common fixed point, whence

$$T_1(x,t) = \lambda(x,t)$$
$$T_2(x,t) = \mu(x + a, t)$$

for some $\lambda, \mu > 0$, $a \neq 0$. By B.4.3 and B.4.5 T_1 and T_2 lie in the same conjugacy class in $\mathrm{II}_1(M)$, i.e. $T_2 = S^{-1} \circ T_1 \circ S$ for some $S \in \mathrm{II}_1(M)$. It easily follows that $\lambda = \mu$; let us assume that $\lambda = \mu > 1$ (otherwise we replace T_1 and T_2 by their inverses). Then for $c = \dfrac{\mu a}{(1 - \mu)}$

$$\lim_{n \to \infty} d\big(T_1^n(0,1), T_2^n(c,1)\big) = \lim_{n \to \infty} d\big((0, \lambda^n), (c, \lambda^n)\big) = 0$$

and hence the sequence $(T_2^{-n} \circ T_1^n)(0,1)$ is not discrete in $\mathbb{IH}^2$, and this is absurd.   □

Proof of B.4.18. We omit the index i everywhere. We set

$$F(\vartheta) = \tilde{\Lambda}^{\Theta_0 \dots \vartheta \dots 0\,(h)}(\beta).$$

Let $\pi : \mathbb{H}^2 \to M_0$ be a fixed universal covering, corresponding to the structure h on M_0. Let $T \in \mathcal{I}(\mathbb{H}^2)$ induce the loop β, let δ be the only T-invariant geodesic line in $\mathbb{H}^2$ and choose a point $x_0 \in \pi^{-1}(\beta \cap \gamma)$ ($\gamma = \gamma_i$ is the i-th h-geodesic loop giving the pant decomposition of M_0, as in B.4.8). Let $\tilde{\gamma}^1$ and $\tilde{\gamma}^3$ be liftings of γ through x_0 and $T(x_0)$ respectively. Since the intersection of β and γ consists of at least two points there exists another lifting $\tilde{\gamma}^2$ of γ meeting the open segment with endpoints x_0 and $T(x_0)$ in a point y_0. We remark that by B.4.19 the lines $\tilde{\gamma}^i$ do not meet each other (not even on the boundary of $\mathbb{H}^2$).

Figure B.27 represents the situation and includes the strips obtained as liftings of the collar where the twist can be performed. We remark that the h-geodesic loop in the free-homotopy class of β is simply $\pi(\delta)$. Let us consider now on M the metric $\Theta_\vartheta(h)$; we shall denote by $\mathbb{H}^2_*$ the hyperbolic plane endowed with the pull-back metric of $\Theta_\vartheta(h)$ with respect to π, and by d^* the distance associated to such a metric. Since $\Theta_\vartheta(h)$ differs from h only in the collar, the restriction of $\pi : \mathbb{H}^2 \to M$ is an isometry outside the strips. Hence the lifting of the $\Theta_\vartheta(h)$-geodesic loop β' in the class of β looks more or less as represented in Fig. B.28.

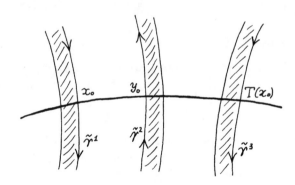

Fig. B.27. Lifting to the hyperbolic plane of the collar and the geodesic in the fixed homotopy class

Let us remark first that $z_1 = T(x_1)$ (the projection of the geodesic must be a loop in the class of β). Moreover we have

$$F(\vartheta) = L^{\Theta_\vartheta(h)}(\beta') = d(x_1, y_1) + d^*(y_1, T(x_1))$$

(remark that $d^*(x_1, y_1) = d(x_1, y_1)$).

For $x \in \tilde{\gamma}^1$ let l_x be the geodesic segment in $\mathbb{H}^2_*$ joining x to $T(x)$, and let $y(x)$ be the intesection of l_x and $\tilde{\gamma}^2$ (this intersection exists and is unique). The shortest of the l_x's is orthogonal to both $\tilde{\gamma}^1$ and $\tilde{\gamma}^3$, so that it induces in the quotient a $\Theta_\vartheta(h)$-geodesic loop in the class of β, i.e. β'. It follows that

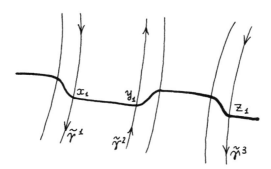

Fig. B.28. The same situation as in Fig. B.27 after the twist

$$F(\vartheta) = \inf_{x \in \tilde{\gamma}^1} \left\{ d\big(x, y(x)\big) + d^*\big(y(x), T(x)\big) \right\}.$$

Moreover it is obvious that

$$d\big(x, y(x)\big) + d^*\big(y(x), T(x)\big) = \inf_{y \in \tilde{\gamma}^2} \left\{ d(x, y) + d^*\big(y, T(x)\big) \right\}.$$

Now, it follows from the definition of Θ that if $y + \vartheta$ denotes the point on $\tilde{\gamma}^2$ at distance ϑ from y towards the positive direction on $\tilde{\gamma}^2$, and $T(x) + \vartheta \in \tilde{\gamma}^3$ is defined in the same way, then

$$d^*\big(y, T(x)\big) = d\big(y + \vartheta, T(x) + \vartheta\big).$$

All these facts imply that if we set for $x \in \tilde{\gamma}^1$, $y \in \tilde{\gamma}^2$ and $\vartheta \in \mathbb{R}$

$$f(x, y, \vartheta) = d(x, y) + d\big(y + \vartheta, T(x) + \vartheta\big)$$

then we have

$$F(\vartheta) = \inf_{x \in \tilde{\gamma}^1, y \in \tilde{\gamma}^2} f(x, y, \vartheta).$$

Suppose a divergent sequence $\{(x_n, y_n, \vartheta_n)\}$ is given; then we have the following possibilities:

(i) either $\{x_n\}$ or $\{y_n\}$ diverge: then, since $\tilde{\gamma}^1$ and $\tilde{\gamma}^2$ do not intersect in the whole $\overline{\mathbb{H}^2}$, we have

$$d(x_n, y_n) \to \infty;$$

(ii) both $\{x_n\}$ and $\{y_n\}$ remain inside a compact set. Then the same holds for $\{T(x_n)\}$, and hence $\{\vartheta_n\}$ diverges, so that

$$d\big(y_n + \vartheta_n, T(x_n) + \vartheta_n\big) \to \infty.$$

It follows that f is a proper mapping.

Moreover A.6.1 implies that f is strictly convex in all its arguments (the convex combination being canonically defined on the lines $\tilde{\gamma}^1$ and $\tilde{\gamma}^2$): in fact

$$x \mapsto d(x,y) \qquad\qquad x \mapsto d\big(y + \vartheta, T(x) + \vartheta\big)$$
$$y \mapsto d(x,y) \qquad\qquad y \mapsto d\big(y + \vartheta, T(x) + \vartheta\big)$$

are strictly convex because of Corollary A.6.2 (we only need to remark, for the second function, that $T : \tilde{\gamma}^1 \to \tilde{\gamma}^3$ preserves convex combinations), while strict convexity of

$$\vartheta \mapsto d\big(y + \vartheta, T(x) + \vartheta\big)$$

exploits the whole A.6.1.

We are now ready for the conclusion: f has a minimum k, so that F has minimum k too. Convexity of f in ϑ implies that F is convex too, while an easy argument based on the fact that $\forall\, \vartheta$ there exists a unique pair (x,y) with

$$F(\vartheta) = f(x,y,\vartheta)$$

proves that convexity is strict. $\square$

Remark B.4.20. We remarked in B.4.2 that τ_g can be naturally endowed with a topology. Almost all the constructions we made for the classification of τ_g can be quite easily proved to be continuous with respect to this topology: the only serious problem arises for the definition in a continuous way of the section $\sigma : \mathbb{R}_+^{3(g-1)} \to \tau_g$ needed in B.4.17. The problem is be solved as soon as an intrinsic (and continuous) choice is made of the points where to start glueing the edges of the pants. This is achieved in the following way: the isotopy classes of the (oriented) loops β_i of B.4.18 are fixed from the beginnig. Then, once we are given a hyperbolic structure on the j-th pant, having ∂_i as boundary component, we double it (the glueing on the edges being given by the identity) and then we consider the geodesic loop δ on the resulting surface corresponding to the free-homotopy class of the double of the intersection of β_i with the pant.

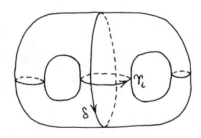

Fig. B.29. A fixed loop allowing an intrinsic choice of the starting point for the glueing between the pants

The intersection of δ with γ_i consists of two points: moreover the pair $(\dot{\gamma}_i, \dot{\delta})$ is positive in one of this points and negative in the other, so that we can choose without ambiguity one of them. This method allows us to fix a point on each boundary component of the pants, and hence we can glue starting from these points.

The above remark (and some more work) allow to prove the following deeper version of Theorem B.4.17:

Theorem B.4.21. The mapping $L : \tau_g \to \mathbb{R}_+^{3(g-1)}$ defined in B.4.8 is the projection associated to a principal fiber bundle structure on τ_g with fiber $\mathbb{R}^{3(g-1)}$.

We refer now to Remark B.4.1. We set

$$\mathrm{Mod}_g = \mathrm{Diff}^+(M_0)\big/_{\mathrm{Diff}^0(M_0)} \qquad \mathcal{M}_g = \tau_g\big/_{\mathrm{Mod}_g}$$

(Mod_g is called the _modular group_, and it operates in a natural way on τ_g: $\mathcal{M}_g$ is the space of moduli of hyperbolic structures on M_0).

As we mentioned at the beginning the structure of $\mathcal{M}_g$ is quite complicated. We state without proof the following result due to Royden [Roy]:

Theorem B.4.22. τ_g can be naturally endowed with a complex structure in such a way that $\mathrm{Aut}(\tau_g) = \mathrm{Mod}_g$.

Moreover it may be shown that Mod_g operates on τ_g in a properly discontinuous way, but not freely. Hence $\mathcal{M}_g$ has singularities corresponding to the points of τ_g which are kept fixed for some non-trivial element of Mod_g, i.e. to the hyperbolic structures on M_0 for which there exists some orientation-preserving auto-isometry non-isotopic to the identity. These singularities cannot be "arbitrarily complicated", as follows from the next result due to Hurwitz (see for instance [Si]): for the sake of completeness we include a sketch of a proof (to be filled in as an exercise) fully in the realm of hyperbolic geometry.

Theorem B.4.23. Every hyperbolic structure on M_0 admits at most $84(g-1)$ non-trivial orientation-preserving auto-isometries.

Sketch of the proof. Given a hyperbolic structure on M_0, we identify as usual $\Pi_1(M_0)$ with a discrete subgroup Γ of $\mathcal{I}(\mathbb{H}^2)$, so that $M_0 = \mathbb{H}^2/\Gamma$. Consider the normalizer $N(\Gamma)$ of Γ in $\mathcal{I}(\mathbb{H}^2)$. Then it is possible to check that $N(\Gamma)$ is discrete (but it need not act freely) and that the index $[N(\Gamma) : \Gamma]$ of Γ in $N(\Gamma)$ is finite. This fact means geometrically that the quotient set

$$M' = \mathbb{H}^2/N(\Gamma)$$

is a "hyperbolic surface with singularities" and the natural projection

$$M_0 = \mathbb{H}^2/\Gamma \to M' = \mathbb{H}^2/N(\Gamma)$$

is a branched covering of finite degree equal to $[N(\Gamma) : \Gamma]$. Then we have that

$$[N(\Gamma) : \Gamma] = \frac{\mathcal{A}(M_0)}{\mathcal{A}(M')}.$$

Then one proves (and this is the most difficult step of the proof) that for any discrete group G of $\mathbb{H}^2$ the following universal inequality holds:

$$\mathcal{A}\left(\mathbb{H}^2/G\right) \geq \frac{\pi}{21}.$$

The conclusion easily follows from B.3.3 and from the fact that $\mathcal{I}(M_0)$ can be naturally identified with $N(\Gamma)/\Gamma$. $\qquad\qquad\square$

Finally, it follows from B.4.22 and B.4.23 that $\mathcal{M}_g$ is a complex analytic space of dimension $3(g-1)$, and hence in particular the following holds:

Corollary B.4.24. M_0 can be endowed with uncountably many pairwise non-isometric hyperbolic structures.

Chapter C.
The Rigidity Theorem
(Compact Case)

In this chapter we are going to prove that there is a very sharp difference between 2-dimensional hyperbolic geometry and higher dimensions (at least for the compact case, but the results we shall prove generalize to the case of finite volume). Namely, we shall prove that for $n \geq 3$ a connected, compact, oriented n-manifold supports at most one (equivalence class of) hyperbolic structure (while it was proved in Chapt. B that a compact surface of genus at least 2 supports uncountably many non-equivalent hyperbolic structures). This is the famous Mostow rigidity theorem: the original proof can be found in [Mos], and others (generalizing the first one) in [Mar] and [Pr]; we shall refer mostly to [Gro3], [Th1, ch. 6] and [Mu]. The core of the proof we present resides in Theorem C.4.2, relating the Gromov norm (introduced in C.3) to the volume of a compact hyperbolic manifold; this result has a deep importance independently of the rigidity theorem: in Chapters E and F we shall meet interesting applications and related ideas.

We shall state the main theorem first, and then we shall carry out the very long proof in rest of the chapter. In the last section we shall also prove a few corollaries of the rigidity theorem and state without proof the generalization to the non-compact case.

We recall that if X and Y are topological spaces, a continuous mapping $f : X \to Y$ is called a <u>homotopy equivalence</u> if there exists a continuous mapping $g : Y \to X$ such that both $f \circ g$ and $g \circ f$ are homotopic to the identity (of the space they are defined on). We shall say that g is a <u>homotopy inverse</u> of f.

Theorem C.0 (Mostow's Rigidity Theorem). Let $n \geq 3$ and let M_1 and M_2 be n-dimensional connected, compact, oriented manifolds endowed with a hyperbolic structure. If $f : M_1 \to M_2$ is a homotopy equivalence then there exists an isometry $\phi : M_1 \to M_2$ homotopic to f.

(For an apparently stronger but actually equivalent statement see also Theorem C.5.2.)

In the sequel we shall keep the notations of the above theorem fixed. Almost all the partial results we shall prove hold even for $n = 2$; only in Sect. 5 we shall need $n \geq 3$. In case the reader wishes to know the plan of the proof before following the separate steps we suggest that he writes down, in order, the statements of Proposition C.1.2, Theorem C.2.1, Proposition C.4.1 and Proposition C.5.1, and then conclude as in Sect. C.5.

C.1 First Step of the Proof: Extension of Pseudo-isometries

Lemma C.1.1. The mapping

$$f_* : \Pi_1(M_1) \to \Pi_1(M_2) \qquad \langle \alpha \rangle \mapsto \langle f \circ \alpha \rangle$$

is a group isomorphism.

Proof. It is well-known that f_* and g_* (defined in the same way) are group homomorphisms, and moreover $f_* \circ g_* = (f \circ g)_*$ and $g_* \circ f_* = (g \circ f)_*$, so that the lemma is deduced from the following assertion:

if X is a topological space, $x \in X$ and $\phi : X \to X$ is a continuous mapping homotopic to the identity, then $\phi_* : \Pi_1(X, x) \to \Pi_1(X, \phi(x))$ is a group isomorphism.

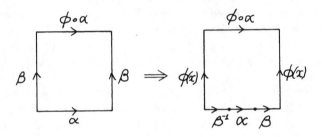

Fig. C.1. Homotopy equivalences induce isomorphisms of fundamental groups

Let F be such a homotopy and define $\beta(t) = F(t, x)$. If α is a loop at x we have the homotopies represented in Fig. C.1 which imply that

$$\phi_*(\langle \alpha \rangle) = \langle \beta^{-1} \cdot \alpha \cdot \beta \rangle,$$

so that ϕ_* is an isomorphism. $\qquad\square$

For $i = 1, 2$ we fix a universal covering of M_i as a projection $\pi_i : \mathbf{H}^n \to M_i$, so that $\Pi_1(M_i)$ is identified with a subgroup Γ_i of $\mathcal{I}^+(\mathbf{H}^n)$. We shall keep calling f_* the isomorphism $\Gamma_1 \to \Gamma_2$ induced by f; let $\tilde{f}$ be a fixed lift of f to the universal cover, *i.e.* a smooth mapping such that the diagram

$$
\begin{array}{ccc}
\mathbf{H}^n & \xrightarrow{\ \tilde{f}\ } & \mathbf{H}^n \\
\Big\downarrow{\pi_1} & & \Big\downarrow{\pi_2} \\
M_1 & \xrightarrow{\ f\ } & M_2
\end{array}
$$

is commutative. We choose the starting point of the lift in such a way that f_* and $\tilde{f}$ are related by the fact that

$$\tilde{f} \circ \gamma = f_*(\gamma) \circ \tilde{f} \qquad \forall \gamma \in \Gamma_1$$

(of course this is possible). Similarly $\tilde{g}$ will be a lift of g such that

$$\tilde{g} \circ \gamma = g_*(\gamma) \circ \tilde{g} \qquad \forall \gamma \in \Gamma_2.$$

If X and Y are metric spaces we shall say a mapping $h : X \to Y$ is a pseudo-isometry if there exist two positive constants c_1, c_2 such that

$$\frac{1}{c_1} \cdot d(x_1, x_2) - c_2 \leq d\big(h(x_1), h(x_2)\big) \leq c_1 \cdot d(x_1, x_2) \qquad \forall x_1, x_2 \in X.$$

This section is devoted to the proof of the following fact:

Proposition C.1.2. We can assume (without changing the homotopy class of f) that $\tilde{f}$ is a pseudo-isometry. If this is the case $\tilde{f}$ can be extended to a continuous mapping $\overline{\mathbf{H}^n} \to \overline{\mathbf{H}^n}$ whose restriction to the boundary is one-to-one, in such a way that the relation

$$\tilde{f} \circ \gamma = f_*(\gamma) \circ \tilde{f} \qquad \forall \gamma \in \Gamma_1$$

holds in the whole $\overline{\mathbf{H}^n}$.

For the first part of C.1.2 we shall need the notion of fundamental domain, which will be used also in Sect. 4. If G is a group operating on a topological space X, a Borel subset D of X is called a fundamental domain for G if

$$\bigcup_{g \in G} g(D) = X \qquad g(D) \bigcap D \subseteq \partial D \qquad \forall g \neq \mathrm{id}.$$

Proposition C.1.3. If $M = \mathbf{H}^n/\Gamma$ is a complete hyperbolic manifold and $x_0 \in \mathbf{H}^n$, the set

$$D = \big\{ x \in \mathbf{H}^n : d(x, x_0) \leq d(x, \gamma(x_0)) \; \forall \gamma \in \Gamma \big\}$$

is a fundamental domain for Γ. Moreover $\mathrm{vol}(D) = \mathrm{vol}(M)$ and if M is compact then D is a compact geodesic polyhedron.

Proof. For $y_1, y_2 \in \mathbf{H}^n$ the set

$$\big\{ y \in \mathbf{H}^n : d(y, y_1) = d(y, y_2) \big\}$$

is easily recognized to be a hyperbolic hyperplane (in $\mathrm{III}^{n,+}$ we can assume y_1 and y_2 have the same height, and the conclusion is immediate).

Since D is the intersection of closed hyperbolic half-spaces it is certainly closed, and hence it is a Borel set. Moreover for $\gamma \in \Gamma$ we have

$$\gamma(D) = \big\{ x \in \mathbf{H}^n : d(x, \gamma(x_0)) \leq d(x, \delta(x_0)) \; \forall \delta \in \Gamma \big\}.$$

As Γ is discrete, for $x \in \mathbf{H}^n$ the set

$$\big\{ d(x, \delta(x_0)) : \delta \in \Gamma \big\}$$

has a minimum $d(x, \delta_0(x_0))$, which implies that $x \in \delta_0(D)$, and hence it is proved that the orbit of D covers $\mathbf{H}^n$.

Finally, for $\gamma \in \Gamma \setminus \{\mathrm{id}\}$

$$\gamma(D) \cap D \subseteq D \cap \{x \in \mathbf{H}^n : d(x, x_0) = d(x, \gamma(x_0))\} \subseteq \partial D.$$

This implies that D is a fundamental domain.

Since the projection $\pi : \mathbf{H}^n \to M$ is a local isometry such that its restriction to $\overset{\circ}{D}$ is one-to-one and its restriction to D is onto, we have obviously $\mathrm{vol}(D) = \mathrm{vol}(M)$.

Assume now that M is compact; we start by proving that D is compact. This is easily checked since M contains a dense open subset isometric to $\overset{\circ}{D}$, and hence D must necessarily have finite diameter, which implies that D is a closed bounded subset of $\mathbf{H}^n$, $i.e.$ a compact set. Let $r = diam(D)$; the set

$$\{\gamma \in \Gamma : d(x_0, \gamma(x_0)) \leq 4r\}$$

is a finite set $\{\gamma_1, ..., \gamma_m\}$. We shall check now that

$$D = \{x \in \mathbf{H}^n : d(x, x_0) \leq d(x, \gamma_i(x_0)) \text{ for } i = 1, ..., m\}.$$

For $\gamma \in \Gamma$ let

$$H_\gamma = \{x \in \mathbf{H}^n : d(x, x_0) = d(x, \gamma(x_0))\}$$
$$S_\gamma = \{x \in \mathbf{H}^n : d(x, x_0) \leq d(x, \gamma(x_0))\}.$$

Since $D \subset B_r(x_0)$ then obviously

$$D = \bigcap \{S_\gamma : H_\gamma \cap B_{2r}(x_0) \neq \phi\}$$

and then it suffices to remark that if $H_\gamma \cap B_{2r}(x_0) \neq \phi$ then $d(x_0, \gamma(x_0)) \leq 4r$.

We have checked that D is the intersection a finite number of closed hyperbolic half-spaces, and then it is a geodesic polyhedron. $\qquad\square$

Lemma C.1.4. We can assume $\tilde{f}$ is a pseudo-isometry.

Proof. Since M_1 and M_2 are compact every continuous mapping $M_1 \to M_2$ is homotopic to a differentiable one (see [Mi2]), hence we can assume both f and its homotopy inverse g are of class C^1. Compactness implies that the norms of df and dg are uniformly bounded by a constant c_1, so that the same holds for $d\tilde{f}$ and $d\tilde{g}$, and hence

$$d(\tilde{f}(x_1), \tilde{f}(x_2)) \leq c_1 d(x_1, x_2) \; \forall x_1, x_2 \in \mathbf{H}^n$$
$$d(\tilde{g}(x_1), \tilde{g}(x_2)) \leq c_1 d(x_1, x_2) \; \forall x_1, x_2 \in \mathbf{H}^n.$$

Now, let us remark that our choices of the basepoints for $\tilde{f}$ and $\tilde{g}$ imply that $g_* = f_*^{-1}$, so that for $\gamma \in \Gamma_1$ we have

$$\tilde{g} \circ \tilde{f} \circ \gamma = \tilde{g} \circ f_*(\gamma) \circ \tilde{f} = (g_* \circ f_*)(\gamma) \circ \tilde{g} \circ \tilde{f} = \gamma \circ \tilde{g} \circ \tilde{f}$$

$i.e.$ $\tilde{g} \circ \tilde{f}$ commutes with all the elements of Γ_1. Since, by C.1.3, Γ_1 admits a compact fundamental domain, it follows that for some $b > 0$

$$d\big(y, (\tilde{g} \circ \tilde{f})(y)\big) \leq b \ \forall y \in \mathbf{H}^n$$

whence

$$d\big((\tilde{g} \circ \tilde{f})(y_1), (\tilde{g} \circ \tilde{f})(y_2)\big) \geq d(y_1, y_2) - 2b \ \forall y_1, y_2 \in \mathbf{H}^n$$

and then

$$d\big(\tilde{f}(x_1), \tilde{f}(x_2)\big) \geq \frac{1}{c_1} d\big((\tilde{g} \circ \tilde{f})(x_1), (\tilde{g} \circ \tilde{f})(x_2)\big) \geq \frac{1}{c_1} d(x_1, x_2) - \frac{2b}{c_1}$$

and the lemma is proved.   □

In the remainder of this section we will use the symbol P instead of $\tilde{f}$: P is a fixed pseudo-isometry of $\mathbf{H}^n$ with constants c_1 and c_2; our aim is to prove that P can be extended to the boundary in the required way.

For $A \subseteq \mathbf{H}^n$ and $r > 0$ we shall set $N_r(A) = \big\{x \in \mathbf{H}^n : d(x, A) < r\big\}$; if γ is a geodesic line in $\mathbf{H}^n$ we shall denote by π_γ the orthogonal projection of $\mathbf{H}^n$ onto γ. For $x, y \in \mathbf{H}^n$ we shall denote by $[x, y]$ the closed geodesic arc joining x to y and by $\gamma(x, y)$ the entire geodesic line passing through them.

Lemma C.1.5. If α is a geodesic line in $\mathbf{H}^n$ and $p, q \in \mathbf{H}^n$ lie at the same distance s from α then

$$d(p, q) \geq \cosh(s) \cdot d\big(\pi_\alpha(p), \pi_\alpha(q)\big).$$

Proof. We consider the hyperboloid model $\mathbf{I}^n$; according to A.5.1 α is the intersection of $\mathbf{I}^n$ with a linear 2-subspace L of $\mathbb{R}^{n+1}$. We shall denote by W the orthogonal space to L (with respect to $\langle . | . \rangle_{(n,1)}$) and by S the unit sphere in W. Moreover we shall denote by $C_s(\alpha)$ the set of all points in $\mathbf{I}^n$ lying at distance s from α; A.5.1 again implies that the mapping

$$\zeta : \alpha \times S \to C_s(\alpha) \qquad (u, w) \mapsto \cosh(s) \cdot u + \sinh(s) \cdot w$$

is a bijection, and it easily follows that it is a diffeomorphism. If $u' \in L$ and $w' \in W$ we have

$$d_{(u,w)}\zeta(u', w') = \cosh(s) \cdot u' + \sinh(s) \cdot w'$$
$$\big\|d_{(u,w)}\zeta(u', w')\big\|_{(n,1)} = \cosh^2(s) \cdot \|u'\| + \sinh^2(s) \cdot \|w'\| \geq \cosh^2(s) \cdot \|u'\|$$

which implies that

$$d\big(\zeta(u_1, w_1), \zeta(u_2, w_2)\big) \geq \cosh(s) \cdot d(u_1, u_2) \qquad \qquad □$$

Lemma C.1.6. We can find a constant $r > 0$ such that for each geodesic line β in $\mathbf{H}^n$ there exists a unique geodesic line $A(\beta)$ with

$$P(\beta) \subset N_r\big(A(\beta)\big).$$

Proof. Let β be a fixed line (we shall check that the constant r we are going to find is independent of β). We start by proving that there exists a constant t depending only on c_1 and c_2 such that

$$P\big([q_1, q_2]\big) \subset N_t\big(\gamma(P(q_1), P(q_2))\big) \quad \forall\, q_1, q_2 \in \mathbb{H}^n \qquad (*)$$

Let q_1, q_2 be fixed and set $p_1 = P(q_1)$, $p_2 = P(q_2)$, $\gamma = \gamma(p_1, p_2)$. Let $s_0 > 0$ be such that $\cosh(s_0) = c_1^2 + 1$. If $[a, b]$ is a connected component of

$$[q_1, q_2] \bigcap P^{-1}\big(N_{s_0}(\gamma)\big)$$

then, by continuity of P we have

$$d\big(P(a), \gamma\big) = d\big(P(b), \gamma\big) = s_0.$$

It follows that

$$\frac{1}{c_1} \cdot d(a, b) - c_2 \leq d(P(a), P(b)) \leq$$
$$\leq d\big(P(a), \pi_\gamma(P(a))\big) + d\big(\pi_\gamma(P(a)), \pi_\gamma(P(b))\big) + d\big(P(b), \pi_\gamma(P(b))\big) \leq$$
$$\leq 2s_0 + \frac{d(P(a), P(b))}{\cosh(s_0)} \leq 2s_0 + \frac{c_1}{c_1^2 + 1} \cdot d(a, b)$$

and hence in conclusion

$$d(a, b) \leq \left(\frac{1}{c_1} - \frac{c_1}{c_1^2 + 1}\right)^{-1} \cdot (c_2 + 2s_0) = \lambda$$

(where λ depends only on c_1 and c_2). Then we can define another constant $t = s_0 + c_1 \cdot \lambda + 1$ depending only on c_1 and c_2. With the above notations, for $x \in [q_1, q_2]$ we have the following possibilities:

(i) $P(x) \in N_{s_0}(\gamma) \subset N_t(\gamma)$.

(ii) $P(x) \notin N_{s_0}(\gamma)$; if $[a, b]$ is the connected component of

$$[q_1, q_2] \bigcap P^{-1}\big(N_{s_0}(\gamma)\big)$$

containing x, then

$$d\big(P(x), \gamma\big) \leq d\big(P(x), P(a)\big) + d\big(P(a), \gamma\big) \leq$$
$$\leq c_1 \cdot d(x, a) + s_0 \leq c_1 \cdot d(a, b) + s_0 \leq c_1 \cdot \lambda + s_0 < t$$
$$\Rightarrow P(x) \in N_t(\gamma)$$

Formula $(*)$ is finally proved.

We complete now the proof of the lemma.

Since P is certainly a proper mapping, if $\{q_\nu\}$ is a sequence in β converging (in $\overline{\mathbb{H}^n}$) to one of the endpoints of β, then $\{P(q_\nu)\}$ is a divergent sequence in $\mathbb{H}^n$: we claim that it converges in $\overline{\mathbb{H}^n}$. If this were not the case we could find two subsequences $\{q'_\nu\}$ and $\{q''_\mu\}$ such that $\{P(q'_\nu)\}$ and $\{P(q''_\mu)\}$ converge in

$\overline{\mathbf{H}^n}$ to different points of the boundary. It follows quite easily (see Fig. C.2) that we can find ν, μ large enough that

$$q'_\nu \in [q''_0, q''_\mu] \qquad P(q'_\nu) \notin N_t\big(\gamma(P(q''_0), P(q''_\mu))\big)$$

and this contradicts (∗).

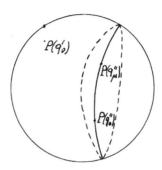

Fig. C.2. The restriction of P to β extends to the endpoints

It follows that we can find two points p^1_∞ and p^2_∞ in $\partial\mathbf{H}^n$ such that if $\{q^1_\nu\}$ and $\{q^2_\nu\}$ are sequences in β converging to the opposite endpoints then

$$P(q^1_\nu) \to p^1_\infty \qquad P(q^2_\nu) \to p^2_\infty \qquad (\text{in } \overline{\mathbf{H}^n}).$$

Let us remark first that $p^1_\infty \neq p^2_\infty$: suppose the converse and take $q \in \beta$. For ν large enough we have (as represented in Fig. C.3)

$$q \in [q^1_\nu, q^2_\nu] \qquad P(q) \notin N_t\big(\gamma(P(q^1_\nu), P(q^2_\nu))\big)$$

and this contradicts (∗).

Since $p^1_\infty \neq p^2_\infty$ we can consider the only geodesic line α having them as endpoints. If we denote by γ_ν the line $\gamma(P(q^1_\nu), P(q^2_\nu))$ we can prove that for compact $K \subset \mathbf{H}^n$

$$\lim_{\nu\to\infty} \sup_{x\in\gamma_\nu \cap K} d(x, \alpha) = 0.$$

In fact it easily follows from the relations $P(q^i_\nu) \to p^i_\infty$ $(i = 1, 2)$ that the endpoints of the γ_ν's converge to the endpoints of α, so that if we confine ourselves to a compact set K the distance between $\gamma_\nu \cap K$ and its projection on α goes to 0 as $\nu \to \infty$ (see Fig. C.4).

Now, given $q \in \beta$, let K be the closed ball of centre $P(q)$ and radius $t + 1$. As soon as ν is large enough that $q \in [q^1_\nu, q^2_\nu]$ we have that

$$d(P(q), \alpha) \le \inf_{x\in\gamma_\nu\cap K} \big(d(P(q), x) + d(x, \alpha)\big) \le$$

$$\le \inf_{x\in\gamma_\nu\cap K} d(P(q), x) + \sup_{x\in\gamma_\nu\cap K} d(x, \alpha) \le t + \sup_{x\in\gamma_\nu\cap K} d(x, \alpha).$$

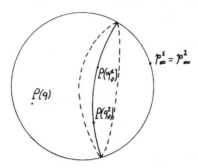

Fig. C.3. The restriction of P to β takes different values at the endpoints

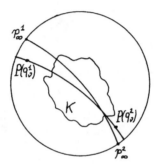

Fig. C.4. The sequence of geodesic lines γ_ν converge uniformly on K to the geodesic line α

As $\nu \to \infty$ we obtain

$$d(P(q), \alpha) \leq t$$

and hence if we set $r = t + 1$ and $A(\beta) = \alpha$ the required property

$$P(\beta) \subset N_r(A(\beta))$$

holds. Uniqueness of $A(\beta)$ is easily verified, as it must necessarily have p_∞^1 and p_∞^2 as endpoints (the details can be filled in as an exercise). □

Lemma C.1.7. Let β be a geodesic line in $\mathbf{H}^n$ and let H be a hyperbolic hyperplane orthogonal to β. Then there exists a constant c depending only on c_1 and c_2 such that

$$diam\big(\pi_{A(\beta)}(P(H))\big) \leq c.$$

Proof. Let $\beta \cap H = \{x\}$, and take $y \in H \setminus \{x\}$. Let l be the geodesic half-line in H starting from x and passing through y. Let l_1 and l_2 be the geodesic lines asymptotically parallel to both β and l, and let x_i be the point on l_i of minimal distance from x, as constructed in Fig. C.5.

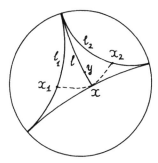

Fig. C.5. Construction proving that the image of a hyperplane ortogonal to β has projection of bounded diameter; first step

Transitivity of $\mathcal{I}(\mathbb{H}^n)$ implies that $d(x, x_1) = d(x, x_2)$ is a constant k independent of the construction (it does not depend even on n).

Let us remark now that A maps asymptotically parallel lines into asymptotically parallel lines (this is easily deduced from the definition of A and from the fact that P is a pseudo-isometry). Moreover if $z_0 = \pi_{A(\beta)}(P(x))$ we have

$$d(z_0, A(l_i)) \leq d(z_0, P(x)) + d(P(x), P(x_i)) + d(P(x_i), A(l_i)).$$

By the choice of A and r (Lemma C.1.6) and by the properties of P, we have

$$d(z_0, A(l_i)) \leq r + c_1 \cdot k + r = d$$

(the constant d depends only on c_1 and c_2). If z_i is the point on $A(l_i)$ of minimal distance from z_0, we have the situation represented in Fig. C.6.

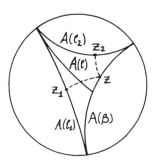

Fig. C.6. Construction proving that the image of a hyperplane ortogonal to β has projection of bounded diameter; second step

By definition of A we have

$$d\big(\pi_{A(\beta)}(P(y)), z_0\big) \leq$$
$$\leq d\big(\pi_{A(\beta)}(P(y)), \pi_{A(\beta)}(A(l))\big) + d\big(\pi_{A(\beta)}(A(l)), z_0\big) \leq$$
$$\leq d(P(y), A(l)) + d\big(\pi_{A(\beta)}(A(l)), z_0\big) \leq$$
$$\leq r + d\big(\pi_{A(\beta)}(A(l)), z_0\big)$$

and Fig. C.6 easily implies that the last number cannot exceed d, so that it suffices to set $c = 2(d + r)$. $\square$

Proposition C.1.8. P extends to a continuous mapping $\overline{P} : \overline{\mathbb{H}^n} \to \overline{\mathbb{H}^n}$ in such a way that $\overline{P}(\partial \mathbb{H}^n) \subset \partial \mathbb{H}^n$ and $\overline{P}|_{\partial \mathbb{H}^n}$ is one-to-one.

Proof. If β is a line in $\mathbb{H}^n$ and β^+ is a half of β, given any point $p_0 \in A(\beta)$ we can choose in a unique way the half-line $A(\beta)^+$ starting at p_0 with the property that as $q \to \infty$ along β^+, $P(q)$ converges to the endpoint of $A(\beta)^+$. As we remarked above, A preserves asymptotical parallelism; since the boundary of $\mathbb{H}^n$ (see A.5.10) can be defined as the set of all the half-lines in $\mathbb{H}^n$ up to asymptotical parallelism, A allows one to define a mapping

$$\partial P : \partial \mathbb{H}^n \to \partial \mathbb{H}^n \qquad \langle \beta^+ \rangle \mapsto \langle A(\beta)^+ \rangle.$$

We define $\overline{P}$ on $\overline{\mathbb{H}^n}$ as the union of P and ∂P. We want to prove that $\overline{P}$ is globally continuous. This is obvious at the points of $\mathbb{H}^n$. Now, let x_∞ be a point of the boundary, and let β be a half-line with endpoint x_∞. A fundamental system of neighborhoods of $\partial P(x_\infty)$ is obtained in the following way: consider a point $y \in A(\beta)$, let K be the hyperbolic hyperplane orthogonal to $A(\beta)$ passing through y and let Q be the connected component of $\overline{\mathbb{H}^n} \setminus K$ containing $\partial P(x_\infty)$; as y varies in $A(\beta)$ these Q's constitute a basis of neighborhoods at $\partial P(x_\infty)$.

Let one of these Q's be fixed; according to the definition of ∂P, if $x \to x_\infty$ along β then $P(x) \to \partial P(x_\infty)$ and moreover $P(x)$ lies in the r-neighborhood of $A(\beta)$ (where r is independent of x and β): it is easily deduced from this that we can find $x_0 \in \beta$ such that if $x \in [x_0, x_\infty)$ then the ball of centre $\pi_{A(\beta)}(P(x))$ and radius c is contained in Q (c is the constant of the above lemma and $[x_0, x_\infty)$ is the half-line starting at x_0 with endpoint x_∞). The corresponding situation is represented in Fig. C.7.

Now let H be the hyperbolic hyperplane orthogonal to β passing through x_0, and Q' be the connected component of $\overline{\mathbb{H}^n} \setminus H$ containing x_∞. If H_1 is a hyperplane orthogonal to β at $x \in [x_0, x_\infty)$ then by the choices we made we obtain $P(H_1) \subset Q$, as represented in Fig. C.8.

Moreover by definition of ∂P we have $\overline{P}(\overline{H_1}) \subset Q$. Since Q' is the union of these H_1's we have

$$\overline{P}(Q') \subset Q$$

and hence the continuity of $\overline{P}$ is proved.

As for the fact that the restriction to the boundary is one-to-one, let us take two points $x_1 \neq x_2$ in $\partial \mathbb{H}^n$; if β is the geodesic line having x_1 and x_2

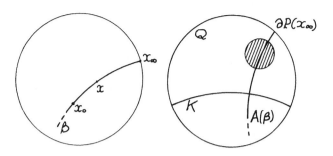

Fig. C.7. Construction proving the continuity of $\overline{P}$: first step

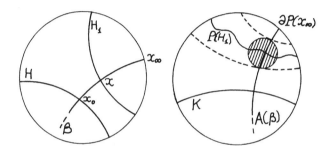

Fig. C.8. Construction proving the continuity of $\overline{P}$: second step

as endpoints, then $\overline{P}(x_1)$ and $\overline{P}(x_2)$ are the endpoints of $A(\beta)$, and therefore they are distinct.   □

Though we are not going to need this fact, we remark that the Jordan-Schönflies theorem (see [Mass2]) implies that *if $f : S^n \to S^n$ is a continuous one-to-one mapping, then f is onto, i.e. it is a homeomorphism*. Since $\partial \mathbf{H}^n$ is canonically homeomeorphic to S^{n-1}, the restriction of $\overline{P}$ to the boundary is a homeomorphism.

We recall that we were considering $P = \tilde{f}$. By simplicity we will keep denoting by $\tilde{f}$ the extension $\overline{P}$ to the boundary. For the conclusion of the proof of C.1.2 we only need the following:

Remark C.1.9. The relation

$$f_*(\gamma) \circ \tilde{f} = \tilde{f} \circ \gamma \qquad \forall \gamma \in \Gamma_1$$

holds in the whole $\overline{\mathbf{H}^n}$.

Proof. For fixed γ all the mappings γ, $f_*(\gamma)$ and $\tilde{f}$ extend to continuous mappings on $\overline{\mathbf{H}^n}$, and we only need to remark that the relation holds in $\mathbf{H}^n$, which is dense in $\overline{\mathbf{H}^n}$.   □

C.2 Second Step of the Proof:
Volume of Ideal Simplices

We shall denote by S_n the set of all n-simplices in $\overline{\mathbb{H}^n}$ having hyperbolic faces (in particular, the edges must be geodesic segments). An element σ of S_n is called <u>ideal</u> if all its vertices lie on $\partial \mathbb{H}^n$, and it is called <u>regular</u> if every permutation of its vertices can be obtained as the restriction of an isometry of $\mathbb{H}^n$ (in particular, if a regular simplex σ has all its vertices in $\mathbb{H}^n$, all its edges have the same length).

It is not difficult to check (using the half-space model and the explicit computation of the metrics A.4.3) that each element σ of S_n has finite volume.

The second step of the proof is expressed by the following:

Theorem C.2.1. The volume function has a maximum v_n on S_n. Moreover v_n is the volume of all and only the regular ideal simplices.

We start by setting $v_n = \sup_{\sigma \in S_n} \mathrm{vol}(\sigma)$.

Remark C.2.2. For the definition of v_n it is enough to confine the supremum to ideal simplices.

Proof. It suffices to show that for $\sigma \in S_n$ we can find an ideal simplex containing σ. Let p belong to the interior of σ; if $p_0, ..., p_n$ are the vertices of σ, we define p_i' to be the endpoint of the geodesic half-line starting from p and passing through p_i. The simplex having vertices $p_0', ..., p_n'$ is ideal and contains σ. $\square$

Lemma C.2.3. $v_n \leq \dfrac{\pi}{(n-1)!}$

Proof. We shall prove this by induction on n. For $n = 2$, since $\mathcal{I}(\mathbb{H}^2)$ operates transitively on the triples of points on the boundary, v_2 is the volume of any arbitrary ideal simplex. According to A.6.4 the volume of a geodesic triangle with inner angles α, β and γ is given by $\pi - \alpha - \beta - \gamma$; in the limit case the triangle is ideal, *i.e.* all its inner angles are 0, the volume becomes π, so that $v_2 = \pi$. The conclusion follows from induction as soon as we show that for $n \geq 3$

$$v_n \leq \frac{v_{n-1}}{n-1}.$$

Let σ be an ideal simplex in $\mathbb{H}^n$; we can assume in the half-space model that one of its vertices is ∞, as represented in Fig. C.9. The other n vertices of σ lie on a (Euclidean) half-sphere Γ of radius r and centre $c \in \mathbb{R}^{n-1} \times \{0\}$. We proved (A.5.6 and A.5.7) that Γ inherits from $\mathbb{H}^n$ the structure of $(n-1)$-dimensional hyperbolic space.

Let τ be the projection of σ on $\mathbb{R}^{n-1} \times \{0\}$ and σ_0 be the face of σ opposite to the vertex ∞ (σ_0 is an ideal simplex in Γ).

Then if $h(x) = \sqrt{r^2 - \|x - c\|^2}$ we have

$$\mathrm{vol}(\sigma) = \int_\tau dx \int_{h(x)}^\infty \frac{dt}{t^n} = \frac{1}{n-1} \int_\tau \frac{dx}{h(x)^{n-1}}.$$

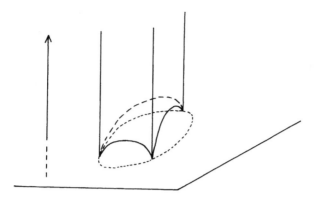

Fig. C.9. An ideal simplex in the half-space model with a vertex at infinity

Now, σ_0 can be parametrized as $\tau \ni x \mapsto (x, h(x))$, which implies that

$$\text{vol}(\sigma_0) = \int_\tau \alpha(x)dx$$

where $\alpha(x)$ is the coefficient of dilation of the volume element with respect to such a parametrization. We claim that

$$\alpha(x) \geq \frac{1}{h(x)^{n-1}} \quad \forall x \in \tau \tag{$*$}$$

which implies

$$\text{vol}(\sigma) \leq \frac{\text{vol}(\sigma_0)}{n-1}$$

and hence the conclusion.

We shall prove $(*)$ with $c = 0$ and $r = 1$ by the explicit computation

$$\alpha(x) = \frac{1}{h(x)^n}$$

(remark that $h(x) \leq 1$). If $\phi(x) = (x, h(x))$ then α is given by the following general formula

$$\alpha(x) = \left[\det\left(\langle d\phi_x(e_i)|d\phi_x(e_j)\rangle_{\phi(x)}\right)_{i,j=1,\ldots,n-1}\right]^{1/2}$$

where $e_1, \ldots, e_{n-1}$ is the canonical basis of $\mathbb{R}^{n-1}$. We have that

$$d\phi_x(v) = \left(v, -\frac{(x|v)}{(1-|x|^2)^{1/2}}\right)$$

$$\Rightarrow \langle d\phi_x(e_i)|d\phi_x(e_j)\rangle_{\phi(x)} = \frac{1}{1-|x|^2} \cdot \left(\delta_j^i + \frac{(x|e_i)(x|e_j)}{1-|x|^2}\right).$$

Let us remark that if $a_1, ..., a_p \in \mathbb{R}$ and a $p \times p$ matrix A is defined by $A^i_j = \delta^i_j + a_i a_j$ then

$$\det(A) = 1 + a_1^2 + ... + a_p^2$$

(this fact can be easily proved as an exercise). It follows that

$$\alpha(x)^2 = \frac{1}{(1 - |x|^2)^{n-1}} \cdot \left(1 + \frac{(x|e_1)^2 + ... + (x|e_{n-1})^2}{1 - |x|^2} \right) =$$

$$= \frac{1}{(1 - |x|^2)^{n-1}} \cdot \left(1 + \frac{|x|^2}{1 - |x|^2} \right) = \frac{1}{(1 - |x|^2)^n} = \frac{1}{h(x)^{2n}}.$$

$\square$

Lemma C.2.4. If $\sigma \in \mathcal{S}_n$ is ideal and if in the half-space model σ has vertices $\infty, p_1, ..., p_n$ then σ is regular if and only if the Euclidean $(n-1)$-simplex τ having vertices $p_1, ..., p_n$ is regular. Moreover an m-simplex in $\mathbb{R}^m$ is regular if and only if all its edges have the same length.

Proof. The second assertion is trivial and we shall assume it.

Let σ be regular. If a is a permutation on the set $\{1, ..., n\}$ there exists $\phi \in \mathcal{I}(\mathbb{H}^{n,+})$ with $\phi(\infty) = \infty$ and $\phi(p_j) = p_{a(j)}$ for $j = 1, ..., n$. By the explicit determination of $\mathcal{I}(\mathbb{H}^{n,+})$ we have that $\phi|_{\mathbb{R}^{n-1} \times \{0\}}$ is a multiple of a Euclidean isometry; moreover it is immediately checked that the multiplying constant must be 1 and hence τ is regular.

Conversely, let τ be regular. Obviously every permutation of the vertices of σ keeping ∞ fixed is induced by an isometry of $\mathbb{H}^{n,+}$. Moreover for $1 \leq j \leq n$ all the p_i's with $i \neq j$ have the same distance r from p_j, which implies that the inversion in $\mathbb{R}^n$ with respect to the sphere of centre p_j and radius r (an isometry of $\mathbb{H}^{n,+}$) induces on the vertices of σ the transposition between ∞ and p_j. Hence σ is regular too. $\square$

In the proof of step 2 we will confine ourselves to the case $n = 3$. The general case requires too much complicated calculation, so we decided to omit it; for a complete proof we refer to [Ha-Mu].

The method for $n = 3$ we will present can be found in [Mi3] and [Mu] (see also [Fe]).

We begin by a parametrization of all ideal simplices in $\mathbb{H}^3$. We shall denote by $\mathcal{T}$ the set of all similarity classes of triangles in $\mathbb{R}^2$. If we consider the canonical action of the symmetric group Σ_3 on $\mathbb{R}^3$ (permutations of coordinates) and we set

$$A = \{ (\alpha, \beta, \gamma) \in \mathbb{R}^3 : \alpha, \beta, \gamma \geq 0, \ \alpha + \beta + \gamma = \pi \}$$

then we have in a natural way $\mathcal{T} = A/\Sigma_3$ (i.e. $\mathcal{T}$ is the set of all non-ordered triples of positive numbers with sum π).

Let $\sigma \in \mathcal{S}_3$ be ideal, and let $p_0, ..., p_3$ be its vertices. For fixed $0 \leq i \leq 3$, realize σ in $\mathbb{H}^{3,+}$ with $p_i = \infty$. If t is large enough then $\sigma \cap (\mathbb{R}^2 \times \{t\})$ is a Euclidean triangle. We shall denote by $T_i(\sigma)$ the class of this triangle in $\mathcal{T}$. Let us remark that $T_i(\sigma)$ is well-defined, since an isometry of $\mathbb{H}^{3,+}$ keeping ∞ fixed induces a conformal diffeomorphism of $\mathbb{R}^2 \times \{t\}$ onto $\mathbb{R}^2 \times \{\lambda t\}$ for a suitable $\lambda > 0$.

Lemma C.2.5. $T_i(\sigma)$ is independent of i.

Proof. We can think of the construction of $T_i(\sigma)$ in the following way: consider a horosphere centred at p_i (close enough to p_i), consider its intersection Δ with σ and then realize the horosphere as a horizontal plane (in such a way that Δ becomes a Euclidean triangle); since the horosphere is mapped onto a horizontal plane by a conformal mapping, the angles we are concerned with for the definition of T_i are just the inner angles of Δ.

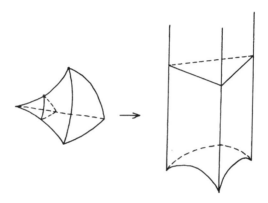

Fig. C.10. Parametrization of ideal simplices. An isometry of an arbitrary ideal simplex onto one with a vertex at infinity

Let $p_j = \infty$ and $i \neq j$ and let p be on the vertical line through p_i close enough to p_i. Since the tangent plane in p to the horosphere centred at p_i is horizontal, the angles considered in Fig. C.11 are equal to each other. Since this construction works for all pairs $i \neq j$, we obtain that the two angles involved in the definition of the $T_i(\sigma)$'s having vertex on the same edge of σ are equal to each other. Hence for suitable angles $\alpha, ..., \gamma'$ we have the situation of Fig. C.12.

Moreover we have the relations:

$$\alpha + \beta + \gamma = \alpha + \beta' + \gamma' = \alpha' + \beta + \gamma' = \alpha' + \beta' + \gamma = \pi$$

and hence

$$\begin{cases} \alpha + \beta = \alpha' + \beta' \\ \alpha + \gamma = \alpha' + \gamma' \\ \beta + \gamma = \beta' + \gamma'. \end{cases}$$

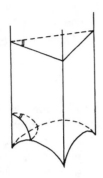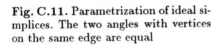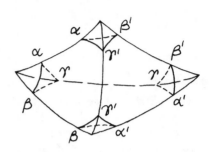

Fig. C.11. Parametrization of ideal simplices. The two angles with vertices on the same edge are equal

Fig. C.12. Parametrization of ideal simplices. In the figure the measure of an angle is given by the symbol near its vertex

Since $\det \begin{pmatrix} 1 & 1 & 0 \\ 1 & 0 & 1 \\ 0 & 1 & 1 \end{pmatrix} \neq 0$ this linear system has the only solution $\alpha = \alpha'$, $\beta = \beta'$, $\gamma = \gamma'$ and the proof is complete. □

The above lemma allows us to define a map T associating to each ideal element of $\mathcal{S}_3$ an element of $\mathcal{T}$. We shall write

$$T(\sigma) = [\alpha(\sigma), \beta(\sigma), \gamma(\sigma)].$$

Lemma C.2.6. $T(\sigma) = T(\sigma') \iff \exists f \in \mathcal{I}(\mathbb{H}^3)$ s.t. $f(\sigma) = \sigma'$.

Proof. $\Rightarrow$. Realize σ and σ' in $\mathbb{H}^{3,+}$ with a vertex at ∞. If $t > 0$ is big enough, $\sigma \cap (\mathbb{R}^2 \times \{t\})$ and $\sigma' \cap (\mathbb{R}^2 \times \{t\})$ are similar triangles, *i.e.* there exists $h \in \text{Conf}(\mathbb{R}^2)$ mapping the first onto the second. By A.3.7 we have $h = \lambda g$ with $g \in \mathcal{I}(\mathbb{R}^2)$ and $\lambda > 0$, and hence by A.4.2, $f : (x, s) \mapsto \lambda(g(x), s)$ is an isometry of $\mathbb{H}^{3,+}$. We can assume $\lambda \geq 1$ (otherwise we interchange σ and σ'), and hence we have quite easily that $f(\sigma) = \sigma'$.

$\Leftarrow$. Realize both σ and σ' in $\mathbb{H}^{3,+}$ with a vertex at ∞; then f keeps ∞ fixed, which implies that f is a similarity of
$$\sigma \cap (\mathbb{R}^2 \times \{t\}) \text{ onto } \sigma' \cap (\mathbb{R}^2 \times \{\lambda t\})$$
(for some $\lambda > 0$) and hence $T(\sigma) = T(\sigma')$. □

C.2.6 implies that in $\mathbb{H}^3$ the isometry classes of ideal simplices are parametrized by $\mathcal{T}$ (of course $\mathcal{T}$ is onto). We are now going to determine the volume of an ideal simplex: this is the only point of the proof where analysis plays an important role, while Mostow's original proof made extensive use of methods from analysis.

We introduce the following <u>Lobachevsky function</u>:

$$\Lambda(\theta) = -\int_0^\theta \log|2\sin t|dt \quad (\theta \in \mathbb{R})$$

(it is quite easily checked that Λ is well-defined, as the integral converges for all θ's).

Proposition C.2.7. (1) Λ is a continuous function.
(2) Λ is an odd function.
(3) $\dot\Lambda(\theta)$ exists for all θ but $k\pi$ $(k \in \mathbb{Z})$.
(4) Λ is π-periodic.
(5) in $[0, \pi]$ Λ has a maximum in the unique point $\pi/6$.
(6) for $m \in \mathbb{Z}$ the following identity holds:

$$\Lambda(m\theta) = m \sum_{k=0}^{m-1} \Lambda\left(\theta + \frac{k\pi}{m}\right).$$

Proof. (1) is obvious.

(2) the function $f : t \mapsto \log|2\sin t|$ is even, and hence Λ is odd.

(3) is obvious.

(4) f has period π, so that Λ is periodic if and only if the mean of f on the period is null, *i.e.* $\Lambda(\pi) = 0$. Since $\sin(2t) = 2\sin t \cos t = 2\sin t \sin(t + \pi/2)$ we have

$$\frac{\Lambda(2\theta)}{2} = -\frac{1}{2}\int_0^{2\theta} \log|2\sin\tau|d(\tau) = -\int_0^\theta \log|2\sin 2t|dt =$$

$$= -\int_0^\theta \log|2\sin t|dt - \int_0^\theta \log|2\sin(t + \pi/2)|dt =$$

$$= \Lambda(\theta) - \int_{\pi/2}^{\pi/2+\theta} \log|2\sin t|dt =$$

$$= \Lambda(\theta) - \int_0^{\pi/2+\theta} \log|2\sin t|dt + \int_0^{\pi/2} \log|2\sin t|dt =$$

$$= \Lambda(\theta) + \Lambda(\theta + \pi/2) - \Lambda(\pi/2).$$

The following identity is then established:

$$\Lambda(2\theta) = 2\big(\Lambda(\theta) + \Lambda(\theta + \pi/2) - \Lambda(\pi/2)\big).$$

If we take $\theta = \pi/2$ we obtain $\Lambda(\pi) = 2\Lambda(\pi)$, whence $\Lambda(\pi) = 0$ and periodicity of Λ is proved.

Let us remark soon that (2) and (4) together imply that $\Lambda(\pi/2) = 0$, so that the above identity reduces to

$$\Lambda(2\theta) = 2\big(\Lambda(\theta) + \Lambda(\theta + \pi/2)\big)$$

which proves (6) for $m = 2$.

(5) Since $\Lambda(0) = \Lambda(\pi) = 0$ and (by (2) and (4)) Λ has both positive and negative values, we confine ourselves to the inner points of the interval, where the derivative exists. The derivative vanishes only for $\theta = \pi/6$ and $\theta = 5/6 \cdot \pi$, and it is easily checked that

$$\Lambda(\pi/6) > 0 > \Lambda(5/6 \cdot \pi),$$

whence the maximum is attained at the only point $\pi/6$.

(6) The case $m = 2$ was already settled above, and it made use of the formula for $\sin(2t)$; for the general case we will need the following generalization of that formula:

$$2\sin(mt) = \prod_{k=0}^{m-1} 2\sin(t + k\pi/m)$$

(to be proved as an exercise). Then we have:

$$
\begin{aligned}
\frac{\Lambda(m\theta)}{m} &= -\frac{1}{m}\int_0^{m\theta} \log|2\sin t|dt = -\int_0^{\theta}\log|2\sin(mt)|dt = \\
&= -\sum_{k=0}^{m-1}\int_0^{\theta}\log|2\sin(t + k\pi/m)|dt = \\
&= -\sum_{k=0}^{m-1}\left(\int_0^{\theta+k\pi/m}\log|2\sin t|dt - \int_0^{k\pi/m}\log|2\sin t|dt\right) = \\
&= \sum_{k=0}^{m-1}\Lambda(\theta + k\pi/m) + C(m)
\end{aligned}
$$

By (2) and (4) we have that for all $m \in \mathbb{N}$ and all $\alpha \in \mathbb{R}$

$$\int_0^{\pi}\Lambda(m\theta)d\theta = \int_0^{\pi}\Lambda(\theta + \alpha)d\theta = 0$$

and hence if we integrate between 0 and π the above identity we obtain that $C(m) = 0$, and the proof is complete.   □

The reason for introducing the function Λ is the following:

Proposition C.2.8. If $\sigma \in \mathcal{S}_3$ then

$$\mathrm{vol}(\sigma) = \Lambda(\alpha(\sigma)) + \Lambda(\beta(\sigma)) + \Lambda(\gamma(\sigma)).$$

Proof. Let us remark first that the formula makes sense, since the sum of the right hand side does not depend on the order of $\alpha(\sigma)$, $\beta(\sigma)$ and $\gamma(\sigma)$.

We realize σ in $\mathbb{II}^{3,+}$ with a vertex at ∞ and the other three vertices on the unit sphere of $\mathbb{R}^2 \times \{0\}$ (of course we can do this by isometries). Let us assume first that α, β and γ are acute. If K is the projection of σ on $\mathbb{R}^2 \times \{0\}$, we can consider the subdivision of K in six triangles as represented in Fig C.13.

If we set

$$\sigma_\alpha = (K_\alpha \times \mathbb{R}_+) \cap \sigma \qquad \sigma_\beta = (K_\beta \times \mathbb{R}_+) \cap \sigma \qquad \sigma_\gamma = (K_\gamma \times \mathbb{R}_+) \cap \sigma$$

we have $\mathrm{vol}(\sigma) = 2\big(\mathrm{vol}(\sigma_\alpha) + \mathrm{vol}(\sigma_\beta) + \mathrm{vol}(\sigma_\gamma)\big)$ and hence it suffices to show that

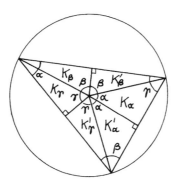

Fig. C.13. Volume of an ideal simplex: subdivision in the case of a projection with acute angles

$$\mathrm{vol}(\sigma_\alpha) = \frac{\Lambda(\alpha)}{2}$$

(of course similar formulae hold for β and γ). If we consider the natural parametrization of σ_α we have

$$\mathrm{vol}(\sigma_\alpha) = \int_0^{\cos\alpha} dx \int_0^{x\,\mathrm{tg}\alpha} dy \int_{\sqrt{1-x^2-y^2}}^{\infty} \frac{dz}{z^3} =$$

$$= \frac{1}{2} \int_0^{\cos\alpha} dx \int_0^{x\,\mathrm{tg}\alpha} \frac{dy}{1-x^2-y^2} = \cdots$$

(using $\dfrac{1}{1-x^2-y^2} = \dfrac{1}{2\sqrt{1-x^2}} \cdot \left(\dfrac{1}{\sqrt{1-x^2}-y} + \dfrac{1}{\sqrt{1-x^2}+y} \right)$)

$$\cdots = \frac{1}{4} \int_0^{\cos\alpha} \log \frac{\sqrt{1-x^2}+y}{\sqrt{1-x^2}-y} \Bigg|_{y=0}^{x\,\mathrm{tg}\alpha} \frac{dx}{\sqrt{1-x^2}}$$

$$= \frac{1}{4} \int_0^{\cos\alpha} \log \frac{\sqrt{1-x^2}\cos\alpha + x\sin\alpha}{\sqrt{1-x^2}\cos\alpha - x\sin\alpha} \frac{dx}{\sqrt{1-x^2}} = \cdots$$

(via the change of coordinate $x = \cos t$)

$$\cdots = -\frac{1}{4} \int_{\pi/2}^{\alpha} \log \frac{\sin t \cos\alpha + \cos t \sin\alpha}{\sin t \cos\alpha - \cos t \sin\alpha} dt = \frac{1}{4} \int_\alpha^{\pi/2} \log \frac{\sin(t+\alpha)}{\sin(t-\alpha)} dt =$$

$$= \frac{1}{4} \int_\alpha^{\pi/2} \log |2\sin(t+\alpha)| dt - \frac{1}{4} \int_\alpha^{\pi/2} \log |2\sin(t-\alpha)| dt =$$

$$= \frac{1}{4} \int_{2\alpha}^{\alpha+\pi/2} \log |2\sin t| dt - \frac{1}{4} \int_0^{\pi/2-\alpha} \log |2\sin t| dt =$$

$$= \frac{1}{4} \left(-\Lambda(\alpha + \pi/2) + \Lambda(2\alpha) + \Lambda(\pi/2 - \alpha) \right).$$

Now, $\Lambda(2\alpha) = 2\Lambda(\alpha) + 2\Lambda(\alpha + \pi/2)$ and

$$\Lambda(\pi/2 - \alpha) = -\Lambda(\alpha - \pi/2) = -\Lambda(\alpha + \pi/2)$$

and then we obtain at last

$$\text{vol}(\sigma_\alpha) = \frac{\Lambda(\alpha)}{2}.$$

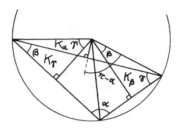

Fig. C.14. Volume of an ideal simplex: subdivision in the case of a projection with an obtuse angle

In case one of the angles is not acute (*e.g.* $\alpha \geq \pi/2$) we can consider the subdivision of σ described in Fig. C.14 and then we have

$$\text{vol}(\sigma) = 2\left(\frac{\Lambda(\beta)}{2} + \frac{\Lambda(\gamma)}{2} - \frac{\Lambda(\pi - \alpha)}{2}\right) = \Lambda(\alpha) + \Lambda(\beta) + \Lambda(\gamma). \qquad \square$$

We are now ready for the conclusion of C.2.1.

Theorem C.2.9. Given $\sigma \in \mathcal{S}_3$ we have $\text{vol}(\sigma) = v_3$ if and only if σ is ideal and regular.

Proof. Since for non-ideal σ we can find another simplex with strictly larger volume, we confine ourselves to ideal simplices. It is immediately checked that $\text{vol}(\sigma) = 0$ if and only if $T(\sigma)$ is a degenerate triangle, and hence, according to C.2.6, and C.2.8, we only have to prove that the maximum of the function

$$(\alpha, \beta, \gamma) \mapsto \Lambda(\alpha) + \Lambda(\beta) + \Lambda(\gamma)$$

on the set $\{\alpha, \beta, \gamma > 0,\ \alpha + \beta + \gamma = \pi\}$ is attained at and only at the point $(\pi/3, \pi/3, \pi/3)$. Equivalently, we must check that the function

$$g : (\alpha, \beta) \mapsto \Lambda(\alpha) + \Lambda(\beta) + \Lambda(\pi - \alpha - \beta)$$

on the set $S = \{\alpha, \beta > 0,\ \alpha + \beta < \pi\}$ is attained at and only at $(\pi/3, \pi/3)$.

Since g is a positive function and it can be continuously extended to 0 on the boundary of S, it certainly has a maximum on S. Moreover it is of class C^1 on S, and hence we must look for the stationary points of its gradient:

$$\begin{cases} \dot{\Lambda}(\alpha) = \dot{\Lambda}(\pi - \alpha - \beta) \\ \dot{\Lambda}(\beta) = \dot{\Lambda}(\pi - \alpha - \beta) \end{cases}$$

$$\Rightarrow \begin{cases} |\sin\alpha| = |\sin(\pi - \alpha - \beta)| \\ |\sin\beta| = |\sin(\pi - \alpha - \beta)|. \end{cases}$$

It is immediately checked that this system has the sole solution $\alpha = \beta = \pi/3$, and the proof is over. □

C.3 Gromov Norm of a Compact Manifold

In this section we shall assume the reader is familiar with the first elements of singular homology theory; several well-established references are available, *e.g.* [Greenb1] and [Mass2], so we will confine ourselves to a brief review of the notations and to the facts we are going to make explicit use of.

If X a is topological space, R is a ring and k is a natural number we shall denote by $C_k(X; R)$ the R-module of the k-chains in X with coefficients in R, *i.e.* the free R-module generated by all the continuous mappings from the k-th standard simplex Δ_k into X. As usual we shall denote the boundary operator $C_k(X; R) \to C_{k-1}(X; R)$ by ∂_k, and we shall call $Z_k(X; R) = Ker(\partial_k)$ the set of the k-cycles, and $B_k(X; R) = Ran(\partial_{k+1})$ the set of the k-boundaries. Since $\partial_k \circ \partial_{k+1} = 0$ we can define the k-th homology R-module on X as the quotient set

$$H_k(X; R) = {}^{Z_k(X;R)}\!/_{B_k(X;R)}.$$

If $A \subset X$ the inclusion $i : A \to X$ induces (by composition) a homomorphism of R-modules

$$i_* : C_k(A; R) \to C_k(X; R)$$

and then we can set $C_k(X, A; R) = {}^{C_k(X;R)}\!/_{i_*(C_k(A;R))}.$

Since for $z \in i_*(C_k(A; R))$ we obviously have $\partial_k(z) \in i_*(C_{k-1}(A; R))$, we can define

$$\hat{\partial}_k : C_k(X, A; R) \to C_{k-1}(X, A; R)$$

by

$$\hat{\partial}_k\big(z + i_*(C_k(A; R))\big) = \partial_k(z) + i_*(C_{k-1}(A; R))$$

and we still have $\hat{\partial}_k \circ \hat{\partial}_{k+1} = 0$, so that we can introduce the k-th homology R-module on X relative to A as the quotient

$$H_k(X, A; R) = {}^{Z_k(X,A;R)}\!/_{B_k(X,A;R)}$$

where $Z_k(X, A; R) = Ker(\hat{\partial}_k)$ and $B_k(X, A; R) = Ran(\hat{\partial}_{k+1})$.

Several facts we shall need about singular homology theory are recalled in the following:

Proposition C.3.1. Let M be a connected compact oriented n-manifold. Then:

(1) $H_n(M; \mathbb{Z}) \cong \mathbb{Z}$ and it has a preferred generator denoted by $[M]$ and called the <u>fundamental</u> <u>class</u> of M;

(2) M can be represented as a compact polyhedron in which each $(n-1)$-simplex is the face of precisely two n-simplices; we shall call this representation a <u>triangulation</u> of M;

(3) the fundamental class of M is canonically represented by the formal sum of the n-dimensional simplices of a triangulation as described above;

(4) $H_n(M; \mathbb{R}) \cong H_n(M; \mathbb{Z}) \otimes \mathbb{R} \cong \mathbb{R}$ (this is a very special case of the universal coefficients theorem); in particular $[M]$ can be viewed as a generator of $H_n(M; \mathbb{R})$ as a real vector space;

(5) given $x \in M$ we have as well $H_n(M, M \setminus \{x\}) \cong \mathbb{Z}$ and the inclusion

$$i : (M, \emptyset) \hookrightarrow (M, M \setminus \{x\})$$

induces an isomorphism

$$i_* : H_n(M; \mathbb{Z}) \to H_n(M, M \setminus \{x\}; \mathbb{Z});$$

the same holds with $\mathbb{R}$ instead of $\mathbb{Z}$.

Now let M and N be connected compact oriented n-manifolds, consider a continuous map $f : M \to N$ and let $f_* : H_n(M; \mathbb{Z}) \to H_n(N; \mathbb{Z})$ be the induced homomorphism; according to C.3.1–(1) there exists an integer $\deg(f)$ called the <u>degree</u> of f such that $f_*([M]) = \deg(f) \cdot [N]$. We state without proof the following result concerning the way the degree of a mapping may be actually computed:

Proposition C.3.2. Let M, N and f be as above;

(1) if there exist triangulations $\mathcal{M}$ of M and $\mathcal{N}$ of N with respect to which f is simplicial then for any $\sigma \in \mathcal{N}$ with $\dim(\sigma) = n$ we have that $\deg(f)$ is given by:

$$\#\{\tau \in \mathcal{M} : f(\tau) = \sigma, \ f : \tau \to \sigma \text{ pos.}\} - \#\{\tau \in \mathcal{M} : f(\tau) = \sigma, \ f : \tau \to \sigma \text{ neg.}\}$$

where pos. (neg.) stands for positive (negative) and means that orientation is preserved (reversed);

(2) if f is smooth then given a regular value y of f we have that:

$$\deg(f) = \#\{x \in f^{-1}(x) : d_x f \text{ pos.}\} - \#\{x \in f^{-1}(x) : d_x f \text{ neg.}\}$$

(for this definition of the degree of a mapping see [Mi2]).

We are now going to define the Gromov norm of a compact oriented manifold. Let us start with an arbitrary topological space X; if $c \in C_k(X; \mathbb{R})$ is a chain in X with real coefficients we can set

$$\|c\| = \inf \left\{ \sum_{i=1}^{p} |a_i| : \exists \sigma_1, ..., \sigma_p : \Delta_k \to X \text{ s.t. } c = \sum_{i=1}^{p} a_i \sigma_i \right\}.$$

$\|.\|$ is a norm on the real vector space $C_k(X; \mathbb{R})$ containing $Z_k(X; \mathbb{R})$, so that the quotient space

$$H_k(X; \mathbb{R}) = \frac{Z_k(X; \mathbb{R})}{B_k(X; \mathbb{R})}$$

can be canonically endowed with the quotient semi-norm:

$$z \in H_k(X; \mathbb{R}) \implies \|z\| = \inf\{\|c\| : c \in Z_k(X; \mathbb{R}), [c] = z\}.$$

Remark that since $B_k(X; \mathbb{R})$ need not be closed in $Z_k(X; \mathbb{R})$ the quotient semi-norm may actually vanish on non-zero elements of $H_k(X; \mathbb{R})$.

Remark C.3.3. If $f : X \to Y$ is continuous and $\alpha \in H_k(X; \mathbb{R})$ then

$$\|f_*(\alpha)\| \leq \|\alpha\|.$$

(In fact if $c = \sum a_i \sigma_i$ is a chain representing α then $f \circ c = \sum a_i (f \circ \sigma_i)$ is a chain representing $f_*(\alpha)$ and then $\|f \circ c\| \leq \|c\|$.)

Assume now M is a compact oriented manifold. According to C.3.1–(4) the semi-norm of $[M] \in H_n(M; \mathbb{R})$ can be calculated; this number will be called the <u>Gromov norm</u> of M, and denoted by $\|M\|$.

Remark that it is not evident from the definition that manifolds with non-zero Gromov norm actually exist. For instance, the next results imply that very nice compact manifolds such as spheres have zero Gromov norm.

Proposition C.3.4. If there exists $f : M \to M$ with $|\deg(f)| \geq 2$ then $\|M\| = 0$.

Proof. By C.3.3 we have

$$\|M\| = \|[M]\| \geq \|f_*([M])\| = \|\deg(f) \cdot [M]\| = |\deg(f)| \cdot \|M\|. \qquad \square$$

Corollary C.3.5. For each $n \geq 2$, $\|S^n\| = 0$.

Proof. Let S^n be viewed as the unit sphere of $\mathbb{C} \times \mathbb{R}^{n-1}$. For integer k the mapping

$$f_k : S^n \ni (z, t) \to (e^{2\pi i k} z, t) \in S^n$$

has degree k, and hence C.3.4 applies. $\qquad \square$

C.4 Third Step of the Proof: the Gromov Norm and the Volume Are Proportional

This section is devoted to the proof of the following fact:

Proposition C.4.1. Let $\tilde{f}$ be the mapping constructed in the first step of the proof (Proposition C.1.2); if $u_0, ..., u_n$ are vertices of a maximal volume simplex then the same holds for $\tilde{f}(u_0), ..., \tilde{f}(u_n)$.

Let us remark that by construction $\tilde{f}(\partial \mathbb{H}^n) \subseteq \partial \mathbb{H}^n$ so that by C.2.1 the above proposition is equivalent to the following assertion: *if $u_0, ..., u_n \in \partial \mathbb{H}^n$ are the vertices of a regular simplex then $\tilde{f}(u_0), ..., \tilde{f}(u_n)$ are the vertices of a regular simplex too.* Our proof will exploit the notion of Gromov norm of a compact oriented manifold.

The basic tool for the proof of C.4.1 is represented by the following result due to Gromov and Thurston:

Theorem C.4.2. If M is an oriented compact hyperbolic manifold then

$$\|M\| = \frac{\mathrm{vol}(M)}{v_n}.$$

Let us remark that the above theorem implies in particular that the hyperbolic volume of M is a topological invariant (as $\|M\|$ is): this fact was easily deduced in the case $n = 2$ by the Gauss-Bonnet formula (B.3.3). Our proof of C.4.1 will require not only C.4.2 but many of the notions we are going to introduce to prove it. We will keep fixed for a while a compact oriented hyperbolic n-manifold M and a universal covering $\pi : \mathbb{H}^n \to M$.

For $u_0, ..., u_k \in \mathbb{H}^n$ ($k \leq n$) we shall denote by $\sigma(u_0, ..., u_k) : \Delta_k \to \mathbb{H}^n$ the barycentric parametrization of the geodesic simplex in $\mathbb{H}^n$ having vertices $u_0, ..., u_k$ (recall that convex combinations are canonically defined in $\mathbb{H}^n$).

We shall say a singular simplex $\varphi : \Delta_k \to M$ is <u>straight</u> if it is expressed as $\varphi = \pi \circ \sigma(u_0, ..., u_k)$ for suitable $u_0, ..., u_k$. A singular chain will be called <u>straight</u> if it can be expressed as the linear combination of straight simplices. Since we will deal only with the n-th homology module of M from now on we will confine ourselves to n-simplices.

We shall say $\sigma(u_0, ..., u_n)$ is <u>degenerate</u> if $u_0, ..., u_n$ belong to some hyperbolic $(n-1)$-subspace; remark as well that

$$\sigma(u_0, ..., u_n) \text{ degenerate } \Leftrightarrow \mathrm{vol}\big(\sigma(u_0, ..., u_n)(\Delta_n)\big) = 0 \Leftrightarrow$$

$$\Leftrightarrow \exists t \in \overset{\circ}{\Delta}_n \text{ s.t. } d_t \sigma(u_0, ..., u_n) \text{ is not an isomorphism.}$$

Moreover each degenerate simplex is homologous to a sum of non-degenerate ones. So we will sometimes suppose without explicit mention that the n-simplices we are considering are non-degenerate.

Lemma C.4.3. Every singular chain in M is naturally homotopic (and hence homologous) to a straight one.

Proof. Let $\varphi : \Delta_n \to M$ be continuous. Since Δ_n is simply connected φ can be globally lifted to a continuous mapping $\tilde{\varphi} : \Delta_n \to \mathbb{H}^n$. Denote $\tilde{\varphi}(e_i)$ (where e_i is the i-th vertex of Δ_n) by v_i and define $\overline{\varphi}$ as $\pi \circ \sigma(v_0, ..., v_n)$. The mapping

$$F : \Delta_n \times [0,1] \ni (t, s) \mapsto \pi\big(s \cdot \tilde{\varphi}(t) + (1-s) \cdot \sigma(v_0, ..., v_n)(t)\big) \in M$$

is a homotopy between φ and $\overline{\varphi}$ with the property that $F(e_i, s) = \varphi(e_i)$ $\forall s \in [0, 1]$, $i = 0, ..., n$. This implies that if $\sum_i a_i \varphi_i$ is a singular cycle in M then $\sum_i a_i \overline{\varphi}_i$ is a straight cycle representing the same homology class. $\quad\square$

Lemma C.4.4. In order to prove that

$$\|M\| \geq \frac{\text{vol}(M)}{v_n}$$

it suffices to check that whenever $\sum_i a_i \sigma_i$ is a straight cycle representing $[M]$ then

$$\sum_i |a_i| \geq \frac{\text{vol}(M)}{v_n}.$$

Proof. We know that

$$\|M\| = \inf \left\{ \sum_{i=1}^h |b_i| : \exists \varphi_1, ..., \varphi_h \ s.t. \ \left[\sum_{i=1}^h b_i \varphi_i\right] = [M] \right\}.$$

Let $\left[\sum_{i=1}^h b_i \varphi_i\right] = [M]$; if $\overline{\varphi}_i$ is constructed as in the above lemma then $\left[\sum_{i=1}^h b_i \overline{\varphi}_i\right] = [M]$ and hence the proof is easily completed. $\quad\square$

Assume $\varphi = \pi \circ \sigma : \Delta_n \to M$ is a straight simplex and σ is non-degenerate. As π is a covering mapping, if for some $t_0 \in \overset{\circ}{\Delta}_n$ we have that $d_{t_0}\varphi$ preserves the orientation then the same holds for all $t \in \overset{\circ}{\Delta}_n$, so that we can define the <u>algebraic volume</u> of φ as

$$\text{algvol}(\varphi) = \begin{cases} +\text{vol}(\sigma(\Delta_n)) & \text{if } d_t\varphi \text{ preserves orientation} \\ -\text{vol}(\sigma(\Delta_n)) & \text{otherwise.} \end{cases}$$

Remark that we simply have

$$\text{algvol}(\varphi) = \int_{\varphi(\Delta_n)} \alpha(x) dv(x)$$

where dv is the volume element on M and $\alpha(x) = \alpha^+(x) - \alpha^-(x)$, with

$$\alpha^+(x) = \#\{t \in \overset{\circ}{\Delta}_n : \varphi(t) = x, \ d_t\varphi \text{ pos.}\}$$
$$\alpha^-(x) = \#\{t \in \overset{\circ}{\Delta}_n : \varphi(t) = x, \ d_t\varphi \text{ neg.}\}$$

This formula explains the name of algebraic volume: in fact this number reprents the measure of the region actually covered by ϕ, in which each point is taken according to its algebraic multiplicity.

Remark as well that by our choice of φ if for some x we have $\alpha^+(x) \neq 0$ then $\alpha^-(y) = 0 \, \forall y$ (which implies the converse: if $\alpha^-(x) \neq 0$ for some x then $\alpha^+(y) = 0 \, \forall y$); in particular $\alpha^+ \cdot \alpha^- = 0$ everywhere.

Of course the definition of the algebraic volume extends by linearity to straight chains: from now on we shall assume chains are always written in their shortest length expression, *i.e.* we shall not accept expressions such as $a\varphi + b\varphi$ instead of $(a+b)\varphi$.

Remark C.4.5. If φ is a straight simplex then $|\text{algvol}(\varphi)| \leq v_n$ (in fact by construction $|\text{algvol}(\varphi)| = \text{vol}(\sigma(\Delta_n))$, and $\sigma(\Delta_n)$ is a geodesic simplex in hyperbolic space).

Proposition C.4.6. $\|M\| \geq \dfrac{\text{vol}(M)}{v_n}$.

Proof. Let $z = \sum_{i=1}^{k} a_i\varphi_i$ be a straight cycle representing $[M]$, and define a subset N of M as $\cup_{i=1}^{k} \varphi_i(\partial\Delta_n)$. As above we introduce (for $x \in M$) a number $\alpha_i(x)$ as

$$\#\{t \in \overset{\circ}{\Delta}_n : \varphi_i(t) = x,\ d_t\varphi_i \text{ pos.}\} - \#\{t \in \overset{\circ}{\Delta}_n : \varphi_i(t) = x,\ d_t\varphi_i \text{ neg.}\}$$

and we set

$$\Phi_z(x) = \sum_{i=1}^{k} a_i\alpha_i(x).$$

We claim that $\Phi_z(x) = 1 \ \forall\, x \in M \setminus N$. Let us remark first that by straightening the simplices of any triangulation of M we obtain a straight triangulation $\mathcal{T}$. Such a triangulation produces the canonical representative of $[M]$ as a straight cycle z_0. Of course if x does not belong to the $(n-1)$-skeleton of $\mathcal{T}$ and if Φ_{z_0} is defined just as Φ_z then $\Phi_{z_0}(x) = 1$. Assume now $\Phi_z(x) \neq 1$ for some $x \in M \setminus N$; since Φ_z is locally constant on $M \setminus N$ and N is closed there is no loss in generality in assuming that x does not belong to the $(n-1)$-skeleton of $\mathcal{T}$. Then we are left to prove that $\Phi_{z_0}(x) - \Phi_z(x) = 0$, or equivalently $\Phi_{z_0-z}(x) = 0$. Recall that by C.3.1–(5) the inclusion

$$i : (M, \phi) \hookrightarrow (M, M \setminus \{x\})$$

induces an isomorphism

$$i_* : H_n(M; \mathbb{R}) \to H_n(M, M \setminus \{x\}; \mathbb{R});$$

moreover it is easily verified that if in the second space we take 1 as generator then

$$i_*([w]) = \Phi_w(x)$$

provided the representative $w = \sum_j b_j w_j$ is chosen to be straight and such that $x \notin \cup_j w_j(\partial\Delta_n)$. This implies that $\Phi_{z_0-z}(x) = 0$ and hence our claim is proved.

Moreover N has null volume, so that

$$\text{vol}(M) = \int_M \Phi(x) dv(x).$$

On the other hand we have that

$$\int_M \alpha_i(x) dv(x) = \int_{\varphi_i(\Delta_n)} \alpha_i(x) dv(x) = \text{algvol}(\sigma_i)$$

so that

$$\text{vol}(M) = \sum_{i=1}^{k} a_i \cdot \text{algvol}(\sigma_i) = \text{algvol}\left(\sum_{i=1}^{k} a_i \sigma_i \right)$$

and then (using C.4.5)

$$\text{vol}(M) = \left| \sum_{i=1}^{k} a_i \text{algvol}(\sigma_i) \right| \le \sum_{i=1}^{k} |a_i| \cdot |\text{algvol}(\sigma_i)| \le v_n \cdot \sum_{i=1}^{k} |a_i|$$

and hence the conclusion follows from C.4.4.   □

C.4.6 establishes one half of Theorem C.4.2, *i.e.* inequality $\ge$; the proof of the opposite inequality is harder. A special technique works for $n = 2$; actually, as it follows from the next proof, Theorem C.4.2 in case $n = 2$ is equivalent to the Gauss-Bonnet formula for hyperbolic surfaces (B.3.3).

Proposition C.4.7. If $n = 2$, $\|M\| \le \dfrac{\text{vol}(M)}{v_2}$.

Proof. Let $g \ge 2$ be the genus of M. By B.3.3 we have that $\text{vol}(M) = 4\pi(g-1)$ and, by C.2.3, $v_2 = \pi$; then we only have to check that $\|M\| \le 4(g-1)$.

Let us realize M as the quotient of a polygon with $4g$ sides and let us cover M by $4g - 2$ triangles $\sigma_1, ..., \sigma_{4g-2}$ (as shown in Fig. C.15 for $g = 2$).

Fig. C.15. A surface of genus g is covered by $4g - 2$ triangles

It is quite evident that $\sum_{i=1}^{4g-2} \sigma_i$ is a cycle representing $[M]$, so that we have $\|M\| \le 4g - 2$.

Since $\Pi_1(M)$ contains some subgroup isomorphic to $\mathbb{Z}$, for $d \in \mathbb{N}$ there exists some d-sheet covering $p : \tilde{M} \to M$ (remark that $\tilde{M}$ is a compact oriented surface, so that the above inequality applies to $\tilde{M}$ too). Since $\chi(\tilde{M}) = d \cdot \chi(M)$ we have $g(\tilde{M}) = 1 + d(g - 1)$; moreover $\deg(p) = d$, which implies that $\|\tilde{M}\| = d\|M\|$ and then

$$d \cdot \|M\| \le 4(1 + d(g - 1)) - 2 = 4d(g - 1) + 2.$$

If we divide by d and pass to the limit as $d \to \infty$ we obtain the required inequality.   □

We are now going to prove this inequality in the general case. We start with the following:

Lemma C.4.8. In order to check that $\|M\| \leq \dfrac{\text{vol}(M)}{v_n}$ it suffices to show that $\forall \varepsilon > 0$ there exists a straight cycle $\sum_{i=1}^{k} a_i \sigma_i$ representing $[M]$ such that

$$\text{sgn}(a_i) \cdot \text{algvol}(\sigma_i) \geq v_n - \varepsilon \ \forall i.$$

Proof. Under these hypotheses we have, by definition,

$$\sum_{i=1}^{k} |a_i| \geq \|M\|.$$

Moreover the same method used in the proof of C.4.6 allows us to see that

$$\text{vol}(M) = \sum_{i=1}^{k} a_i \cdot \text{algvol}(\sigma_i)$$

and hence

$$\text{vol}(M) = \sum_{i=1}^{k} |a_i| \text{sgn}(a_i) \cdot \text{algvol}(\sigma_i) \geq \sum_{i=1}^{k} |a_i| \cdot (v_n - \varepsilon) \geq \|M\| \cdot (v_n - \varepsilon)$$

and the conclusion follows from the arbitrariness of ε. $\qquad\qquad\square$

In the sequel we shall use Proposition C.1.3 proving the existence of a compact geodesic polyhedron in $\mathbf{H}^n$ being a fundamental domain for $\Pi_1(M)$ (which is identified with a group of isometries of $\mathbf{H}^n$).

We recall now that if G is a (real finite-dimensional) Lie group a Borel measure μ on G is called a <u>left-invariant</u> <u>Haar</u> <u>measure</u> if
$- \forall A \subset G$ Borel set, $\forall g \in G$ we have $\mu(g \cdot A) = \mu(A)$;
$-$ if A is open and non-empty then $\mu(A) > 0$;
$-$ if A is compact then $\mu(A) < \infty$.
It is rather easily checked that such measures do exist in any case and that they are multiples of each other.

A similar definition and the same existence and uniqueness property can be given for a <u>right-invariant</u> <u>Haar</u> <u>measure</u>. We shall say G is <u>unimodular</u> if a left-invariant Haar measure on G is right-invariant too.

Lemma C.4.9. Compact Lie groups are unimodular.

Proof. Let μ be left-invariant, let $g \in G$ and define another measure μ_g by the relation $\mu_g(A) = \mu(A \cdot g)$. Since it is a left-invariant Haar measure we have $\mu_g = k \cdot \mu$. But now

$$k \cdot \mu(G) = \mu_g(G) = \mu(G \cdot g) = \mu(G) < \infty$$

so that $k = 1$ and the lemma is proved. $\qquad\qquad\square$

Now let $\mathcal{G}$ be the Lie algebra of G. We recall that for $X \in \mathcal{G}$ an endomorphism of $\mathcal{G}$ called the <u>adjoint</u> of X is defined by $\text{ad}(X) : Y \mapsto [X, Y]$. The following elementary characterization of unimodular groups can be found *e.g.* in [He]; tr denotes the trace of a linear map.

Proposition C.4.10. G is unimodular if and only if

$$tr(\mathrm{ad}(X)) = 0 \ \forall X \in \mathcal{G}.$$

We come now to the fact these notions were introduced for; we recall that $\mathcal{I}(\mathbf{H}^n)$ was proved to be a closed subgroup of $\mathrm{Gl}(n+1)$ and then it is naturally endowed with a Lie group structure.

Proposition C.4.11. $\mathcal{I}(\mathbf{H}^n)$ is a unimodular Lie group.

Proof. Let $G = \mathrm{O}(\mathbb{R}^{n+1}, \langle .|.\rangle_{(n,1)})$ and recall that in A.2.4 we have checked that $\mathcal{I}(\mathbf{H}^n) = \mathrm{O}(\mathbb{I}^n)$ consists of "half" of G: precisely, it consists of two of the four connected components of G. Then it suffices to show that G is unimodular.

Let $\mathcal{G}$ be the Lie algebra of G: according to general facts (see [He] again) $\mathcal{G}$ is the set of all $(n+1) \times (n+1)$-matrices X such that $\exp(tX) \in G \ \forall t \in \mathbb{R}$. Then $\mathcal{G}$ is the direct sum of the following two subspaces:

$$W = \left\{ \begin{pmatrix} A & 0 \\ 0 & 0 \end{pmatrix} : A \ n \times n\text{-matrix}, \ {}^t\!A + A = 0 \right\}$$

$$V = \left\{ \begin{pmatrix} 0 & v \\ {}^t v & 0 \end{pmatrix} : v \in \mathbb{R}^n \right\}.$$

(Hint: write down the relations defining G, and check that it has dimension $n(n+1)/2$; remark that $\exp(W) \subset G$ and by direct computation of the exponential mapping prove also that $\exp(V) \subset G$, and then conclude using dimensions.)

Since the mappings $X \mapsto \mathrm{ad}(X)$ and $Y \mapsto trY$ are linear, it suffices to prove that $tr(\mathrm{ad}(X)) = 0 \ \forall X \in W \cup V$.

Let $X = \begin{pmatrix} A & 0 \\ 0 & 0 \end{pmatrix}$. Then $\mathrm{ad}(X)(W) \subset W$ and $\mathrm{ad}(X)$ must have null trace if restricted to W: in fact the space of skew-symmetric $n \times n$-matrices is the algebra of the compact group $\mathrm{O}(n)$, and Lemma C.4.9 applies. Moreover for $v \in \mathbb{R}^n$ we have

$$\mathrm{ad}(X)\begin{pmatrix} 0 & v \\ {}^t v & 0 \end{pmatrix} = \begin{pmatrix} 0 & Av \\ {}^t(Av) & 0 \end{pmatrix}$$

so that in a natural way $tr(\mathrm{ad}(X)) = tr(A) = 0$.

Let $X = \begin{pmatrix} 0 & v \\ {}^t v & 0 \end{pmatrix}$; then $\mathrm{ad}(X)(W) \subset V$ and $\mathrm{ad}(X)(V) \subset W$, so that $\mathrm{ad}(X)$ certainly has null trace. $\qquad\square$

Remark C.4.12. Given $x \in \mathbf{H}^n$ and a Borel subset A of $\mathcal{I}(\mathbf{H}^n)$ we can set

$$\mu_x(A) = \mathrm{vol}(A(x));$$

it is quite easily checked that μ_x is a left-invariant Haar measure: the above result implies that it is right-invariant too, and then (by the transitivity of $\mathcal{I}(\mathbf{H}^n)$) we have that it is actually independent of x.

In the sequel we shall consider a fixed bi-invariant Haar measure μ on $\mathcal{I}(\mathbf{H}^n)$; we shall often implicitly make use of the concrete characterization of μ suggested by C.4.12. All the subsets of $\mathcal{I}(\mathbf{H}^n)$ we are going to consider are "very nice" compact subsets, namely finite intersections of closures of open subsets with smooth boundary. We shall often use without explicit mention the fact that the measure of such a nice set equals the measure of its interior.

We will also need the following result (to be compared with C.2.4):

Lemma C.4.13. Let $u_0, ..., u_n \in \mathbf{H}^n$ and let σ be their convex hull; then σ is regular if and only if $d(u_i, u_j)$ is constant for $i \neq j$.

Proof. The "only if" part is quite obvious: if i, j, k are different from each other there exists $\phi \in \mathcal{I}(\mathbf{H}^n)$ keeping u_k fixed and interchanging u_i and u_j, so that

$$d(u_k, u_i) = d(u_k, u_j)$$

and the conclusion follows at once.

As for the "if" part, it suffices to show that each transposition of the vertices is induced by an isometry of $\mathbf{H}^n$. Let $i \neq j$ and choose k different from both of them. In the disc model assume $u_k = 0$ and define H as the (linear) hyperplane in $\mathbb{R}^n$ spanned by $\{u_l : l \neq i, j, k\}$ and by the middle point of the segment $[u_i, u_j]$. Since $d(0, u_i) = d(0, u_j)$ the segment $[u_i, u_j]$ is orthogonal to H and hence the symmetry with respect to H interchanges u_i and u_j; moreover it is an isometry of $\mathbb{D}^n$ and it keeps u_l fixed for all $l \neq i, j$. $\square$

Now, for $R > 0$ we set

$$\mathcal{S}(R) = \left\{ (u_0, ..., u_n) \in \left(\mathbf{H}^n\right)^{n+1} : d(u_i, u_j) = R \; \forall i \neq j \right\}.$$

According to the above lemma, if $(u_0, ..., u_n) \in \mathcal{S}(R)$ then the geodesic simplex with vertices $u_0, ..., u_n$ is regular.

Lemma C.4.14. If $(u_0^R, ..., u_n^R) \in \mathcal{S}(R)$ is fixed, the mapping

$$\Psi : \mathcal{I}(\mathbf{H}^n) \ni g \mapsto (g(u_0^R), ..., g(u_n^R)) \in \mathcal{S}(R)$$

is a bijection.

Proof. Assume $\Psi(g_1) = \Psi(g_2)$; in the disc model we can assume $u_0^R = 0$, so that $g_1^{-1} \circ g_2$ is a linear mapping keeping n linearly independent points fixed, which implies that it is the identity, and therefore Ψ is one-to-one.

Let $(u_0, ..., u_n) \in \mathcal{S}(R)$. By application of an isometry we can reduce to the case $u_0 = u_0^R = 0 \in \mathbb{D}^n$. Since both u_1 and u_1^R lie on the same sphere centred at 0 we can find $A \in O(n)$ such that $Au_1^R = u_1$ and hence we can assume $u_1 = u_1^R$. Similarly we can continue, and surjectivity of Ψ is proved. $\square$

In the sequel we shall assume that a point $(u_0^R, ..., u_n^R) \in \mathcal{S}(R)$ is fixed for each $R > 0$. According to the above lemma the Haar measure μ we fixed on $\mathcal{I}(\mathbf{H}^n)$ induces a measure m on $\mathcal{S}(R)$ by

$$m(A) = \mu\big(\{g \in \mathcal{I}(\mathbf{H}^n) : (g(u_0^R), ..., g(u_n^R)) \in A\}\big) \quad (A \subset \mathcal{S}(R)).$$

Remark as well that right-invariance of μ implies that the induced measure on $\mathcal{S}(R)$ is independent of the starting point $(u_0^R, ..., u_n^R)$.

We shall say an element $(u_0, ..., u_n)$ of $\mathcal{S}(R)$ is positive if the mapping

$$\sigma(u_0, ..., u_n) : \Delta_n \ni (t_0, ..., t_n) \mapsto \sum_{i=0}^{n} t_i u_i \in \mathbf{H}^n$$

is orientation-preserving (remark that if $(u_0, ..., u_n) \in \mathcal{S}(R)$ then $\sigma(u_0, ..., u_n)$ is non-degenerate). Then we can divide $\mathcal{S}(R)$ in a natural way into $\mathcal{S}_+(R)$ and $\mathcal{S}_-(R)$. Moreover if in the above lemma we assume the starting point $(u_0^R, ..., u_n^R)$ is positive then $\mathcal{S}_+(R)$ corresponds to $\mathcal{I}^+(\mathbf{H}^n)$.

We define now

$$\tilde{\mathcal{S}}(R) = \big\{\sigma(u_0, ..., u_n) : (u_0, ..., u_n) \in \mathcal{S}(R)\big\}$$

(we will often confuse a simplex with its parametrization, so that $\tilde{\mathcal{S}}(R)$ will be viewed as a subset of $\mathcal{S}_n$, the set of all the simplices we introduced in Sect. C.2).

Let us remark that by C.4.14 if $\tau, \tau' \in \tilde{\mathcal{S}}(R)$ then there exists $g \in \mathcal{I}(\mathbf{H}^n)$ such that $g(\tau) = \tau'$ so that $\mathrm{vol}(\tau) = \mathrm{vol}(\tau')$. It follows that the function

$$V : \mathbb{R}_+ \to \mathbb{R}_+ \qquad R \mapsto \mathrm{vol}(\tau) \quad (\text{for } \tau \in \tilde{\mathcal{S}}(R))$$

is well-defined.

Lemma C.4.15. $\lim_{R \to \infty} V(R)$ exists and equals v_n.

Proof. V is obviously non-decreasing and bounded by v_n. Moreover in $\mathbb{D}^n$ given a regular ideal simplex τ we can find a function $R \mapsto \tau_R \in \tilde{\mathcal{S}}(R)$ (as suggested by Fig. C.16) such that

$$\lim_{R \to \infty} \mathrm{vol}(\tau_R) = \mathrm{vol}(\tau)$$

and hence C.2.1 implies the conclusion. $\square$

We are finally ready to conclude the proof of Theorem C.4.2.

Proposition C.4.16. $\|M\| \leq \dfrac{\mathrm{vol}(M)}{v_n}$.

Proof. Let $\Gamma \cong \Pi_1(M)$ be such that $M = \mathbf{H}^n/\Gamma$ (we recall that a covering $\pi : \mathbf{H}^n \to M$ was fixed at the beginning). We define an action of Γ on Γ^{n+1} as left translation on each component:

$$\Gamma \ni \gamma \mapsto \Big(\Gamma^{n+1} \ni (\gamma_0, ..., \gamma_n) \mapsto (\gamma \cdot \gamma_0, ..., \gamma \cdot \gamma_n)\Big).$$

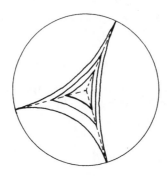

Fig. C.16. A sequence of regular simplices whose volume converges to the volume of a regular ideal simplex

We define Ω as the quotient set Γ^{n+1}/Γ with respect to this action.

Let us fix a compact geodesic polyhedron D in $\mathbb{H}^n$ as a fundamental domain for Γ and a point $u \in \overset{\circ}{D}$. We shall denote by d the (finite) diameter of D.

For $\omega = [(\gamma_0, ..., \gamma_n)] \in \Omega$ we define the simplex

$$\sigma_\omega = \pi \circ \sigma(\gamma_0(u), ..., \gamma_n(u)), \qquad i.e.$$

$$\sigma_\omega : \Delta_n \ni t \mapsto \pi \left(\sum_{i=0}^{n} t_i \gamma_i(u) \right) \in M$$

(of course it is well-defined); we define as well for $R > 0$ a number $a_R^+(\omega)$ in the following way

$$a_R^+(\omega) = m(\{(u_0, ..., u_n) \in \mathcal{S}_+(R) : u_j \in \gamma_j(D) \forall j\});$$

we must check that this is well-defined: for $\gamma \in \Gamma$ we have

$$m(\{(u_0, ..., u_n) \in \mathcal{S}_+(R) : u_j \in \gamma\gamma_j(D) \forall j\}) =$$
$$= \mu(\{g \in \mathcal{I}^+(\mathbb{H}^n) : g(u_j^R) \in \gamma\gamma_j(D) \forall j\}) =$$
$$= \mu(\{\gamma g : g \in \mathcal{I}^+(\mathbb{H}^n), \ g(u_j^R) \in \gamma_j(D) \forall j\}) =$$
$$= \mu(\gamma\{g \in \mathcal{I}^+(\mathbb{H}^n) : g(u_j^R) \in \gamma_j(D) \forall j\}) =$$
$$= m(\{(u_0, ..., u_n) \in \mathcal{S}_+(R) : u_j \in \gamma_j(D) \forall j\})$$

(left-invariance of μ and the fact that $\Gamma \subset \mathcal{I}^+(\mathbb{H}^n)$ were used here).

Remark also that since D is compact a set of the form

$$\{\delta \in \mathcal{I}(\mathbb{H}^n) : \delta(x_0) \in D\}$$

is compact in $\mathcal{I}(\mathbb{H}^n)$ and hence has finite measure. This implies immediately that $a_R^+(\omega) < \infty$.

Similarly we introduce the number

$$a_R^-(\omega) = m\big(\{(u_0, ..., u_n) \in \mathcal{S}_-(R) : u_j \in \gamma_j(D) \,\forall j\}\big) < \infty.$$

We define at last

$$a_R(\omega) = a_R^+(\omega) - a_R^-(\omega).$$

We consider now the formal sum

$$z_R = \sum_{\omega \in \Omega} a_R(\omega) \cdot \sigma_\omega.$$

We shall prove that z_R is indeed a cycle representing a non-zero multiple of $[M]$ and that it allows us to conclude the proof as suggested by C.4.8; this fact will require several partial results.

(i) *z_R is expressed by a finite sum.*
Let us remark that each $\omega \in \Omega$ has one and only one representative of the form

$$(\text{id}, \gamma_1, ..., \gamma_n).$$

Assume now $a_R(\omega) \neq 0$; then there exists at least one simplex $\sigma(u_0, ..., u_n)$ in $\tilde{\mathcal{S}}(R)$ (*i.e.* a regular simplex with edges of length R) such that

$$u_0 \in D \qquad u_i \in \gamma_i(D) \ (1 \leq i \leq n).$$

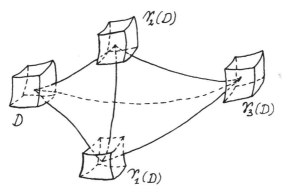

Fig. C.17. Proof that z_R is a finite sum: a regular R-simplex as represented exists only for finitely many $\gamma_1, \gamma_2, ...$

This implies that $d(u, \gamma_i(u)) \leq 2d + R$ for $i = 1, ..., n$ (we recall that u is a fixed point in the interior of D). Since Γ is discrete only finitely many choices of the γ_i's are possible, and hence $a_R(\omega) \neq 0$ for finitely many ω's.

(ii) *z_R is a cycle.*
We shall consider an $(n-1)$-face τ of one of the σ_ω's appearing in z_R and we shall prove that its coefficient in ∂z_R is zero (which implies that z_R is a cycle). By construction τ is obtained as projection of a geodesic $(n-1)$-simplex having vertices in the Γ-orbit of u, that is

$$\tau : \Delta_{n-1} \ni (t_0, ..., t_{n-1}) \mapsto \pi \left(\sum_{i=0}^{n-1} t_i \gamma_i(u) \right)$$

for suitable $\gamma_0, ..., \gamma_{n-1}$.

We want to prove now that the coefficient of τ in ∂z_R is zero. Such a coefficient is explicitly given by

$$\sum_{j=0}^{n} (-1)^{n-j} \sum_{\gamma \in \Gamma} a_R \big([(\gamma_0, ..., \gamma_{j-1}, \gamma, \gamma_j, ..., \gamma_{n-1})] \big)$$

and we shall prove that for each j the second sum vanishes. We take by simplicity $j = n$. Then the second sum is given by

$$\sum_{\gamma_n \in \Gamma} a_R \big([(\gamma_0, ..., \gamma_n)] \big) =$$
$$= \sum_{\gamma_n \in \Gamma} a_R^+ \big([(\gamma_0, ..., \gamma_n)] \big) - \sum_{\gamma_n \in \Gamma} a_R^- \big([(\gamma_0, ..., \gamma_n)] \big).$$

Since D is a fundamental domain for Γ we have

$$\sum_{\gamma_n \in \Gamma} a_R^+ \big([(\gamma_0, ..., \gamma_n)] \big) =$$
$$= \sum_{\gamma_n \in \Gamma} m \big(\{ (u_0, ..., u_n) \in \mathcal{S}_+(R) : u_i \in \gamma_i(D) \, \forall i \} \big) =$$
$$= m \Big(\bigcup_{\gamma_n \in \Gamma} \{ (u_0, ..., u_n) \in \mathcal{S}_+(R) : u_i \in \gamma_i(D) \, \forall i \} \Big) =$$
$$= m \big\{ (u_0, ..., u_n) \in \mathcal{S}_+(R) : u_i \in \gamma_i(D) \, \forall i \le n - 1,$$
$$\exists \gamma_n \in \Gamma \ s.t. \ u_n \in \gamma_n(D) \big\}$$
$$= m \big\{ (u_0, ..., u_n) \in \mathcal{S}_+(R) : u_i \in \gamma_i(D) \, \forall i \le n - 1 \big\}.$$

A similar calculation for the other sum proves that if for $s \in \{+, -\}$ we set

$$A_s = \big\{ (u_0, ..., u_n) \in \mathcal{S}_s(R) : u_i \in \gamma_i(D) \, \forall i \le n - 1 \big\}$$

then the coefficient we are considering is given by

$$m(A_+) - m(A_-).$$

Now let $g_0 \in \mathcal{I}(\mathbb{H}^n)$ be the reflection with respect to the hyperbolic hyperplane containing $u_0^R, ..., u_{n-1}^R$; then we easily get

$$\{ g \in \mathcal{I}^-(\mathbb{H}^n) : g(u_i^R) \in \gamma_i(D) \, \forall i < n \} =$$
$$= \{ g \in \mathcal{I}^+(\mathbb{H}^n) : g(u_i^R) \in \gamma_i(D) \, \forall i < n \} \cdot g_0$$

so that by definition of m and right-invariance of μ we have $m(A_+) = m(A_-)$ and hence the proof of (ii) is complete.

(iii) if $R > 2d$ then $a_R^+(\omega) \cdot a_R^-(\omega) = 0 \ \forall \omega$.

It is easily checked (see also Fig. C.17 again) that if an element σ_0 of $\tilde{\mathcal{S}}(R)$ has the first vertex in D, the second in $\gamma_1(D)$, ... and the n-th in $\gamma_n(D)$, then any other element of $\tilde{\mathcal{S}}(R)$ with this property has the same orientation as σ_0.

(iv) $\forall\, \varepsilon > 0$ *if* R *is big enough then* $|\text{algvol}(\sigma_\omega)| \geq v_n - \varepsilon$ *whenever* $a_R(\omega)$ *is not zero.*

Let us assume for $\omega = [(\gamma_0, ..., \gamma_n)]$ we have $a_R(\omega) \neq 0$; then there exists $\sigma(u_0, ..., u_n) \in \tilde{\mathcal{S}}(R)$ such that $u_i \in \gamma_i(D) \forall\, i$, and hence $d(\gamma_i(u), u_i) \leq d \forall\, i$. Moreover by definition

$$\big|\text{algvol}(\sigma_\omega)\big| = \text{vol}\big(\sigma(\gamma_0(u), ..., \gamma_n(u))\big).$$

Since the vertices of $\sigma(\gamma_0(u), ..., \gamma_n(u)))$ are d-close to the vertices of an element of $\tilde{\mathcal{S}}(R)$, Lemma C.4.15 rather easily implies that the volume of $\sigma(\gamma_0(u), ..., \gamma_n(u)))$ goes to v_n as $R \to \infty$. (We have to prove that the function

$$\theta(R) = \sup \big\{ |\text{vol}(\tau) - \text{vol}(\tau_0)| : \tau_0 = \sigma(v_0, ..., v_n) \in \tilde{\mathcal{S}}(R),$$
$$\tau = \sigma(w_0, ..., w_n) \; s.t. \; d(v_i, w_i) \leq d \,\forall\, i \big\}$$

has limit 0 as $R \to \infty$, and this is almost immediate.)

(v) *if* $R > 2d$ *and* $a_R(\omega) \neq 0$ *then* $a_R(\omega) \cdot \text{algvol}(\sigma_\omega) > 0$.

Assume $a_R(\omega) \neq 0$ and define for $x \in M$ the number $\alpha_\omega(x)$ as

$$\#\big\{ t \in \mathring{\Delta}_n : \sigma_\omega(t) = x, d_t \sigma_\omega \text{ pos.} \big\} - \#\big\{ t \in \mathring{\Delta}_n : \sigma_\omega(t) = x, d_t \sigma_\omega \text{ neg.} \big\}.$$

Since

$$\text{algvol}(\sigma_\omega) = \int_M \alpha_\omega(x) dv(x)$$

we only have to check that $a_R(\omega) \alpha_\omega(x) \geq 0 \;\forall\, x$.

We shall prove that if $a_R^+(\omega) \neq 0$ and $\alpha_\omega(x) \neq 0$ then $\alpha_\omega(x) > 0$ (the case $a_R^-(\omega) \neq 0$ being settled analogously). Let $\tilde{x} \in \pi^{-1}(x)$ and consider the lift $\tilde{\sigma}_\omega$ of σ_ω starting at $\tilde{x}$; $\tilde{\sigma}_\omega$ is given by $\sigma(\gamma_0(u), ..., \gamma_n(u))$, with $[(\gamma_0, ..., \gamma_n)] = \omega$. Since $a_R^+(\omega) \neq 0$ we can find $(u_0, ..., u_n) \in \mathcal{S}_+(R)$ such that $u_i \in \gamma_i(D) \forall\, i$. In particular $d(u_i, \gamma_i(u)) \leq d$ and since $R > 2d$ this implies that $\sigma(\gamma_0(u), ..., \gamma_n(u))$ is positively oriented and hence $\alpha_\omega(x) > 0$.

(vi) *if* $R > 2d$ *then* $\text{algvol}(z_R) > 0$, *so that in particular* z_R *is a non-trivial cycle.*

This fact will follow from *(v)* if we show that not all the $a_R(\omega)$'s vanish.

Pick $(u_0^0, ..., u_n^0) \in \mathcal{S}(R)$; since $\Gamma(D) = \mathbb{H}^n$ we can find $\gamma_0, ..., \gamma_n \in \Gamma$ such that $u_i^0 \in \gamma_i(D) \forall\, i$. By perturbing a little $(u_0^0, ..., u_n^0)$ and changing if necessary the γ_i's we can assume as well

$$u_i^0 \in \gamma_i(\mathring{D}) \,\forall\, i.$$

It follows that

$$m\big(\big\{ (u_0, ..., u_n) \in \mathcal{S}(R) : u_i \in \gamma_i(D) \,\forall\, i \big\}\big) \neq 0$$

and hence for $\omega = [(\gamma_0, ..., \gamma_m)]$ either $a_R^+(\omega) \neq 0$ or $a_R^-(\omega) \neq 0$, which implies $a_R(\omega) \neq 0$.

Conclusion of the proof of C.4.16.

Given $\epsilon > 0$ we take $R > 2d$ such that *(iv)* applies. Then, according to *(i)* and *(ii)*, z_R is a singular cycle. *(iv)* and *(v)* imply that

$$\mathrm{sgn}(a_R(\omega)) \cdot \mathrm{algvol}(\sigma_\omega) \geq v_n - \varepsilon$$

whenever $a_R(\omega) \neq 0$.

By *(vi)* for some real number $k \neq 0$ we have $[z_R] = k[M]$, *i.e.* $1/k \cdot z_R$ is a representative of $[M]$, and hence $\mathrm{algvol}(z_R) = k \cdot \mathrm{vol}(M)$ so that $k > 0$. Then the conclusion follows from Lemma C.4.8 as

$$1/k \cdot z_R = \sum_{\omega \in \Omega} 1/k \cdot a_R(\omega) \cdot \sigma_\omega$$

is a straight cycle and

$$\mathrm{sgn}\left(1/k \cdot a_R(\omega)\right) \cdot \mathrm{algvol}(\sigma_\omega) \geq v_n - \varepsilon.$$

$\square$

Theorem C.4.2 is proved at last.

Though we will not need it for the proof of C.4.1, we record the following consequence of C.4.2 (and C.3.4):

Corollary C.4.17. If M is a compact oriented hyperbolic manifold and $f \in C^0(M, M)$ then $|\deg(f)| \leq 1$.

We are now ready for the conclusion of the third step of the proof of the rigidity theorem. We recall that for $u_0, ..., u_k \in \mathbb{H}^n$ we denoted by $\sigma(u_0, ..., u_k)$ the (barycentric parametrization of) the geodesic simplex with vertices $u_0, ..., u_k$. If we forget the parametrization and we refer to a simplex merely as a subset of $\overline{\mathbb{H}^n}$, this notation generalizes in a natural way to the case $u_0, ..., u_k \in \overline{\mathbb{H}^n}$.

Proof of C.4.1. Assume by contradiction that there exists a geodesic simplex $\tau = \sigma(w_0, ..., w_n)$ in $\overline{\mathbb{H}^n}$ such that $\mathrm{vol}(\tau) = v_n$ and
$$\mathrm{vol}(\sigma(\tilde{f}(w_0), ..., \tilde{f}(w_n))) = v_n - 2\varepsilon < v_n.$$

Of course we can assume the vertices are ordered in such a way that τ is positive (*i.e.* the orientation given by the ordering of the vertices matches the orientation of $\mathbb{H}^n$).

Since $\tilde{f}$ is continuous on $\overline{\mathbb{H}^n}$ for $j = 0, ..., n$ we can choose a neighborhood U_j of w_j in $\overline{\mathbb{H}^n}$ such that if $u_j \in U_j$ (for $j = 0, ..., n$) then
$$\mathrm{vol}(\sigma(\tilde{f}(u_0), ..., \tilde{f}(u_n))) \leq v_n - \varepsilon.$$

Let the notations of the proof of C.4.2 be fixed for $M = M_1$, $\pi = \pi_1$, $\Gamma = \Gamma_1$. We set now:

$$c_R = \sum \left\{ a_R(\omega) \cdot \sigma_\omega : \omega = [(\gamma_0, ..., \gamma_n)] \in \Omega \ s.t. \ \gamma_i(u) \in U_i \forall i \right\}.$$

We claim that there exist two positive numbers α_1 and α_2 such that if R is big enough: (a) $\|z_R\| = \alpha_1$; (b) $\|c_R\| \geq \alpha_2$. Let us assume this claim for a moment and let us conclude the proof.

Since f is a homotopy equivalence we have $f_*([M_1]) = \pm[M_2]$ and hence $\|M_1\| = \|M_2\|$ so that by C.4.2 it is $\text{vol}(M_1) = \text{vol}(M_2)$. We also recall that $z_R = k[M_1]$ for some $k > 0$, which implies $f_*([z_R]) = \pm k[M_2]$. As a representative of $f_*([z_R])$ we consider the straightening z'_R of the cycle $f \circ z_R$, explicitly given by

$$z'_R = \sum_{\omega=[(\gamma_0,\ldots,\gamma_n)]\in\Omega} a_R(\omega) \cdot \Big(\pi_2 \circ \sigma\big(\tilde{f}(\gamma_0(u)),\ldots,\tilde{f}(\gamma_n(u))\big)\Big).$$

(In fact the straightening of $f \circ \sigma_\omega = f \circ \pi_1 \circ \sigma(\gamma_0(u),\ldots,\gamma_n(u))$ is easily recognized to be $\pi_2 \circ \sigma\big(\tilde{f}(\gamma_0(u)),\ldots,\tilde{f}(\gamma_n(u))\big)$.)

Since $[z_R] = k[M_1]$ and $[z'_R] = \pm k[M_2]$ we have

$$\text{algvol}(z_R) = k \cdot \text{vol}(M_1) \qquad \text{algvol}(z'_R) = \pm k \cdot \text{vol}(M_2)$$

(this fact was already used several times); but $\text{vol}(M_1) = \text{vol}(M_2)$ and hence

$$\text{algvol}(z_R) = \pm\text{algvol}(z'_R).$$

Let us recall now that during the proof of C.4.16, point *(iv)*, it was checked that if $\omega = [(\gamma_0,\ldots,\gamma_n)]$ is such that $a_R(\omega) \neq 0$ then the volume of $\sigma(\gamma_0(u),\ldots,\gamma_n(u))$ differs from v_n by a quantity that goes to 0 as R goes to ∞. It follows that

$$|\text{algvol}(z_R)| \geq \|z_R\| \cdot \inf\Big\{ \text{vol}(\sigma(\gamma_0(u),\ldots,\gamma_n(u))) :$$
$$\omega = [(\gamma_0,\ldots,\gamma_n)]\in\Omega,\ a_R(\omega)\neq 0\Big\}$$

and the right hand side converges to $\alpha_1 \cdot v_n$ as $R \to \infty$. Moreover:

$$|\text{algvol}(z'_R)| = \sum_{\omega=[(\gamma_0,\ldots,\gamma_n)]\in\Omega} |a_R(\omega)| \cdot \text{vol}\big(\sigma\big(\tilde{f}(\gamma_0(u)),\ldots,\tilde{f}(\gamma_n(u))\big)\big) =$$

$$= \sum\Big\{|a_R(\omega)| \cdot \text{vol}\big(\sigma\big(\tilde{f}(\gamma_0(u)),\ldots,\tilde{f}(\gamma_n(u))\big)\big) :$$
$$: \omega = [(\gamma_0,\ldots,\gamma_n)] \in \Omega,\ \exists i\ s.t.\ \gamma_i(u) \notin U_i\Big\} +$$

$$+ \sum\Big\{|a_R(\omega)| \cdot \text{vol}\big(\sigma\big(\tilde{f}(\gamma_0(u)),\ldots,\tilde{f}(\gamma_n(u))\big)\big) :$$
$$: \omega = [(\gamma_0,\ldots,\gamma_n)] \in \Omega,\ \gamma_i(u)\in U_i\,\forall i\Big\} \leq$$

$$\leq v_n \cdot \sum\Big\{|a_R(\omega)| : \omega = [(\gamma_0,\ldots,\gamma_n)] \in \Omega,\ \exists i\ s.t.\ \gamma_i(u) \notin U_i\Big\} +$$

$$+ (v_n - \varepsilon) \cdot \sum\Big\{|a_R(\omega)| : \omega = [(\gamma_0,\ldots,\gamma_n)] \in \Omega,\ \gamma_i(u) \in U_i\,\forall i\Big\} \leq$$

$$\leq v_n \cdot (\|z_R\| - \|c_R\|) + (v_n - \varepsilon) \cdot \|c_R\| \leq \alpha_1 v_n \left(1 - \frac{\varepsilon\alpha_1}{v_n\alpha_2}\right)$$

and this is absurd since

$$\liminf |\mathrm{algvol}(z'_R)| \geq \alpha_1 v_n.$$

We are left to prove the claim. We can assume the points $\{(u_0^R, ..., u_n^R)\}_{R>0}$ used in C.4.14 for the identification between $\mathcal{I}^+(\mathbb{H}^n)$ and $\mathcal{S}_+(R)$ are chosen in such a way that $u_i^R \to w_i$ as $R \to \infty$ (compare with the construction described in Fig. C.16). Then, as soon as R is such that $a_R^+(\omega) \cdot a_R^-(\omega) = 0 \ \forall \omega$, we have

$$\|z_R\| = \sum_{\omega \in \Omega} |a_R(\omega)| =$$

$$= \sum_{\omega \in \Omega,\ a_R^+(\omega) \neq 0} a_R^+(\omega) + \sum_{\omega \in \Omega,\ a_R^-(\omega) \neq 0} a_R^-(\omega) = \sum_{\omega \in \Omega} \left(a_R^+(\omega) + a_R^-(\omega) \right) =$$

$$= \sum_{\omega = [(\mathrm{id}, \gamma_1, ..., \gamma_n)] \in \Omega} \left(m\left\{ (u_0, ..., u_n) \in \mathcal{S}_+(R) : u_0 \in D, u_i \in \gamma_i(D) \right\} + \right.$$

$$\left. + m\left\{ (u_0, ..., u_n) \in \mathcal{S}_-(R) : u_0 \in D, u_i \in \gamma_i(D) \right\} \right) =$$

$$= \sum_{\omega = [(\mathrm{id}, \gamma_1, ..., \gamma_n)] \in \Omega} m\left\{ (u_0, ..., u_n) \in \mathcal{S}(R) : u_0 \in D, u_i \in \gamma_i(D) \right\} =$$

$$= \sum_{\omega = [(\mathrm{id}, \gamma_1, ..., \gamma_n)] \in \Omega} \mu\left\{ \delta \in \mathcal{I}(\mathbb{H}^n) : \delta(u_0^R) \in D, \delta(u_i) \in \gamma_i(D) \right\} =$$

$$= \mu\left(\bigcup_{\gamma_1, ..., \gamma_n \in \Gamma} \left\{ \delta \in \mathcal{I}(\mathbb{H}^n) : \delta(u_0^R) \in D, \delta(u_i) \in \gamma_i(D) \right\} \right) =$$

$$= \mu\left\{ \delta \in \mathcal{I}(\mathbb{H}^n) : \delta(u_0^R) \in D \right\}$$

and right-invariance of μ combined with transitivity of $\mathcal{I}(\mathbb{H}^n)$ implies that this number is a constant α_1 independent of u_0^R (and hence of R); part (a) of the claim is proved. As for part (b), we obtain as above that:

$$\|c_R\| = \sum_{\omega = [(\gamma_0, ..., \gamma_n)] \in \Omega,\ \gamma_i(u) \in U_i} \mu\left\{ \delta \in \mathcal{I}(\mathbb{H}^n) : \delta(u_i^R) \in \gamma_i(D) \right\}$$

and then

$$\|c_R\| \geq \sum_{\omega = [(\gamma_0, ..., \gamma_n)] \in \Omega,\ \gamma_i(D) \subset U_i} \mu\left\{ \delta \in \mathcal{I}(\mathbb{H}^n) : \delta(u_i^R) \in \gamma_i(D) \right\} \geq$$

$$\geq \mu\left(\bigcup_{\gamma_0, ..., \gamma_n \in \Gamma,\ \gamma_i(D) \subset U_i} \left\{ \delta \in \mathcal{I}(\mathbb{H}^n) : \delta(u_i^R) \in \gamma_i(D), \delta(u) \in D \right\} \right) \geq$$

$$\geq \mu\left\{ \delta \in \mathcal{I}(\mathbb{H}^n) : \delta(u_i^R) \in U_i, \delta(u) \in D \right\}.$$

(In the passage before the last one condition $\delta(u) \in D$ was added in order to consider the fact that each $\omega \in \Omega$ has more than one representative of the form $[(\gamma_0, ..., \gamma_n)]$.) Now let us consider another small neighborhood U'_i of w_i in $\mathbb{H}^n$ such that $\overline{U'_i} \subset \mathring{U}_i$ and set

$$\mathcal{M} = \left\{ \delta \in \mathcal{I}(\mathbf{H}^n) : \delta(U_i') \subset U_i \, \forall i, \delta(u) \in D \right\};$$

then $\mathcal{M}$ is a neighborhood of the identity in $\mathcal{I}(\mathbf{H}^n)$: in fact $\mathcal{I}(\mathbf{H}^n)$ can be viewed as a set of continuous functions of $\overline{\mathbf{H}^n}$ onto itself, and $\mathcal{M}$ is a neighborhood of the identity with respect to the compact-open topology (on $\overline{\mathbf{H}^n}$); moreover it is easily checked that the Lie group structure we are considering on $\mathcal{I}(\mathbf{H}^n)$ induces a topology not coarser than this compact-open topology.

If R is big enough we have $u_i^R \in U_i'$ and then

$$\mathcal{M} \subset \left\{ \delta \in \mathcal{I}(\mathbf{H}^n) : \delta(u_i^R) \in U_i, \delta(u) \in D \right\}$$

and then it suffices to set $\alpha_2 = \mu(\mathcal{M}) > 0$. $\qquad\qquad\square$

C.5 Conclusion of the Proof, Corollaries and Generalizations

The following result is the only step in the proof of the rigidity theorem requiring n to be at least three.

Proposition C.5.1. Let $n \geq 3$ and let $P : \partial \mathbb{H}^n \to \partial \mathbb{H}^n$ be a continuous one-to-one mapping such that if an ideal geodesic simplex with vertices $u_0, ..., u_n$ has volume v_n then the ideal geodesic simplex with vertices $P(u_0), ..., P(u_n)$ has volume v_n too. Then P is the trace of an element of $\mathcal{I}(\mathbf{H}^n)$.

Proof. We shall carry out the proof with $n = 3$, the generalization being obvious. According to C.2.1 P has the property that it maps the vertices of every (ideal) regular simplex into the vertices of a (ideal) regular simplex. Since all regular ideal simplices are conjugate in $\mathcal{I}(\mathbf{H}^n)$, we can assume P keeps the vertices of one of these simplices fixed. In the half-space model, recalling C.2.4, we can assume P keeps fixed ∞ and the vertices v_1, v_2, v_3 of an equilateral triangle T in $\mathbb{R}^2 \times \{0\}$ (we denote by l the length of the edges of T). We shall prove that P is indeed the identity on the whole $\mathbb{R}^2 \times \{0\}$, which implies the conclusion.

Let us consider the point v_1' symmetric to v_1 with respect to the segment $[v_2, v_3]$; since the simplex with vertices v_1', v_2, v_3, ∞ is regular, the same must hold for the simplex with vertices $P(v_1'), v_2, v_3, \infty$, so that $P(v_1') \in \{v_1, v_1'\}$. Injectivity implies that $P(v_1') = v_1'$. It follows from this argument that every vertex of the tesselation of $\mathbb{R}^2 \times \{0\}$ associated to T is kept fixed by P.

Now, let v be the barycentre of T. Since v is the image of ∞ under an isometry of $\mathbb{H}^{3,+}$ keeping v_1, v_2, v_3 fixed (an inversion centred at v) the simplex with vertices v, v_1, v_2, v_3 is regular too, and hence as above we have $P(v) \in \{v, \infty\} \Rightarrow P(v) = v$. Of course the same holds for the barycentre of each triangle in the tesselation in question.

Now, let z_1 be the middle point of the segment $[v_2, v_3]$, and let v and v_1' be as above. The inversion i centred at v and keeping v_1, v_2, v_3 fixed is such that $i(v_1') = z_1$; since the simplex with vertices v_1', v_2, v_3, ∞ is regular, the

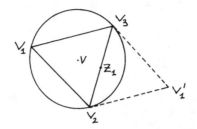

Fig. C.18. Construction of a dense set of fixed points starting from the fixed vertices of a regular ideal simplex

same holds for the simplex with vertices z_1, v_2, v_3, v; once again the properties of P imply that $P(z_1) \in \{z_1, v_1\} \Rightarrow P(z_1) = z_1$. It follows that the middle point of each edge in the tesselation we are considering is kept fixed by P too. Then we can consider a new tesselation of $\mathbb{R}^2$ by triangles of edge $^1/_2$ whose vertices are kept fixed by P: if we iterate this procedure we obtain at last a dense subset of $\mathbb{R}^2 \times \{0\}$ on which P is the identity, and then the continuity implies that P is the identity everywhere. $\square$

Proof of C.0. According to the partial results we have obtained by now, we can assume that there exists a lift $\tilde{f} : \mathbb{H}^n \to \mathbb{H}^n$ such that:
(1) $\tilde{f}$ extends in a continuous way to $\partial\mathbb{H}^n$;
(2) the relation $\tilde{f} \circ \gamma = f_*(\gamma) \circ \tilde{f}$ $\forall \gamma \in \Gamma_1$ holds in the whole of $\overline{\mathbb{H}^n}$;
(3) there exists $q \in \mathcal{I}(\mathbb{H}^n)$ such that $\tilde{f}|_{\partial\mathbb{H}^n} = q|_{\partial\mathbb{H}^n}$.
These facts imply that

$$q \circ \gamma = f_*(\gamma) \circ q \ \forall \gamma \in \Gamma_1 \ \text{ on } \partial\mathbb{H}^n;$$

however this identity involves only elements of $\mathcal{I}(\mathbb{H}^n)$, and hence A.5.13 implies that it holds in the whole of $\overline{\mathbb{H}^n}$.

We set now for $x \in \mathbb{H}^n$, $F(\pi_1(x)) = \pi_2(q(x))$; the conclusion of the proof is deduced from the following facts:
(i) $F : M_1 \to M_2$ is well-defined and one-to-one.
In fact we have that

$$\pi_1(x) = \pi_1(x') \Leftrightarrow \exists \gamma \in \Gamma_1 \ s.t. \ x' = \gamma(x) \Leftrightarrow$$
$$\Leftrightarrow \exists \gamma \in \Gamma_1 \ s.t. \ q(x') = (q \circ \gamma)(x) \Leftrightarrow$$
$$\Leftrightarrow \exists \gamma \in \Gamma_1 \ s.t. \ q(x') = (f_*(\gamma) \circ q)(x) \Leftrightarrow$$
$$\Leftrightarrow \exists \delta \in \Gamma_2 \ s.t. \ \delta(q(x)) = q(x') \Leftrightarrow \pi_2(q(x)) = \pi_2(q(x')).$$

(ii) F is onto.
In fact $\pi_2(y) = F(\pi_1(q^{-1}(y)))$.
(iii) F is an isometry of M_1 onto M_2.
In fact π_1, π_2 and q are local isometries, and bijectivity was checked above.
(iv) F is homotopic to f.

In fact if we set

$$H(t, \pi_1(x)) = \pi_2\big(t \cdot \tilde{f}(x) + (1 - t) \cdot q(x)\big),$$

then (2) and the analogous relation for q easily imply that H is a well-defined homotopy between f and F.   □

The proof of the rigidity theorem is now complete; we prove now an equivalent statement which is apparently stronger:

Theorem C.5.2 (Rigidity Theorem – Sharp Formulation). For $i = 1, 2$ let $M_i = \mathbf{H}^n/\Gamma_i$ be a compact oriented hyperbolic manifold with $n \geq 3$. If there exists a group isomorphism ϕ of Γ_1 onto Γ_2 then we can find $q \in \mathcal{I}(\mathbf{H}^n)$ such that

$$q \circ \gamma = \phi(\gamma) \circ q \ \forall \gamma \in \Gamma_1,$$

so that q induces an isometry g of M_1 onto M_2 with $g_* = \phi$.

Proof. It is sufficient to show that under these assumptions there exists a homotopy equivalence $f : M_1 \to M_2$ such that $f_* = \phi$ (with the basepoints suitably chosen).

Let $v \in \mathbf{H}^n$ and let $v_i = \pi_i(v)$ be the basepoint for $\Pi_1(M_i)$. According to general facts about compact manifolds (see [Greenbl]) M_i admits a presentation as a finite CW-complex where $\{v_i\}$ is the only 0-cell; since cells are contractible such a presentation can be lifted (starting from the fixed point v) to a presentation of $\mathbf{H}^n$. We are now going to define a mapping $\tilde{f} : \mathbf{H}^n \to \mathbf{H}^n$ such that

$$\tilde{f} \circ \gamma = \phi(\gamma) \circ \tilde{f} \ \forall \gamma \in \Gamma_1.$$

$\tilde{f}$ will be defined recursively on the k-th skeletons of the two CW-structures considered on $\mathbf{H}^n$. Since the 0-skeleton is respectively given by $\Gamma_1(v)$ and $\Gamma_2(v)$ for the first step it suffices to set $\tilde{f}(\gamma(v)) = \phi(\gamma)(v) \ \forall \gamma \in \Gamma_1$.

Assume now $\tilde{f}$ is defined on the $(k - 1)$-skeleton and let c be a k-cell in the first presentation of $\mathbf{H}^n$ as a CW-complex, with the associated function

$$h : \overline{D^k} \to \mathbf{H}^n \quad s.t. \quad h\big|_{D^k} : D^k \xrightarrow{\sim} c.$$

If $p_0 \in S^{k-1}$ is such that $h(p_0)$ belongs to the 0-skeleton we consider the mapping

$$j : S^{k-1} \times [0, 1] \to \overline{D^k} \qquad (p, t) \mapsto (1 - t) \cdot p_0 + t \cdot p;$$

let us remark that the restriction of j to $j^{-1}(D^k) = \big(S^{k-1} \setminus \{p_0\}\big) \times (0, 1)$ is one-to-one; then if we set for $(p, t) \in j^{-1}(D^k)$

$$\tilde{f}\big(h(j(p, t))\big) = (1 - t) \cdot \tilde{f}\big(h(p_0)\big) + t \cdot \tilde{f}\big(h(p)\big)$$

(the convex combination being canonically defined in $\mathbf{H}^n$), the function $\tilde{f}$ is well-defined on c, and it easily follows from the very same formula that $\tilde{f}$ is continuous on $\bar{c}$. If c' is another k-cell conjugate to c, *i.e.* such that for some $\gamma \in \Gamma_1$ we have $c' = \gamma(c)$, then we can associate to c' the function $h' = \gamma \circ h$ and perform the same construction as above on c' using h'. This method applied to all the conjugacy classes of k-cells allows one to define continuously $\tilde{f}$ on the k-skeleton. Moreover the relation

$$\tilde{f} \circ \gamma = \phi(\gamma) \circ \tilde{f} \quad \forall \gamma \in \Gamma_1$$

holds by construction.

Now we can set $f(\pi_1(x)) = \pi_2(\tilde{f}(x))$, and the mapping $f : M_1 \to M_2$ is well-defined and continuous. The same construction applied to the isomorphism $\phi^{-1} : M_2 \to M_1$ leads to the construction of a continuous mapping $\tilde{f}_1 : \mathbf{H}^n \to \mathbf{H}^n$ such that

$$\tilde{f}_1 \circ \gamma = \phi^{-1}(\gamma) \circ \tilde{f}_1 \quad \forall \gamma \in \Gamma_2$$

inducing a mapping $f_1 : M_2 \to M_1$. The mapping

$$F(\pi_1(x), t) = \pi_1\big((1 - t) \cdot x + t \cdot (\tilde{f}_1 \circ \tilde{f})(x)\big)$$

is a well-defined homotopy between $f_1 \circ f$ and the identity of M_1; similarly we can prove that $f \circ f_1$ is homotopic to the identity of M_2, so that f is a homotopy equivalence and the proof is over.   □

During the above proof a special case of the following general fact was checked:

Theorem C.5.3. If X and Y are finite CW-complexes and for $i \geq 2$ the i-th homotopy group of Y is trivial, then each homomorphism

$$\phi : \Pi_1(X) \to \Pi_1(Y)$$

is induced by a continuous mapping $f : X \to Y$.

The following generalization (due to [Pr]) of Theorems C.0 and C.5.2 holds:

Theorem C.5.4. If M_1 and M_2 are finite-volume complete connected hyperbolic n-manifolds with $n \geq 3$ and there exists a group isomorphism ϕ of $\Pi_1(M_1)$ onto $\Pi_1(M_2)$, then there exists an isometry f of M_1 onto M_2 such that $f_* = \phi$.

The proof of such a generalization is based essentially on the same methods used above for the compact case.

The use of the methods of Gromov and Thurston in all their power allows to establish the following very deep generalization of Theorem C.0, which could be considered as the "final version" of the rigidity theorem (see [Th1, ch. 6]):

Theorem C.5.5. If M_1 and M_2 are compact connected hyperbolic n-manifolds, with $n \geq 3$, and if $f : M_1 \to M_2$ is such that

$$\mathrm{vol}(M_1) = |\deg(f)| \cdot \mathrm{vol}(M_2)$$

then f is homotopic to a locally isometric covering of degree $|\deg(f)|$ of M_1 onto M_2.

(It is easily checked that this fact together with C.4.2 implies C.0: in fact if f is a homotopy equivalence then $|\deg(f)| = 1$ and $\|M_1\| = \|M_2\|$, which implies that $\mathrm{vol}(M_1) = |\deg(f)| \cdot \mathrm{vol}(M_2)$, and we only have to remark that a locally isometric covering of degree 1 is just an isometry.)

We remark that this generalization is not immediate. The main difficulty is that, in general, the hypothesis of the theorem does not allow a continuous extension of the function to the boundary of $\mathbf{H}^n$ as we saw in the first step of our proof. The extension one can obtain is just measurable, and then the remainder of the proof must be modified according to this weaker fact. In [Th1, ch. 6] the existence of such a measurable extension is proved by a direct geometric argument, but it is also possible to deduce it from Furstenberg's work about random walks in hyperbolic space: "a random continuous path hits a definite point of $\partial \mathbf{H}^n$ with probability 1". The complete proof of Theorem C.5.5 is beyond the aims of the present book, so we refer the reader to the original source [Th1].

We shall prove now an interesting corollary of the rigidity theorem. If G is a group we shall denote by $\mathrm{Out}(G)$ the quotient group

$$\mathrm{Aut}(G)\big/_{\mathrm{Int}(G)}$$

where $\mathrm{Aut}(G)$ denotes the group of all the automorphisms of G and $\mathrm{Int}(G)$ denotes the normal subgroup of $\mathrm{Aut}(G)$ consisting of all inner automorphisms.

Theorem C.5.6. Let M be an oriented compact connected hyperbolic n-manifold, with $n \geq 3$; then $\mathrm{Out}(\Pi_1(M))$ is a finite group and it is isomorphic to $\mathcal{I}(M)$.

Proof. Given $f \in \mathcal{I}(M)$ we have that

$$f_* : \Pi_1(M, x) \to \Pi_1(M, f(x))$$

is a group isomorphism: since the change of basepoint induces a conjugacy on the fundamental group, a homomorphism

$$\mathcal{K} : \mathcal{I}(M) \to \mathrm{Out}(\Pi_1(M)) \qquad f \mapsto \langle f_* \rangle$$

is well-defined.

C.5.2 implies that $\mathcal{K}$ is onto.

We claim now that $\mathcal{K}$ is one-to-one: we must check that if $f \in \mathcal{I}(M)$ and $\langle f_* \rangle = \langle \mathrm{id} \rangle$ then $f = \mathrm{id}$. Assume $M = \mathbf{H}^n\big/_\Gamma$ and let $\tilde{f} \in \mathcal{I}(\mathbf{H}^n)$ be a lift of f; the lift can be chosen in such a way that $\tilde{f} \circ \gamma = \gamma \circ \tilde{f} \; \forall \gamma \in \Gamma$ (for a general lift we have

$$\tilde{f} \circ \gamma = \left(\delta^{-1} \circ \gamma \circ \delta\right) \circ \tilde{f} \; \forall \gamma \in \Gamma$$

and hence we can consider $\delta \circ \tilde{f}$ instead of $\tilde{f}$).

It follows that we only need to prove that the centralizer of Γ in $\mathcal{I}(\mathbf{H}^n)$ is trivial. Assume the converse, *i.e.* there exists $\delta \neq \mathrm{id}$ such that $[\gamma, \delta] = \mathrm{id}$

$\forall \gamma \in \Gamma$. Let $\gamma \in \Gamma \setminus \{\text{id}\}$; by B.4.4 γ is of hyperbolic type; if l is the only γ-invariant geodesic line we have

$$\gamma(\delta(l)) = \delta(\gamma(l)) = \delta(l) \Rightarrow \delta(l) = l.$$

It follows that δ cannot be of parabolic type; now we define H as

$$\begin{cases} \text{the only } \delta\text{-invariant line} & \text{if } \delta \text{ is hyperbolic} \\ \{x \in \mathbb{H}^n : \delta(x) = x\} & \text{if } \delta \text{ is elliptic.} \end{cases}$$

Remark that H is a hyperbolic proper subspace of $\mathbb{H}^n$, and for all $\gamma \in \Gamma \setminus \{\text{id}\}$ we have that H contains the only γ-invariant line and moreover $\gamma(H) = H$. Let $x_0 \in H$ and let l_0 be any line orthogonal to H passing through x_0; if $\varepsilon > 0$ is such that

$$\varepsilon < \frac{1}{2} \cdot \min \left\{ d(x_0, \gamma(x_0)) : \gamma \in \Gamma \setminus \{\text{id}\} \right\}$$

then the closed ε-neighborhood $\overline{N_\varepsilon(l_0)}$ of l_0 does not meet its $(\Gamma \setminus \{\text{id}\})$-orbit. Then M contains a closed subset homeomorphic to $\overline{N_\varepsilon(l_0)}$ which is not compact, and this is absurd.

Our claim is proved.

As for finiteness, it follows from above that we only have to prove that $\mathcal{I}(M)$ contains only finitely many homotopy classes. Since M is compact we can find $\rho > 0$ such that if $x, y \in M$ and $d(x, y) < \rho$ then there exists (one and) only one geodesic loop of length $d(x, y)$ joining x to y; in particular if $d(x, y) < \rho$ the convex combination of x and y is well-defined. Moreover the group $\mathcal{I}(M)$ is compact with respect to the topology induced by the distance

$$d(f_1, f_2) = \sup_{x \in M} d\big(f_1(x), f_2(x)\big)$$

so that $\mathcal{I}(M)$ is covered by a finite number of ρ-balls, and we only have to remark that by the choice of ρ if $d(f_1, f_2) < \rho$ then f_1 and f_2 are homotopic. $\square$

We remark that we have already proved in B.4.23 that half of the above result (*i.e.* finiteness of $\mathcal{I}(M)$) holds also for $n = 2$. On the contrary, it is possible to check that for a surface M of genus $g \geq 2$ the group $\mathrm{Out}(\Pi_1(M))$ is canonically identified with the modular group Mod_g, which is definitely not finite.

Remark C.5.7. It is possible to prove that C.5.6 holds also for finite-volume hyperbolic manifolds and not only for compact ones. The proof works in the same way except for finiteness of $\mathcal{I}(M)$; this is obtained by exploiting a notion we shall introduce in the following chapter: M must be replaced by its ε-thick part $M_{[\varepsilon, \infty)}$, where ε is so small that the ε-thin part contains no tube components.

Proposition C.4.6 extends to the following result, to be found in [Gro3]:

Proposition C.5.8. $\forall n \in \mathbb{N} \; \exists c_n \in \mathbb{R}_+$ with $0 < c_n \leq \dfrac{\pi}{(n-1)}$ such that if M is a compact connected oriented Riemannian n-manifold whose sectional curvatures do not exceed -1, then

$$\mathrm{vol}(M) \leq c_n \cdot \|M\|.$$

(The proof works substantially as that of C.4.6: the hypothesis about the curvature allow one to define the straightening of a cycle and to prove that there exists a bound c_n for the volume of a straight simplex.)

Though it is very difficult in general to compute the Gromov norm of a manifold, the following result may be useful:

Proposition C.5.9. (1) $\forall n \in \mathbb{N} \; \exists b(n) \in \mathbb{R}_+$ such that if for $i = 1,2$ M_i is a connected compact oriented n_i-manifold, then

$$b(n_1 + n_2) \cdot \|M_1\| \cdot \|M_2\| \leq \|M_1 \times M_2\| \leq \frac{\|M_1\| \cdot \|M_2\|}{b(n_1 + n_2)}.$$

(2) If $M_1 \# M_2$ denotes the connected sum of two n-dimensional connected compact oriented n-manifolds M_1 and M_2, with $n \geq 3$, then

$$\|M_1 \# M_2\| = \|M_1\| + \|M_2\|.$$

(For the proof we refer to [Gro3] again; we just remark that (1) is not very complicated, while (2) requires refined methods from the theory of bounded cohomology: we shall sketch this theory and give a proof of (1) in Chapt. F.)

We can prove now as an application of C.4.2 also for $n \geq 3$ an analogue of Theorem B.3.5 concerning the uniqueness of the geometric structure of a manifold.

Proposition C.5.10. Let M be a compact oriented manifold. Then M supports at most one constant sectional curvature Riemannian structure (*i.e.* the fact that it is flat or hyperbolic or elliptic makes the other two cases impossible).

Proof. The elliptic case is distinguished from the other two becase the universal covering S^n of an elliptic manifold is not diffeomorphic to $\mathbf{H}^n$ or $\mathbb{R}^n$, which cover respectively hyperbolic and flat manifolds.

If M is flat it follows from one of Bieberbach's theorems (we shall state in D.3.15) that it is covered by a torus, and hence it easily follows from the same argument presented for C.3.5 that the Gromov norm of M vanishes, while we know (C.4.2) that for hyperbolic M we have $\|M\| = \dfrac{\mathrm{vol}(M)}{v_n} \neq 0$. $\square$

As a conclusion of this chapter we prove some results about hyperbolic manifolds fibering over S^1 (see [Jo]). We are going to use some definitions the reader will find in Sect. F.3 (namely, the notion of flat fiber bundle and the relation of weak equivalence between flat bundles).

Given a manifold V (without boundary) of dimension $n - 1$ and a diffeomorphism ϕ of V onto itself we define an n-manifold V_ϕ as the quotient of $V \times [0,1]$ with the respect to the natural equivalence relation identifying $V \times \{0\}$ with $V \times \{1\}$ via ϕ. It is easily checked that the mapping $(x,t) \mapsto \exp(2\pi i \cdot t) \in S^1$ is well-defined on V_ϕ, and it is the projection associated to a structure of a flat fiber bundle with base S^1, fiber F and structure group generated by ϕ (moreover, the holonomy of such a flat bundle maps the canonical generator of $\Pi_1(S^1)$ onto ϕ).

Example C.5.11. (1) For $V = \mathbb{R}$ and $\phi(x) = -x$ we obtain the (infinite) Möbius strip;
(2) for $V = S^1$ and $\phi(x) = -x$ we obtain the Klein bottle.

Remark C.5.12. If ϕ and ψ are isotopic diffeomorphisms of V then V_ϕ and V_ψ are weakly equivalent fiber bundles (in particular, they are diffeomorphic manifolds).

Proof. If $F : V \times [0,1] \to V$ is the isotopy between ϕ and ψ, the mapping $(x,t) \mapsto F(x,t)$ is well-defined from V_ϕ to V_ψ and it gives an equivalence. $\square$

We are going to compute now the fundamental group of V_ϕ; let us recall that if G and H are groups and p is a homomorphism of H into $\mathrm{Aut}(G)$ the semi-direct product of the groups G and H with respect to p is denoted by $G \coprod_p H$. Moreover if $H = \mathbb{Z}$ and $a \in \mathrm{Aut}(G)$ we can define $G \coprod_a \mathbb{Z}$ as $G \coprod_p \mathbb{Z}$ where p is defined by $p(k) = a^k$. Let us remark that the diffeomorphism ϕ of V induces an automorphism ϕ_* of $\Pi_1(V)$.

Proposition C.5.13. $\Pi_1(V_\phi) \cong \Pi_1(V) \coprod_{\phi_*} \mathbb{Z}$.

Proof. Let $\tilde{V}$ be the universal cover of V. If we fix basepoints in V and $\tilde{V}$, the group $\Pi_1(V)$ can be identified with a group of diffeomorphisms of $\tilde{V}$ acting freely and properly discontinuously, in such a way that $V = \tilde{V} \big/ \Pi_1(V)$. If $\tilde{\phi} : \tilde{V} \to \tilde{V}$ is a lift of ϕ (with respect to the fixed basepoints) we have

$$\phi_*(\gamma) = \tilde{\phi} \circ \gamma \circ \tilde{\phi}^{-1} \qquad \forall \gamma \in \Pi_1(V).$$

Let us define the action of $\Pi_1(V)$ and $\mathbb{Z}$ on $\tilde{V} \times \mathbb{R}$ in the following way:

$$\Pi_1(V) \ni \gamma : (\tilde{x}, t) \mapsto (\gamma(\tilde{x}), t)$$
$$\mathbb{Z} \ni k : (\tilde{x}, t) \mapsto (\tilde{\phi}^k(\tilde{x}), t + k)$$

and let G be the group of diffeomorphisms of $\tilde{V} \times \mathbb{R}$ generated by them. It is quite obvious that G operates freely and properly discontinuously; moreover we have

$$V_\phi \cong \left(\tilde{V} \times \mathbb{R} \right) \big/ G$$

(in fact if we first consider the action of $\Pi_1(V)$ we reduce to $V \times \mathbb{R}$, and the induced action we still have to consider is the action of $\mathbb{Z}$ defined by

$$k : (x,t) \mapsto (\phi^k(t), t + k)$$

and the quotient of $V \times \mathbb{R}$ with respect to it is V_ϕ). Since $\tilde{V} \times \mathbb{R}$ is connected and simply connected we obtain that $\Pi_1(V_\phi) \cong G$. By direct computation it is possible to prove that in G we have

$$1 \circ \gamma = \phi_*(\gamma) \circ 1$$

and it is easily deduced from this that we actually have $G \cong \Pi_1(V) \coprod_{\phi_*} \mathbb{Z}$.

$\square$

We shall denote from now on by $[\phi]$ the isotopy class of a diffeomorphism ϕ of V. The problem we are going to consider is the determination of the pairs $(V, [\phi])$ for which the manifold V_ϕ can be endowed with a complete hyperbolic structure (briefly: V_ϕ is hyperbolic). We will confine ourselves to the case of a compact, connected and oriented manifold V of dimension $n - 1$ and an orientation-preserving diffeomorphism ϕ (and these assumptions will be implicit).

Remark C.5.14. For $n = 2$ there exists no pair $(V, [\phi])$ such that V_ϕ is hyperbolic.

Proof. By our assumptions we have $V \cong S^1$ and $[\phi] = [\text{id}]$ and then V_ϕ is the (one-hole) torus, and we know (B.3.5) that it is not hyperbolic. $\square$

Proposition C.5.15. For $n = 3$, if V_ϕ is hyperbolic then V has genus at least 2 (and hence it is hyperbolic too) and moreover none of the following cases can occur:

(1) ϕ has finite order;

(2) there exists a hyperbolic structure on V with the property that the isotopy class of ϕ contains an isometry with respect to such a structure;

(3) $\exists \gamma \in \Pi_1(V) \setminus \{1\}$ such that $\phi_*(\gamma) \in \{\gamma, \gamma^{-1}\}$;

(4) there exists a ϕ-invariant simple loop in V not homotopic to 0.

Proof. Let us start by proving that the genus of V cannot be 0; otherwise we have that $V = S^2$ and then necessarily $[\phi] = [\text{id}]$ (see [Mi2]). Then we have that V_ϕ is diffeomorphic to $S^2 \times S^1$, whose universal cover $S^2 \times \mathbb{R}$ is not homeomorphic to $\mathbb{H}^3$.

Assume now V has genus 1; then V is the torus and $\Pi_1(V)$ is $\mathbb{Z}^2$. Then by C.5.13 we have that $\Pi_1(V_\phi)$ has a subgroup isomorphic to $\mathbb{Z}^2$. V_ϕ is hyperbolic and then (by B.4.4) $\Pi_1(V_\phi)$ is isomorphic to a discrete group of isometries of hyperbolic type. If a and b are the generators of the subgroup isomorphic to $\mathbb{Z}^2$, since $a \cdot b = b \cdot a$, we must have that a and b have the same fixed points on the boundary (otherwise b would interchange the fixed points of a and then it would have a fixed point in the interior of the geodesic line joining them, which is absurd). In the half-space model we can assume these points are 0 and ∞, and hence we have for suitable $\lambda, \mu > 0$ and $\theta, \tau \in \mathbb{R}$

$$a : (z, t) \mapsto \lambda(e^{i\theta} \cdot z, t) \qquad b : (z, t) \mapsto \mu(e^{i\tau} \cdot z, t).$$

Since we have

$$a^n \cdot b^m = \text{id} \quad \Leftrightarrow \quad n = m = 0$$

we have in particular $\lambda^n \cdot \mu^m = 1 \Leftrightarrow n = m = 0$ and this implies that

$$\left\{ \lambda^n \cdot \mu^m : n, m \in \mathbb{Z} \right\}$$

is not discrete in $\mathbb{R}_+$, and this is absurd.

The first assertion is proved, and we turn to the proof that the conditions (1)...(4) are absurd.

(1) Let us assume $\phi^k = \text{id}$. Then $V \times S^1$ is a k-fold covering of V_ϕ, which implies, for the Gromov norms

$$\|V_\phi\| = k \cdot \|V \times S^1\| = 0$$

and this is absurd according to the Gromov-Thurston theorem (C.4.2).

(2) We can assume ϕ itself is an isometry, and then according to B.4.23 it has finite order, so that (1) applies.

(3) According to B.4.4 we have that γ corresponds in $\mathbb{H}^3$ to an isometry of hyperbolic type, and hence it cannot have finite order. In the two cases $\phi_*(\gamma) = \gamma$ and $\phi_*(\gamma) = \gamma^{-1}$ we have respectively that $\Pi_1(V_\phi)$ contains a subgroup isomorphic to $\mathbb{Z}^2$ and $\mathbb{Z} \coprod_{-1} \mathbb{Z}$. (From a geometric viewpoint these conditions correspond to the fact that V_ϕ contains respectively a torus and a Klein bottle whose inclusion is one-to-one at the Π_1-level.) We have already proved that $\Pi_1(V_\phi)$ does not have subgroups isomorphic to $\mathbb{Z}^2$. In the other case let $a, b \in \mathcal{I}^+(\mathbb{H}^3)$ correspond to the canonical generators of $\mathbb{Z} \coprod_{-1} \mathbb{Z}$, so that $aba = b$. In the half-space model assume a keeps 0 and ∞ fixed; this implies that $b(\{0, \infty\}) = \{0, \infty\}$, and hence, as above, that b fixes 0 and ∞ too. Then both a and b have the form

$$(z, t) \mapsto \lambda(e^{i\theta} \cdot z, t)$$

and hence they commute, which is absurd.

(4) This is a special case in which (3) holds.   □

The above proposition gives some information about ϕ in order that V_ϕ is hyperbolic, but it is not evident at all that any hyperbolic V_ϕ does exist. However there do exist examples of such pairs $(V, [\phi])$, due to Jorgensen and Riley (we shall meet one in E.7, in the non-compact case). It is quite interesting to remark that these examples historically induced Thurston to formulate his famous geometrization conjecture (we will not fully discuss this conjecture as it would take us too far from our aims: we address the reader to the end of E.7 for a partial statement and to [Th2] for a complete one). We also recall that Thurston has classified all the pairs $(V, [\phi])$ in dimension three such that V_ϕ is hyperbolic.

For the case of dimension larger than three we have the following negative result:

Proposition C.5.16. For $n \geq 4$ there exist no pair $(V, [\phi])$ such that both V and V_ϕ are hyperbolic.

Proof. Assume V is hyperbolic. According to the rigidity theorem in the homotopy class of ϕ we can find an isometry ψ. Since V_ϕ and V_ψ have the same homotopy type (easy verification) then they have the same Gromov norm. Since $\mathcal{I}(V)$ is finite (C.5.6) the same proof as for point (1) in C.5.15 allows one to obtain that $V \times S^1$ is a covering of V_ψ with a finite numbers of leaves, which implies that V_ψ have 0 Gromov norm and then the conclusion follows from C.4.2. $\square$

We do not know whether the condition that V be hyperbolic is actually necessary for the above result to hold. In other words we do not know in dimension larger than three whether there exist or not pairs $(V, [\phi])$ with the property that V_ϕ is hyperbolic (maybe a negative answer can be obtained assuming Thurston's conjecture).

Chapter D.
Margulis' Lemma and its Applications

In this chapter we begin the study of complete hyperbolic manifolds which are not necessarily compact. The essential tool of our investigations is Margulis' lemma, which is proved in the first section, while the second section is devoted to the basic properties of the thin-thick decomposition of a hyperbolic manifold. In the third section several facts are deduced about the shape of the ends, in particular in the case of finite volume.

D.1 Margulis' Lemma

Let M be a Riemannian manifold. If σ is a (piecewise differentiable) path in M, we shall denote by $L(\sigma)$ its length. Remark that each loop in M is homotopic to a piecewise differentiable loop based at the same point, so that we can think of $\Pi_1(M)$ as the set of all piecewise differentiable loops up to homotopy. For $\varepsilon > 0$ we set

$$M_{(0,\varepsilon]} = \{x \in M : \exists \langle \sigma \rangle \in \Pi_1(M, x) \setminus \{1\} \ s.t. \ L(\sigma) \leq \varepsilon\}$$

$$M_{[\varepsilon,\infty)} = \{x \in M : \forall \langle \sigma \rangle \in \Pi_1(M, x) \setminus \{1\}, \ L(\sigma) \geq \varepsilon\}.$$

Of course if M is a compact manifold we have $M_{(0,\varepsilon]} = \phi$ whenever ε is small enough.

We shall say that $M_{(0,\varepsilon]}$ is the <u>ε-thin part</u> of M, and $M_{[\varepsilon,\infty)}$ is the <u>ε-thick part</u> of M (when a constant ε is fixed we will omit its specification, so we shall speak of thin and thick part of M).

Margulis' lemma, which we are going to prove, can be heuristically expressed in the following way: *for every natural number n there exists a constant ε_n such that if M is a complete oriented hyperbolic n-manifold and $x \in M_{(0,\varepsilon_n]}$, the subgroup of $\Pi_1(M, x)$ generated by ε_n-short loops at x is not "very complicated" (which implies, more geometrically, that the ε_n-thin part of M is not "very complicated").*

At the end of this section we shall sketch a generalization of Margulis' lemma which applies to manifolds of bounded non-positive sectional curvature, and not only to hyperbolic manifolds.

Before stating the main result we recall some elementary notions in group theory; let G be a fixed group;

–if H and K are subgroups of G, we shall denote by $[H, K]$ the subgroup of G generated by all the elements $[h, k] = hkh^{-1}k^{-1}$, for $h \in H$ and $k \in K$;
– the m-th <u>commutator</u> G^m of G is recursively defined by

$$\begin{cases} G^1 = [G, G] \\ G^{m+1} = [G, G^m] \quad (m \geq 1) \, ; \end{cases}$$

– G is said to be <u>nilpotent</u> if for some integer m, $G^m = \{1\}$;
– a subgroup H of G is said to have <u>finite index</u> if the quotient set $G/_H$ is finite;
– if G admits a subgroup of finite index enjoying a property $\mathcal{P}$, G is said to enjoy the property <u>almost-$\mathcal{P}$</u>.

Margulis' lemma will be formulated by means of properly discontinuous subgroups of $\mathcal{I}(\mathbf{H}^n)$, the relation with complete hyperbolic manifolds being given as usual by the representation of the manifold as a quotient of $\mathbf{H}^n$; remark that the lemma holds without assuming the group to act freely. For the original proof we refer to [Ka-Ma] and [Ra].

We shall denote by d the hyperbolic distance in $\mathbf{H}^n$.

Theorem D.1.1 (Margulis' Lemma). $\forall n \in \mathbb{N} \; \exists \varepsilon_n \geq 0$ such that for any properly discontinuous subgroup $\Gamma < \mathcal{I}(\mathbf{H}^n)$ and for any $x \in \mathbf{H}^n$, the group $\Gamma_{\varepsilon_n}(x)$ generated by the set

$$F_{\varepsilon_n}(x) = \big\{ \gamma \in \Gamma : d(x, \gamma(x)) \leq \varepsilon_n \big\}$$

is almost-nilpotent.

Proof. The proof consists of several steps. From now on we shall consider Γ and x fixed; we shall say a constant is "universal" when it does not depend on the choice of Γ and x, but only on the integer n; Theorem D.1.1 is then formulated in the following way: *there exists a universal constant ε_n such that $\Gamma_{\varepsilon_n}(x)$ is almost-nilpotent.* This is what we are going to prove.

We start by fixing some notations. For $x \in \mathbf{H}^n$, $g \in \mathcal{I}(\mathbf{H}^n)$ and $a > 0$ we set

$$\|g\|_{a,x} = \max \big\{ a \cdot d(x, g(x)), \mathrm{dist}_x(\mathrm{I}, P_{g(x),x} \circ D_x g) \big\}$$

where:
– for $y, z \in \mathbf{H}^n$, $P_{y,z}$ is the parallel transport from $T_y \mathbf{H}^n$ to $T_z \mathbf{H}^n$ along the only geodesic arc joining y to z;
– $D_x g$ is the differential of g in x (we are not going to use the previous notation $d_x g$ in order to avoid confusion with the hyperbolic distance d);
– dist_x is the distance on $O(T_x \mathbf{H}^n)$ defined by

$$\mathrm{dist}_x(A, B) = \max\{ \sphericalangle (Aw, Bw) : w \in T_x \mathbf{H}^n \}$$

in which $\sphericalangle (v_1, v_2)$ is the measure of the angle between two vectors v_1, v_2.

Now, for $\varepsilon > 0$, we set

$$F'_\varepsilon(x)_a = \{\gamma \in \Gamma : \|\gamma\|_{a,x} \le \varepsilon\}$$

and we denote by $\Gamma'_\varepsilon(x)_a$ the group generated by $F'_\varepsilon(x)_a$.

Lemma D.1.2. There exist universal constants ε'_n and a_n such that for all g and h in $\mathcal{I}(\mathbf{H}^n)$ we have

$$\|[g,h]\|_{a_n,x} \le 1/2 \cdot \max \{\|g\|_{a_n,x}, \|h\|_{a_n,x}\}$$

whenever $\|g\|_{a_n,x}$ and $\|h\|_{a_n,x}$ do not exceed ε'_n.

Proof. We shall prove that it suffices to choose $a_n = 1$; the lemma was formulated in this way because of the generalization we are going to sketch at the end of the section.

We start by remarking that for $k \in \mathcal{I}(\mathbf{H}^n)$, $y, z \in \mathbf{H}^n$ we have

(*) $$P_{k(y),k(z)} = D_z k \circ P_{y,z} \circ (D_y k)^{-1}$$

(it suffices to recall that k preserves geodesics and its differential preserves angles). We deduce from this that for $k \in \mathcal{I}(\mathbf{H}^n)$ and $y \in \mathbf{H}^n$

(**) $$\|k^{-1}gk\|_{a,y} = \|g\|_{a,k(y)};$$

in fact $d(y, k^{-1}gk(y)) = d(k(y), gk(y))$ and moreover by (*)

$$\max_{v \in T_y} \sphericalangle \left(v, P_{k^{-1}gk(y),y} \circ D_y(k^{-1}gk)(v)\right) =$$

$$= \max_{v \in T_y} \sphericalangle \left(v, (D_y k)^{-1} \circ P_{gk(y),k(y)} \circ \left(D_{gk(x)}k^{-1}\right)^{-1} \circ \right.$$

$$\left. \circ D_{gk(x)}k^{-1} \circ D_{k(y)}g \circ D_y k(v)\right) =$$

$$= \max_{v \in T_y} \sphericalangle \left(D_y k(v), P_{gk(y),k(y)} \circ D_{k(y)}g\left(D_y k(v)\right)\right) =$$

$$= \max_{w \in T_{k(y)}} \sphericalangle \left(w, P_{gk(y),k(y)} \circ D_{k(y)}g(w)\right).$$

Using (**) and the obvious relation $[k^{-1}gk, k^{-1}hk] = k^{-1}[g,h]k$, it is readily verified that the lemma can be proved for an arbitrary fixed point $\bar{x}$ of $\mathbf{H}^n$. We shall refer to the disc model and we shall choose $\bar{x} = 0$; we simplify the notation by setting $\|\cdot\| = \|\cdot\|_{1,0}$.

Now for $x_0 \in \mathbf{D}^n$ we consider the only element $f^{(x_0)} \in \mathcal{I}(\mathbf{D}^n)$ with the property that $f^{(x_0)}(0) = x_0$ and $D_0 f^{(x_0)}$ is a positive multiple of the identity operator on $\mathbb{R}^n$ (the tangent bundle to $\mathbf{D}^n$ being canonically identified with $D^n \times \mathbb{R}^n$). $f^{(x_0)}$ is obviously unique, $f^{(0)} = \text{id}$, and for $x_0 \ne 0$ $f^{(x_0)}$ is easily calculated by the following method: consider the only inversion $i_{(x_0)}$ in $\mathcal{I}(\mathbf{D}^n)$ such that $i_{(x_0)}(0) = x_0$ and the reflection $s_{(x_0)}$ with respect to the hyperplane $x_0^\perp$; then $f^{(x_0)}$ is given by $i_{(x_0)} \circ s_{(x_0)}$. This method gives the explicit expression for $f^{(x_0)}$:

$$f^{(x_0)}(x) = \frac{\left(1 - \|x_0\|^2\right) \cdot x + \left(1 + \|x\|^2 + 2(x_0|x)\right) \cdot x_0}{1 + 2(x_0|x) + \|x_0\|^2 \cdot \|x\|^2}.$$

It follows that the mapping

$$\mathbf{D}^n \times \mathbf{D}^n \ni (x, x_0) \mapsto f^{(x_0)}(x) \in \mathbf{D}^n$$

is rational (in particular, it is analytic).

Now, let us consider the bijective mapping

$$\phi : O(n) \times \mathbf{D}^n \to \mathcal{I}(\mathbf{D}^n) \qquad (A, x_0) \mapsto f^{(x_0)} \circ A.$$

It is easily verified that this one-chart atlas endows $\mathcal{I}(\mathbf{D}^n)$ with a Lie group structure (operations are rational functions); moreover, it can be checked that this structure equals the one $\mathcal{I}(\mathbf{H}^n)$ inherits from $\mathrm{Gl}(n+1, \mathbb{R})$ (see A.2.4).

Now, let us remark that we have $P_{0,x_0} = D_0 f^{(x_0)}$; in fact, since hyperbolic geodesics passing through the origin are obtained by change of parameter from Euclidean geodesics, P_{0,x_0} must be a positive multiple of the identity operator; the multiplying constant is uniquely determined by the requirement that P_{0,x_0} be an isometry, and hence it coincides with the constant appearing in $D_0 f^{(x_0)}$. This implies that

$$\max_{v \in T_0} \sphericalangle \left(v, P_{x_0,0} \circ D_0 \phi(A, x_0) \right) =$$

$$= \max_{v \in T_0} \sphericalangle \left(v, \left(D_0 f^{(x_0)} \right)^{-1} \circ D_0 f^{(x_0)} \circ D_0 A(v) \right) =$$

$$= \max_{v \in \mathbf{R}^n} \sphericalangle (v, Av).$$

So $\|.\|$ corresponds in $O(n) \times \mathbf{D}^n$ under ϕ to the natural distance from the point $(\mathrm{I}, 0) = \phi^{-1}(\mathrm{id})$:

$$\|\phi(A, x_0)\| = \max \left\{ \mathrm{dist}(\mathrm{I}, A), d(0, x_0) \right\}.$$

Now it suffices to recall that for any Lie group G the differential of the mapping $G \times G \ni (g, h) \mapsto [g, h] \in G$ is null in $(1, 1)$; this implies that

$$\|[g, h]\| = o\left(\max\{\|g\|, \|h\|\} \right)$$

and the lemma is proved.   □

From now on we shall fix $a = 1$ (and omit all subscripts a) and we shall always mean by ε'_n the number appearing in the above lemma.

Lemma D.1.3. $\Gamma'_{\varepsilon'_n}(x)$ is a nilpotent group.

Proof. It follows from the proper discontinuity of Γ that if $\alpha > 0$ is small enough then $\Gamma'_\alpha(x) = \{\mathrm{id}\}$. Now we can choose $m \in \mathbb{N}$ such that $\varepsilon'_n \cdot (1/2)^m \le \alpha$. We claim that $\left(\Gamma'_{\varepsilon'_n}(x) \right)^m = \{\mathrm{id}\}$. Let $g_1, ..., g_m \in F'_{\varepsilon'_n}(x)$; Lemma D.1.2 yields

$$\|[g_1, [g_2, ... [g_{m-1}, g_m]...]]\|_x \le \varepsilon'_n \cdot (1/2)^m \le \alpha \implies [g_1, [g_2, ... [g_{m-1}, g_m]...] = \mathrm{id}.$$

Hence every m-commutator of the generators of $\Gamma'_{\varepsilon'_n}(x)$ is trivial; by the general formula (true in any group)

$$[a, b \cdot c] = [a, b] \cdot [b, [a, c]] \cdot [a, c]$$

we obtain that every element of $\left(\Gamma'_{\varepsilon'_n}(x)\right)^m$ can be expressed as the product of commutators of order at least m of the generators (remark that, by D.1.2, the commutator of two element of $F'_{\varepsilon'_n}(x)$ belongs to $F'_{\varepsilon'_n}(x)$ again) and therefore it is trivial.   □

Lemma D.1.4. There exists a universal constant $m_n \in \mathbb{N}$ such that for any subset $\mathcal{S}$ of $O(T_x \mathbf{H}^n)$ containing at least m_n elements it is possible to find $A, B \in \mathcal{S}$ such that $A \neq B$ and $\mathrm{dist}_x(A, B) \leq \varepsilon'_n/2$.

Proof. It suffices to recall the following facts:
- $O(T_x \mathbf{H}^n)$ is canonically isomorphic to $O(n)$;
- $O(n)$ is compact;
- dist_x is a distance inducing the natural topology on $O(T_x \mathbf{H}^n)$.   □

Our next step consists in finding a universal constant $k_n > 0$ such that, if we set $\varepsilon_n = \varepsilon'_n/k_n$, the group G generated by

$$\left\{ \gamma \in \Gamma_{\varepsilon_n}(x) : \|\gamma\|_x \leq \varepsilon'_n \right\}$$

has finite index in $\Gamma_{\varepsilon_n}(x)$. After this, the proof of Theorem D.1.1 will be complete, since G is contained in $\Gamma'_{\varepsilon'_n}(x)$ and hence it is nilpotent.

Instead of choosing k_n *a priori*, we shall look for the conditions it must satisfy. Since Γ is properly discontinuous $F_{\varepsilon_n}(x)$ is finite; let

$$F_{\varepsilon_n}(x) = \{\gamma_1, ..., \gamma_h\}.$$

We want to determine k_n in such a way that for all $\gamma \in \Gamma_{\varepsilon_n}(x)$ there exists $\tilde{\gamma} = \gamma_{j_1} \cdot ... \cdot \gamma_{j_l}$ with $l \leq m_n$ and $\gamma \cdot G = \tilde{\gamma} \cdot G$ (if this is the case of course G has finite index).

Let $\gamma = \gamma_{i_1} \cdot ... \cdot \gamma_{i_l}$ be a reduced expression of γ in the generators (*i.e.* l is the least possible length), and assume $l \geq m_n + 1$. For $s = 0, 1, ..., m_n$ we set

$$\theta_s = \gamma_{i_{(l-s)}} \cdot \gamma_{i_{(l-s+1)}} \cdot ... \cdot \gamma_{i_l}$$

and we consider the set

$$\left\{ P_{\theta_s(x),x} \circ D_x \theta_s : s = 0, 1, ..., m_n \right\} \subset O(T_x(\mathbf{H}^n)).$$

By the choice of m_n (Lemma D.1.4) it follows that there exist $s, t \in \{0, ..., m_n\}$, $s \neq t$, such that

$$\max_{w \in T_x} \sphericalangle \left(P_{\theta_s(x),x} \circ D_x \theta_s(w), P_{\theta_t(x),x} \circ D_x \theta_t(w) \right) \leq \varepsilon'_n/2.$$

We assume for instance $s < t$ and we set

$$\alpha = \gamma_{i_1} \cdot ... \cdot \gamma_{i_{(l-t-1)}} = \gamma \cdot \theta_t^{-1}$$
$$\eta = \gamma_{i_{(l-t)}} \cdot ... \cdot \gamma_{i_{(l-s-1)}} = \theta_t \cdot \theta_s^{-1}$$
$$\beta = \gamma_{i_{(l-s)}} \cdot ... \cdot \gamma_{i_l} = \theta_s.$$

We have found a decomposition $\gamma = \alpha \cdot \eta \cdot \beta$ with the following properties:

$-\eta \neq \mathrm{id}$, $\beta \neq \mathrm{id}$ (otherwise l would not be minimal);

$-\eta \cdot \beta$ has length at most $m_n + 1$ in the generators;

$-\max_{w \in T_x} \not< \big(P_{\eta\beta(x),x} \circ D_x \eta\beta(w), P_{\beta(x),x} \circ D_x \beta(w)\big) \leq \varepsilon'_n/2$.

Now, let us assume that there exists a universal choice of k_n such that under these hypotheses

$$\|\beta^{-1} \cdot \eta \cdot \beta\|_x \leq \varepsilon'_n$$

i.e. $\beta^{-1} \cdot \eta \cdot \beta \in G$; then we have $\gamma \cdot G = \alpha \cdot \eta \cdot \beta \cdot G = \alpha \cdot \beta \cdot G$ and the proof is easily completed by induction on l (we just proved that if the length of γ is bigger than m_n, then γ can be replaced by an element whose length is strictly smaller).

We are left to prove the following:

Lemma D.1.5. There exists a universal constant k_n such that if we set $\varepsilon_n = \varepsilon'_n/k_n$, if $F_{\varepsilon_n}(x) = \{\gamma_1, ..., \gamma_h\}$, and if we are given $\eta = \gamma_{i_1} \cdot ... \cdot \gamma_{i_p}$ and $\beta = \gamma_{j_1} \cdot ... \cdot \gamma_{j_q}$ with the properties:

$-\eta \neq \mathrm{id}$, $\beta \neq \mathrm{id}$;

$-p + q \leq m_n + 1$;

$-\max_{w \in T_x} \not< \big(P_{\eta\beta(x),x} \circ D_x \eta\beta(w), P_{\beta(x),x} \circ D_x \beta(w)\big) \leq \varepsilon'_n/2$;

then $\|\beta^{-1} \cdot \eta \cdot \beta\|_x \leq \varepsilon'_n$.

Proof. We must give upper bounds for the numbers

(1) $$d(x, (\beta^{-1}\eta\beta)(x))$$

(2) $$\max_{w \in T_x} \not< \big(P_{\beta^{-1}\eta\beta(x),x} \circ D_x(\beta^{-1}\eta\beta)(w), w\big).$$

For (1) we have

$$d(x, (\beta^{-1}\eta\beta)(x)) = d(\beta(x), \eta\beta(x)) \leq d(x, \beta(x)) + d(x, \eta\beta(x));$$
$$d(x, \beta(x)) = d(x, \gamma_{j_1} \cdot ... \cdot \gamma_{j_q}(x)) \leq$$
$$\leq d(x, \gamma_{j_1}(x)) + d(\gamma_{j_1}(x), \gamma_{j_1} \cdot ... \cdot \gamma_{j_q}(x)) \leq$$
$$\leq \varepsilon'_n/k_n + d(x, \gamma_{j_2} \cdot ... \cdot \gamma_{j_q}(x)) \leq ... \leq q \cdot \varepsilon'_n/k_n;$$
$$d(x, \eta\beta(x)) \leq ...(\text{similarly})... \leq (p + q) \cdot \varepsilon'_n/k_n;$$
$$\Rightarrow \quad d(x, (\beta^{-1}\eta\beta)(x)) \leq (p + 2q) \cdot \varepsilon'_n/k_n \leq 2(m_n + 1)\varepsilon'_n/k_n$$

and hence every choice of $k_n \geq 2(m_n + 1)$ gives $d(x, (\beta^{-1}\eta\beta)(x)) \leq \varepsilon'_n$.

Now we turn to (2). As in the proof of D.1.2 we obtain

$$\max_{w \in T_x} \not< \big(P_{\beta^{-1}\eta\beta(x),x} \circ D_x(\beta^{-1}\eta\beta)(w), w\big) =$$
$$= \max_{v \in T_{\beta(x)}} \not< \big(P_{\eta\beta(x),\beta(x)} \circ D_{\beta(x)}\eta(v), v\big).$$

The third hypothesis about η and β yields

$$\varepsilon_n'/2 \geq \max_{w \in T_x} \sphericalangle \left(P_{\eta\beta(x),x} \circ D_x\eta\beta(w), P_{\beta(x),x} \circ D_x\beta(w)\right) =$$

$$= \max_{w \in T_x} \sphericalangle \left(P_{x,\beta(x)} \circ P_{\eta\beta(x),x} \circ D_{\beta(x)}\eta\big(D_x\beta(w)\big), D_x\beta(w)\right) =$$

$$= \max_{v \in T_{\beta(x)}} \sphericalangle \left(P_{x,\beta(x)} \circ P_{\eta\beta(x),x} \circ D_{\beta(x)}\eta(v), v\right).$$

Now, let us consider in $\mathbf{H}^n$ the geodesic triangle T having vertices x, $\beta(x)$ and $\eta\beta(x)$; A.6.5 implies that if $Z \subset T_{\beta(x)}\mathbf{H}^n$ is the the hyperbolic 2-subspace containing T, the parallel transport

$$\varphi = P_{\eta\beta(x),\beta(x)} \circ P_{x,\eta\beta(x)} \circ P_{\beta(x),x} \in O(T_{\beta(x)}\mathbf{H}^n),$$

along the boundary of T is given by

$$\varphi|_Z = \text{rot}(\pm\mathcal{A}(T)) \qquad\qquad \varphi|_{Z^\perp} = \text{id}$$

where $\mathcal{A}(T)$ denotes the area of T and $\text{rot}(\vartheta)$ is the rotation by an angle ϑ on Z. Then, using the above formulas, for $v \in T_{\beta(x)}\mathbf{H}^n$ we have

$$\sphericalangle \left(P_{\eta\beta(x),\beta(x)} \circ D_{\beta(x)}\eta(v), v\right) =$$

$$= \sphericalangle \left(\varphi \circ P_{x,\beta(x)} \circ P_{\eta\beta(x),x} \circ D_{\beta(x)}\eta(v), v\right) \leq$$

$$\leq \sphericalangle \left(\varphi\big(P_{x,\beta(x)} \circ P_{\eta\beta(x),x} \circ D_{\beta(x)}\eta(v)\big), P_{x,\beta(x)} \circ P_{\eta\beta(x),x} \circ D_{\beta(x)}\eta(v)\right) +$$

$$+ \sphericalangle \left(P_{x,\beta(x)} \circ P_{\eta\beta(x),x} \circ D_{\beta(x)}\eta(v), v\right) \leq$$

$$\leq \mathcal{A}(T) + \varepsilon_n'/2.$$

Now, it is obvious that there exists a universal constant λ such that for any geodesic triangle Δ in $\mathbf{H}^n$ the area of Δ does not exceed λ times the maximal length of the sides of Δ. If we apply this to T we obtain

$$\mathcal{A}(T) \leq \lambda \cdot 2(m_n + 1) \cdot \varepsilon_n'/k_n$$

and hence every choice of $k_n \geq 4 \cdot (m_n + 1) \cdot \lambda$ gives

$$\max_{w \in T_x} \sphericalangle \left(P_{\beta^{-1}\eta\beta(x),x} \circ D_x(\beta^{-1}\eta\beta)(w), w\right) \leq \varepsilon_n'.$$

□

The proof of Theorem D.1.1 is now complete.   □

Margulis' lemma D.1.1 admits the following general version:

Theorem D.1.6. $\forall n \in \mathbb{N} \; \exists \varepsilon_n > 0$ such that, given
– a connected and simply connected complete Riemannian n-manifold X whose sectional curvatures k satisfy $-1 \leq k \leq 0$ everywhere,
– a point x in X,
– a properly discontinuous subgroup Γ of $\mathcal{I}(X)$,
then the subgroup of Γ generated by $\{\gamma \in \Gamma : d(x, \gamma(x)) \leq \varepsilon_n'\}$ is almost-nilpotent.

We will not present a complete proof here, refering to [Bu-Ka, p. 25] and [Ba-Gr-Sc, pp. 105-109]. However, we remark that the proof follows the same scheme presented above for the hyperbolic case. After giving the very same definition for $\| \cdot \|_{a,x}$, one proves the following analogue of Lemma D.1.2:

Lemma D.1.7. There exist universal constants $a_n > 0$, $\varepsilon'_n > 0$, $\delta \in (0,1)$ such that for any $g, h \in \mathcal{I}(X)$ we have

$$\| [g,h] \|_{a_n,x} \leq \delta \cdot \max \left\{ \|g\|_{a_n,x}, \|h\|_{a_n,x} \right\}$$

whenever $\|g\|_{a_n,x}$ and $\|h\|_{a_n,x}$ do not exceed ε'_n.

Then, one repeats the very same steps we followed above until Lemma D.1.5: this lemma can be generalized too, the essential tool being the fact that the assumption about the curvature implies that the parallel transport along the boundary of a geodesic triangle Δ in X can be controlled in a universal way by the diameter of Δ.

D.2 Local Geometry of a Hyperbolic Manifold

This section will be devoted to the study of some geometric consequences of Margulis' Lemma D.1.1. The results can be summarized in the following way: if M is an n-dimensional oriented complete hyperbolic manifold, every $\varepsilon_n/2$-ball in M is isometric to a ball of another hyperbolic manifold whose fundamental group is very special.

Proposition D.2.1. Let M be an n-dimensional oriented complete hyperbolic manifold, let x belong to M and let ε be a positive number; if we identify $\Pi_1(M,x)$ with a subgroup Γ of $\mathcal{I}(\mathbb{H}^n)$, and we denote by Γ_ε the subgroup of Γ generated by loops at x of length less than or equal to ε, then the $\varepsilon/2$-ball centred at x is isometric to a ball of the hyperbolic manifold $\mathbb{H}^n/\Gamma_\varepsilon$.

Proof. Let us consider the commutative diagram of coverings:

$$\mathbb{H}^n \xrightarrow{\omega} \mathbb{H}^n/\Gamma_\varepsilon$$
$$\downarrow \pi \quad \swarrow p$$
$$M = \mathbb{H}^n/\Gamma$$

where p is a local isometry (since ω and π are). Let U be the $\varepsilon/2$-ball centred at x; for $y \in U$, if α and β are geodesic arcs joining x and y in U, then $\langle \beta^{-1} \circ \alpha \rangle \in \Gamma_\varepsilon$; it follows that U can be globally lifted to $\mathbb{H}^n/\Gamma_\varepsilon$ by an isometric mapping. $\square$

The following theorem gives a complete description of all the possibilities for the group Γ_ε, in case ε does not exceed the n-th Margulis constant.

Theorem D.2.2. Let the symbols of Proposition D.2.1 be fixed, and assume $\varepsilon \leq \varepsilon_n$. The following mutually exclusive possibilities are given:

1) $\Gamma_\varepsilon = \{\mathrm{id}\}$;

2) $\Gamma_\varepsilon \cong \mathbb{Z}$, and it is generated by an isometry of hyperbolic type;

3) Γ_ε consists of isometries of parabolic type, all having the same fixed point p at infinity; every horosphere centred at p is Γ_ε-invariant, and the action of Γ_ε is isometric with respect to the Euclidean structure defined on such horospheres; in particular Γ_ε can be identified with a discrete subgroup of $\mathcal{I}^+(\mathbb{R}^{n-1})$ acting freely.

The proof requires a few preliminaries. We begin with the following:

Lemma D.2.3. If $\phi, \psi \in \mathcal{I}(\mathbb{H}^n)$ are not of elliptic type and $[\phi, \psi] = \mathrm{id}$, then one of the following holds:

i) ϕ and ψ are both of hyperbolic type with the same fixed points at infinity;

ii) ϕ and ψ are both of parabolic type with the same fixed point at infinity.

Proof. If ϕ is hyperbolic, and p, q are its fixed points at infinity, then (by commutativity) $\psi(\{p, q\}) = \{p, q\}$; if $\psi(p) = q$ (whence $\psi(q) = p$) then ψ would have a fixed point on the geodesic line with endpoints p and q, and this is absurd since ψ is not elliptic. It follows that ψ is hyperbolic, $\psi(p) = p$ and $\psi(q) = q$.

If ϕ is parabolic then ψ must be parabolic too, and commutativity implies that ϕ and ψ have the same fixed point. $\qquad\square$

Lemma D.2.4. Let G be a nilpotent subgroup of $\mathcal{I}(\mathbb{H}^n)$ containing no isometry of elliptic type; then one of the following holds:

1) $G = \{\mathrm{id}\}$;

2) $G \setminus \{\mathrm{id}\}$ consists of hyperbolic isometries, all having the same fixed points at infinity;

3) $G \setminus \{\mathrm{id}\}$ consists of parabolic isometries, all having the same fixed point at infinity.

Proof. Since a non-trivial nilpotent group has a non-trivial centre, this result is a straight-forward consequence of the previous lemma. $\qquad\square$

Lemma D.2.5. Every element of finite order in $\mathcal{I}(\mathbb{H}^n)$ is of elliptic type. *Hint.* Consider the model $\mathbb{II}^{n,+}$ and choose the fixed points at infinity in a suitable way.)

Proof of Theorem D.2.2. By Margulis' Lemma D.1.1, there exists a nilpotent subgroup G of Γ_ε such that Γ_ε/G is finite. We discuss separately the possibilities for G given by Lemma D.2.4:

1) assume $G = \{\mathrm{id}\}$; by Lemma D.2.5, Γ_ε is trivial too.

2) Let G consist of isometries of hyperbolic type having fixed points p and q at infinity. For $\gamma \in \Gamma_\varepsilon$ there exists an integer k such that $\gamma^k \in G$. Since $[\gamma, \gamma^k] = \mathrm{id}$, γ is an isometry of hyperbolic type having p and q as fixed points. In $\mathbb{II}^{n,+}$ we can assume $\{p, q\} = \{0, \infty\}$. It follows that every element γ of Γ_ε can be written in the form:

$$\gamma(y, t) = \lambda(Ay, t)$$

for suitable $\lambda > 0$ ($\lambda \neq 1$ if $\gamma \neq \mathrm{id}$) and $A \in SO(n-1)$.

Let us define a subgroup K of $\mathrm{Diff}(\mathbb{R}_+)$ by

$$K = \left\{ (t \mapsto \lambda t) : \exists A \in SO(n-1) \ s.t. \ \big((y,t) \mapsto \lambda(Ay,t)\big) \in \Gamma_\varepsilon \right\}.$$

Since Γ_ε is discrete, K is discrete too, and therefore there exists $\lambda_0 > 1$ such that

$$K = \left\{ (t \mapsto \lambda_0^n t) : n \in \mathbb{Z} \right\}.$$

Let $\gamma_0 \in \Gamma_\varepsilon$ be such that $\gamma_0(y,t) = \lambda_0(A_0 y, t)$.

Now, for $\gamma \in \Gamma_\varepsilon$ let $m \in \mathbb{Z}$ be such that $\gamma(y,t) = \lambda_0^m (Ay, t)$. As we remarked above, the only element of Γ_ε of the form $(y,t) \mapsto (Ay, t)$ is the identity; it follows that $\gamma = \gamma_0^m$ and hence Γ_ε is isomorphic to $\mathbb{Z}$ and it is generated by the hyperbolic isometry γ_0.

3) Let G consist of isometries of parabolic type having a common fixed point p at infinity; as above, it is easily verified that the same holds for Γ_ε. Now, in $\mathbb{II}^{n,+}$, we assume $p = \infty$. Every element of Γ_ε can be written in the form

$$(y,t) \mapsto (Ay + b, t)$$

for suitable $A \in SO(n-1)$ and $b \in \mathbb{R}^{n-1}$. Horospheres centred at ∞ are affine hyperplanes $\mathbb{R}^{n-1} \times \{t_0\}$, and then they are Γ_ε-invariant; all other statements about Γ_ε are now obvious. $\qquad\square$

The following global result will be used several times in the sequel.

Proposition D.2.6. *If M is a complete oriented hyperbolic manifold having finite volume and if $\varepsilon > 0$ then $M_{[\varepsilon,\infty)}$ is compact.*

Proof. Of course we can assume $\varepsilon \leq \varepsilon_n$. Since $M_{[\varepsilon,\infty)}$ is closed it suffices to prove that it is covered by a finite number of compact sets. We remark that by D.2.1 and D.2.2 for $x \in M_{[\varepsilon,\infty)}$ the ball $B_{\varepsilon/2}(x)$ of radius $\varepsilon/2$ centred at x is isometric to an $\varepsilon/2$-ball of $\mathbb{H}^n$, so that its volume is a constant ν (depending only on ε). Let us consider the set

$$\mathcal{S} = \left\{ Y \subset M_{[\varepsilon,\infty)} : d(y_1, y_2) \geq \varepsilon \ \forall y_1, y_2 \in Y, \ y_1 \neq y_2 \right\}$$

which is partially ordered by inclusion and inductive, so that it contains a maximal element Y_0. Since the balls

$$\left\{ B_{\varepsilon/2}(y) : y \in Y_0 \right\}$$

are pairwise disjoint, we have

$$\#Y_0 \leq \frac{\mathrm{vol}(M)}{\nu} < \infty.$$

Maximality implies that

$$M_{[\varepsilon,\infty)} \subset \bigcup \left\{ \overline{B_{2\varepsilon}(y)} : y \in Y_0 \right\}$$

and then we only have to recall that by completeness closed balls in M are compact. $\square$

D.3 Ends of a Hyperbolic Manifold

By Proposition D.2.1 and Theorem D.2.2 the local geometry of a complete n-dimensional oriented hyperbolic manifold M falls into a specified set of possibilities. We are now to globalize this fact for the "ends" of the manifold. The result is much more precise for the case $n \leq 3$, but it is quite relevant in the case $n \geq 4$ too. Also, in the case of finite volume our description of the thin-thick decomposition is somewhat sharper. We shall omit the case $n = 2$ since it can be treated as an exercise following the method we shall use for the case $n = 3$.

We begin with a (quite rough) general definition. Let X be a Hausdorff locally compact topological space. For compact $K \subset X$ define

$$\mathcal{E}(K) = \big\{\overline{E} : E \text{ connected component of } X \setminus K, \ \overline{E} \text{ not compact}\big\}.$$

If there exists a compact set $K \subset X$ such that for any other compact set $K' \supset K$ the inclusion induces a bijective mapping $\mathcal{E}(K') \to \mathcal{E}(K)$, then we shall call each element of $\mathcal{E}(K)$ a <u>topological end</u> of X relative to K; of course this definition is essentially independent of K, *i.e.* if H is another compact set having the same property as K then there exists a natural bijective mapping $\mathcal{E}(K) \to \mathcal{E}(H)$. We can think heuristically of the topological ends of X as those parts of X which "go to infinity".

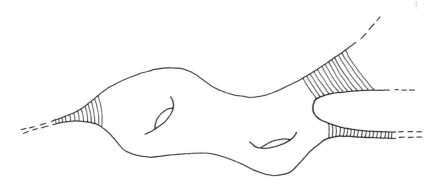

Fig. D.1. Topogical ends of a manifold

If M is a complete oriented hyperbolic n-manifold we shall make use of a completely different definition: for $\varepsilon \leq \varepsilon_n$ (the n-th Margulis constant) we shall call <u>ε-end</u> of M the closure of a connected component of $M \setminus M_{[\varepsilon,\infty)}$ (for a technical reason we shall explain we do not define an ε-end as a connected component of $M_{(0,\varepsilon]}$.

Remark D.3.1. Of course these definitions are related to each other, but they are actually different. In fact we may have that:

(1) an ε-end is not a topological end, in case it is a region where the manifold becomes thin without going to infinity (see Fig. D.2).

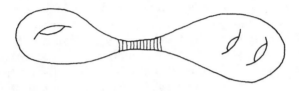

Fig. D.2. A ε-end needs not be a topological end

(2) a topological end is not an ε-end, in case it does not become thin as it goes to infinity (see Fig D.3).

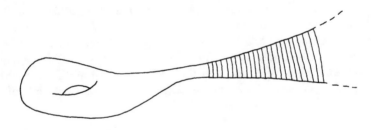

Fig. D.3. A topological end needs not be a ε-end

In the following result we shall assume without proof that the topological ends of a hyperbolic manifold can be defined. (This fact could be deduced quite easily from the other results we are going to prove, independently of the next one, in the present section.)

Proposition D.3.2. If M is a complete oriented hyperbolic manifold having finite volume then each topological end of M is an ε-end, and if ε is small enough the converse is true too.

Proof. Let $K \subset M$ be as in the definition of the topological ends. Since by D.2.6 $M_{[\varepsilon,\infty)}$ is compact, $H = M_{[\varepsilon,\infty)} \cup K$ is compact too, so that we can take as generic topological end the closure of a connected component E of $M \setminus H$; then E is contained in a unique connected component of $M \setminus M_{[\varepsilon,\infty)}$ and the first part is proved.

As for the second part we remark that if ε is small enough we have that $K \subset M_{[\varepsilon,\infty)}$, and then $M_{[\varepsilon,\infty)}$ itself can be used for the definition of the topological ends of M.   □

For the remainder of the section we shall keep fixed a number $\varepsilon > 0$ such that $\varepsilon \leq \varepsilon_n$; when speaking of thin part, thick part and end of a hyperbolic manifold we shall refer to this ε. We define μ to be the half of ε; we recall that every μ-ball in M is isometric to a μ-ball in $\mathbf{H}^n/_{\Gamma_\varepsilon}$, where Γ_ε falls into one of the cases given in Theorem D.2.2; in particular, if x belongs to the thick part $M_{[\varepsilon,\infty)}$ of M, then the μ-ball centred at x is isometric to a μ-ball in $\mathbf{H}^n$.

We begin with the determination of the ends of M up to homeomorphism; further information about the ends shall be given in D.3.11 and D.3.12. However, all geometric arguments are contained in the the long proof of the following theorem. For some illustrations clarifying the constructions we refer to the three-dimensional case D.3.13 described in Figg. D.5 to D.10.

Theorem D.3.3. The thin part $M_{(0,\varepsilon]}$ of M is the union of pieces homeomorphic to one of the following types:
(1) $\overline{D^{n-1}} \times S^1$;
(2) $V \times [0,\infty)$, where V is a differentiable oriented $(n-1)$-manifold without boundary supporting a Euclidean structure;
(3) S^1.
Moreover:
– these pieces have positive distance from each other;
– the ε-ends of M are the pieces of type (1) and (2);
– the pieces of type (3) are closed geodesics of length precisely ε;
– if M has finite volume the pieces of type (1) and (2) are finitely many and in those of type (2) the manifold V is compact.

Proof. Our argument will require several steps.

First of all we choose a point x in $M_{(0,\varepsilon]}$, identify $\Pi_1(M,x)$ with a subgroup Γ of $\mathcal{I}(\mathbf{H}^n)$, denote the quotient projection from $\mathbf{H}^n$ onto M by π, choose $\tilde{x}$ in $\pi^{-1}(x)$ and define $\Gamma_\varepsilon(\tilde{x})$ as usual. By Theorem D.2.2, $\Gamma_\varepsilon(\tilde{x})$ is either infinite cyclic generated by an isometry of hyperbolic type, or it consists of isometries of parabolic type all having the same fixed point at infinity. We distinguish between the two cases. The first one will give "tube" pieces (homeomorphic to $\overline{D^{n-1}} \times S^1$ or to S^1) and the second one will give "cusp" pieces (homeomorphic to $V \times [0,\infty)$).

FIRST CASE. In $\Pi^{n,+}$ assume that $\Gamma_\varepsilon(\tilde{x})$ is generated by a hyperbolic isometry γ with fixed points 0 and ∞ and consider the subgroup Γ_0 of Γ consisting of the identity and of all hyperbolic isometries with fixed points 0 and ∞. The same method as that used for Theorem D.2.2 proves that Γ_0 is infinite cyclic generated by an isometry

$$\gamma_0 : (y,t) \mapsto \lambda_0(A_0 y, t)$$

where $\lambda_0 > 1$ and $A_0 \in SO(n-1)$. Remark that $\Gamma_0 = \Gamma_\varepsilon(\tilde{x}_0)$, where $\tilde{x}_0$ is any point on the geodesic line $\{0\} \times \mathbb{R}_+$.

If d denotes the hyperbolic distance in $\mathrm{III}^{n,+}$, we define the set

$$\tilde{N} = \{(y,t)\in \mathrm{III}^{n,+} : \exists k\in \mathbb{Z}\setminus\{0\} \ s.t. \ d((y,t),\gamma_0^k(y,t)) \le \varepsilon\}.$$

Since γ is a power of γ_0, by definition of $\Gamma_\varepsilon(\tilde{x})$, $\tilde{N}$ contains $\tilde{x}$ and hence it is non-empty; moreover it is easily verified that it is kept invariant under dilations and that it consists either of the geodesic line $\{0\}\times \mathbb{R}_+$ or of a closed neighborhood of that line.

Lemma D.3.4. If $g\in\Gamma$ is such that $g(\tilde{N})\cap \tilde{N} \ne \phi$, then g is a power of γ_0.

Proof. Let $\xi\in g(\tilde{N})\cap \tilde{N}$, $\eta = g^{-1}(\xi)$. For some $k,h\in\mathbb{Z}\setminus\{0\}$ we have:

$$d(\xi,\gamma_0^k(\xi)) \le \varepsilon$$
$$d(\xi,g\gamma_0^h g^{-1}(\xi)) = d(\eta,\gamma_0^h(\eta)) \le \varepsilon.$$

It follows that γ_0^k and $g\gamma_0^h g^{-1}$ both belong to the group $\Gamma_\varepsilon(\xi)$. Theorem D.2.2 applies to this group: since it contains the hyperbolic isometry γ_0^k it must be infinite cyclic generated by a hyperbolic isometry γ_1. As γ_0^k is a power of γ_1, γ_1 must have the same fixed points of γ_0^k at infinity, and hence (by definition of Γ_0) it must be a power of γ_0. It follows that $g\gamma_0^h g^{-1}$ is a power of γ_0 too. We have proved that for some integer m

$$g\gamma_0^h g^{-1} = \gamma_0^m \ \Rightarrow \ g\gamma_0^h = \gamma_0^m g.$$

Remark that $h \ne 0$, whence $m \ne 0$.

Now let l denote the geodesic line $\{0\}\times \mathbb{R}_+$; by the preceding formula, since l is γ_0-invariant, we obtain that $g(l) = \gamma_0^m g(l)$. But l is the only γ_0^m-invariant geodesic line, so $g(l) = l$. Since g cannot be elliptic, g is hyperbolic with fixed points 0 and ∞, and hence it is a power of γ_0.   □

Since $\tilde{N}$ is obviously Γ_0-invariant, Lemma D.3.4 yields

$$\pi(\tilde{N}) = \tilde{N}/\Gamma_0;$$

moreover it easily follows from the definition of $\tilde{N}$ that $\pi(\tilde{N}) \subset M_{(0,\varepsilon]}$.

Since $\tilde{N} \ne \phi$, we have $d((0,1),(0,\lambda_0)) \le \varepsilon$; if equality holds, $\tilde{N}$ reduces to the geodesic line $\{0\}\times \mathbb{R}_+$ and the quotient space $\pi(\tilde{N})$ is a closed geodesic arc of length precisely ε. If inequality holds, the shape of $\pi(\tilde{N})$ is described by the following:

Lemma D.3.5. Assume $d((0,1),(0,\lambda_0)) < \varepsilon$; then $\pi(\tilde{N})$ is homeomorphic to $\overline{D^{n-1}} \times S^1$.

Proof. We claim that $\tilde{N}\cap(\mathbb{R}^{n-1}\times\{1\})$ is homeomorphic to $\overline{D^{n-1}}$. For k in $\mathbb{Z}\setminus\{0\}$ we set

$$E_k = \{y\in\mathbb{R}^{n-1} : d((y,1),\gamma_0^k(y,1)) \le \varepsilon\}.$$

We must prove that the union of all the E_k's is homeomorphic to a closed disc; let us remark first that only finitely many E_k's are non-empty. Moreover it easily follows from the explicit determination of the distance in $\mathbb{I}\mathbb{I}^{n,+}$ (A.5.8) that the equation of E_k is

$$(\star) \qquad\qquad \|\lambda_0^k A_0^k y - y\| \leq \delta_k,$$

for a suitable constant δ_k. It follows that E_k is a closed ellipsoid centred at the origin. (Though we are not going to need it we remark that all the E_k's are co-axial. Hint: consider a normal form of A_0.) Moreover the finite union of closed ellipsoids centred at the origin is certainly homeomorphic to a closed disc, as suggested by Fig. D.4.

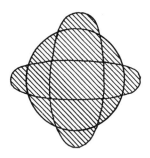

Fig. D.4. A union of concentric coaxial ellipsoids

Our claim is proved. Let us also remark that $\tilde{N} \cap (\mathbb{R}^{n-1} \times \{1\})$ is A_0-invariant. The homeomorphism $\tilde{A}_0$ of $\overline{D^{n-1}}$ corresponding to A_0 under the identification $\tilde{N} \cap (\mathbb{R}^{n-1} \times \{1\}) \cong \overline{D^{n-1}}$ is (continuously) isotopic to the identity: in fact $A_0 \in SO(n-1)$ which is connected, and hence an easy argument (based on the fact that $\tilde{N} \cap (\mathbb{R}^{n-1} \times \{1\})$ is not only homeomorphic to $\overline{D^{n-1}}$, but also starlike) leads to the conclusion.

Now, it follows from dilation-invariance that $\tilde{N}$ is the infinite positive cone with vertex at the origin based on $\tilde{N} \cap (\mathbb{R}^{n-1} \times \{1\})$, and therefore $\tilde{N} \cong \overline{D^{n-1}} \times \mathbb{R}_+$. The action of γ_0 under this identification is given by

$$\gamma_0 : \overline{D^{n-1}} \times \mathbb{R}_+ \ni (w,t) \mapsto (\tilde{A}_0 w, \lambda_0 t).$$

It follows that $\pi(\tilde{N})$ is fibered over S^1 with fiber $\overline{D^{n-1}}$; the isotopy class of $\tilde{A}_0$ being trivial, the fibration is trivial, and hence

$$\pi(\tilde{N}) \cong \overline{D^{n-1}} \times S^1.$$

(Remark that the assumption of the orientability of M is essential here; in fact connectedness of $SO(n-1)$ was used, and it is well-known that $O(n)$ is not connected.) □

SECOND CASE. We refer to the notations we introduced at the beginning of the proof of Theorem D.3.3, and we assume that $\Gamma_\varepsilon(\tilde{x})$ consists of parabolic isometries of $\mathbb{I}\mathbb{I}^{n,+}$ having ∞ as fixed point. We consider the subgroup Γ_1 of

Γ consisting of the identity and of all the parabolic isometries in Γ having ∞ as fixed point, and we set

$$\tilde{L} = \{(y,t)\in \mathbb{II}^{n,+} : \exists \gamma \in \Gamma_1 \setminus \{\text{id}\} \ s.t. \ d((y,t),\gamma(y,t)) \leq \varepsilon\}.$$

Since $\Gamma_\varepsilon(\tilde{x}) \subseteq \Gamma_1$, by definition we obtain that $\tilde{x} \in \tilde{L}$, and hence $\tilde{L}$ is not empty; we are now going to describe the shape of $\pi(\tilde{L})$.

Our first step will be the proof of an analogue of Lemma D.3.4 in this case. We shall need the following technical result:

Lemma D.3.6. A properly discontinuous subgroup of $\mathcal{I}(\mathbf{H}^n)$ cannot contain an isometry of hyperbolic type and one of parabolic type having a common fixed point.

Proof. Consider the half-space model and let $\delta_1, \delta_2 \in \mathcal{I}(\mathbb{II}^{n,+})$ be such that δ_1 is parabolic, δ_2 is hyperbolic, $\delta_1(\infty) = \delta_2(\infty) = \infty$, $\delta_2(0) = 0$. Then δ_1 and δ_2 can be written in the following form:

$$\delta_1(y,t) = (Ay + b, t)$$
$$\delta_2(y,t) = \lambda(By, t)$$

with $A, B \in O(n - 1)$, $b \in \mathbb{R}^{n-1} \setminus \{0\}$, $\lambda \neq 1$; we assume that $\lambda > 1$ (if this is not the case we replace δ_2 by its inverse).

For $n \in \mathbb{N}$, by direct computation we obtain that

$$(\delta_2^{-n} \circ \delta_1 \circ \delta_2^n)(0,1) = (\lambda^{-n} \cdot B^{-n}b, 1).$$

These points are different from each other and converge to $(0,1)$ as n goes to infinity, therefore the group generated by δ_1 and δ_2 is not properly discontinuous. $\square$

Lemma D.3.7. If $g \in \Gamma$ is such that $g(\tilde{L}) \cap \tilde{L} \neq \phi$ then $g \in \Gamma_1$.

Proof. Let $\xi \in g(\tilde{L}) \cap \tilde{L}$, $\eta = g^{-1}(\xi)$. For some $\gamma_1, \gamma_2 \in \Gamma_1 \setminus \{1\}$ we have

$$d(\xi, \gamma_1(\xi)) \leq \varepsilon$$
$$d(\xi, g\gamma_2 g^{-1}(\xi)) = d(\eta, \gamma_2(\eta)) \leq \varepsilon.$$

Hence $\gamma_1, g\gamma_2 g^{-1} \in \Gamma_\varepsilon(\xi)$; since this group contains γ_1, it must consist of parabolic isometries having ∞ as fixed point. In particular

$$g\gamma_2 g^{-1}(\infty) = \infty \quad \Rightarrow \quad \gamma_2(g^{-1}(\infty)) = g^{-1}(\infty);$$

since ∞ is the only fixed point for γ_2 we obtain $g(\infty) = \infty$; by D.3.6, g is necessarily parabolic and the lemma is proved. $\square$

As in the first case, Lemma D.3.7 and Γ_1-invariance of $\tilde{L}$ imply that

$$x \in \pi(\tilde{L}) = {}^{\tilde{L}}\!/_{\Gamma_1} \subset M_{(0,\varepsilon]}.$$

Lemma D.3.8. $\pi(\tilde{L})$ is homeomorphic to $V \times [0, \infty)$ for a suitable oriented Euclidean $(n-1)$-manifold without boundary V.

Proof. For $y \in \mathbb{R}^{n-1}$ we define the continuous function

$$W_y : \mathbb{R}_+ \to \mathbb{R}_+$$
$$t \mapsto \min \{ d((y, t), \gamma(y, t)) : \gamma \in \Gamma_1 \setminus \{\mathrm{id}\} \}.$$

It is easily verified (using A.5.8) that W_y is strictly decreasing and

$$\lim_{t \to 0} W_y(t) = +\infty \qquad\qquad \lim_{t \to \infty} W_y(t) = 0.$$

Moreover the mapping $(y, t) \mapsto W_y(t)$ is jointly continuous; it follows that the function

$$Q : \mathbb{R}^{n-1} \to \mathbb{R}_+ \qquad y \mapsto (W_y)^{-1}(\varepsilon)$$

is well-defined and continuous. The set $\tilde{L}$ is then given by

$$\tilde{L} = \{(y, t) : t \geq Q(y)\}$$

and therefore it is homeomorphic to $\mathbb{R}^{n-1} \times [0, \infty)$. Under this identification the group Γ_1 operates as a group of isometries of $\mathbb{R}^{n-1}$, and keeps the second coordinate fixed. This implies that $\pi(\tilde{L})$ is homeomorphic to

$$\left(\mathbb{R}^{n-1} / \Gamma_1 \right) \times [0, \infty),$$

and the first factor is a differentiable oriented manifold supporting a Euclidean structure. □

We have proved that $M_{(0,\varepsilon]}$ is covered by pieces of the types (1), (2) and (3) (and moreover that those of type (3) are closed geodesics of length precisely ε). For the conclusion of the proof of Theorem D.3.3 we shall need the following technical result:

Lemma D.3.9. A properly discontinuous subgroup of $\mathcal{I}(\mathbf{H}^n)$ cannot contain two isometries of hyperbolic type having only one common fixed point.

Proof. In the half-space model let us have two isometries

$$\delta_1 : (y, t) \mapsto \lambda_1(A_1 y, t)$$
$$\delta_2 : (y, t) \mapsto \lambda_2(A_2 y + b, t)$$

with $\lambda_i \neq 1$, $A_i \in O(n-1)$, $b \neq 0$.

Let $(c, 0)$ be the fixed point of δ_2. We assume $\lambda_i > 1$ (otherwise we replace δ_i by its inverse). For $m \in \mathbb{N}$ we can find (in a unique way) $k'_m \in \mathbb{N}$ and $y'_m \in [1, \lambda_2)$ such that $\lambda_1^m = y'_m \cdot \lambda_2^{k'_m}$. Let us consider a subsequence $\{y'_{m_i}\}$ of $\{y'_m\}$ such that $y'_{m_i} \to y \in [1, \lambda_2]$. We set $k_i = k'_{m_i}$ and $y_i = y'_{m_i}$. We have

$$d\big((\delta_2^{-k_i} \delta_1^{m_i})(0, 1), (c, y_i)\big) = d\big(\delta_1^{m_i}(0, 1), \delta_2^{k_i}(c, y_i)\big) =$$
$$= d\big((0, \lambda_1^{m_i}), (c, \lambda_1^{m_i})\big) \to 0 \quad (\text{as } i \to \infty)$$

whence $d\big((\delta_2^{-k_i}\delta_1^{m_i})(0,1),(c,y)\big) \to 0$ as $i \to \infty$. Moreover

$$d\big((\delta_2^{-k_i}\delta_1^{m_i})(0,1),(c,y)\big) = d\big((0,\lambda_1^{m_i}),(c,\lambda_2^{k_i}\cdot y)\big) > 0 \ \forall i.$$

This implies that the group generated by δ_1 and δ_2 is not properly discontinuous. $\qquad\square$

Lemma D.3.10. Let $C_1, C_2 \subseteq M_{(0,\varepsilon]}$ be sets of the form $\pi(\tilde{N})$ or $\pi(\tilde{L})$; if $C_1 \neq C_2$ then they have positive distance from each other.

Proof. Let J_1 and J_2 be subsets of $\mathbb{H}^{n,+}$ of the form $\tilde{N}$ or $\tilde{L}$; we shall prove that if $J_1 \neq J_2$ then $\pi(J_1)$ has positive distance from $\pi(J_2)$; by Lemmas D.3.4 and D.3.7 this is the same as proving that J_1 has positive distance from J_2, or, equivalently, that the closures of J_1 and J_2 in $\overline{\mathbb{H}^n}$ are disjoint.

For $i = 1, 2$ let Γ_i be the subgroup of Γ used for the definition of J_i (Γ_i is either the group of all hyperbolic isometries in Γ having a specified pair of fixed points, or the group of all parabolic isometries in Γ having a specified fixed point).

Assume J_1 and J_2 have a common point ξ in $\mathbb{H}^n$. By construction of the $\tilde{N}$'s and the $\tilde{L}$'s, the group $\Gamma_\varepsilon(\xi)$ is a (non-trivial) subgroup of both Γ_1 and Γ_2: this implies (by the properties of the Γ_i's) that $\Gamma_1 = \Gamma_2$ and hence $J_1 = J_2$.

If J_1 and J_2 have a common point in $\partial \mathbb{H}^n$, this point is a fixed point at infinity for the elements of both Γ_1 and Γ_2. By Lemma D.3.6, Γ_1 and Γ_2 are of the same type: if they are both parabolic they obviously coincide. If they are both hyperbolic Lemma D.3.9 applies, and once again $\Gamma_1 = \Gamma_2$. $\qquad\square$

The last assertions of D.3.3 left to prove are deduced from the following facts:

– $M \setminus M_{[\varepsilon,\infty)}$ consists precisely of the interior of the pieces of type (1) and (2); the interiors of such pieces are (obviously) open; moreover they are closed as the pieces are closed and pairwise disjoint;

– if M has finite volume then $M_{[\varepsilon,\infty)}$ is a compact topological manifold with boundary; its boundary consists of the boundaries of the pieces of type (1) and (2), so that these boundaries are compact and finitely many. $\qquad\square$

Theorem D.3.3 explains why we defined an ε-end as the closure of a component of $M \setminus M_{[\varepsilon,\infty)}$ and not as a component of $M_{(0,\varepsilon]}$: first of all we do not want to consider a closed geodesic as an end, and moreover it is not obvious (not even in the finite-volume case) that the pieces of type (1), (2) and (3) are actually the components of $M_{(0,\varepsilon]}$. However from now on we shall often confuse the two notions and assume that when speaking of a component of $M_{(0,\varepsilon]}$ we actually mean the closure of a component of its interior.

Theorem D.3.3 settles the determination of the topological structure of the ends of M, but it says nothing about the metric structure. As announced we shall illustrate now two subproducts of the proof of Theorem D.3.3 which give interesting informations about the geometry of the ends. We shall call a tube an end of M homeomorphic to $\overline{D^{n-1}} \times S^1$, and a cusp an end of M homeomorphic to $V \times [0,\infty)$.

Proposition D.3.11. The interior of a tube end C of M is diffeomorphic (and not only homeomorphic) to $D^{n-1} \times S^1$. C contains a unique geodesic loop. If $\delta > 0$ is small enough the closed δ-neighborhood C' of such a loop is contained in C and for some $r > 0$ and $\rho > 0$ it is isometric to the Riemannian manifold $B_r \times S_\rho$, where B_r is a closed ball of radius r in $\mathbb{R}^{n-1}$, S_ρ is a circle of length ρ and the metric is given by

$$ds^2_{(y,z)}(v, l) = \|v + ly\|^2 + l^2$$

(for $y \in B_r$, $z \in S_\rho$, $v \in \mathbb{R}^{n-1}$, $l \in \mathbb{R}$).

Proof. The first assertion is proved by the same construction made for D.3.5; it suffices to remark that the union of a finite number of open ellipsoids centred at the origin in $\mathbb{R}^{n-1}$ is diffeomorphic to an open disc, and that if we confine ourselves to the interior all mappings are differentiable.

The quotient of the geodesic line $\{0\} \times \mathbb{R}_+$ is surely a geodesic loop in C; conversely, let $\xi \in \tilde{N} \setminus \{0\} \times \mathbb{R}_+$ and $k \in \mathbb{Z} \setminus \{0\}$, and consider the geodesic line l passing through ξ and $\gamma_0^k(\xi)$; it is easily verified that the unit tangent vector to l in $\gamma_0^k(\xi)$ is not the image with respect to $D_\xi \gamma_0^k$ of the unit tangent vector to l in ξ: it follows that the quotient of l is not a closed geodesic.

By A.5.8 it is readily checked that for suitable $r > 0$ the closed δ-neighborhood of $\{0\} \times \mathbb{R}_+$ is the infinite positive cone with vertex in the origin based on $B_r \times \{1\}$, where B_r denotes the closed r-ball centred at the origin in $\mathbb{R}^{n-1}$; we shall denote this cone by C_r. Since $r \to 0$ as $\delta \to 0$, if δ is small enough $C_r \subseteq \tilde{N}$; obviously $\gamma_0(C_r) = C_r$ and if $g \in \Gamma$ and $g(C_r) \cap C_r \neq \phi$ then $g \in \Gamma_0$. It follows that the closed δ-neighborhood of the geodesic loop is isometric to C_r/Γ_0. Let us consider the diffeomorphism

$$f : B_r \times \mathbb{R} \to C_r \qquad (y, \tau) \to (e^\tau y, e^\tau).$$

If we set $\rho = \log(\lambda_0)$, γ_0 corresponds via f to

$$\tilde{\gamma}_0 : B_r \times \mathbb{R} \ni (y, \tau) \mapsto (A_0 y, \tau + \rho)$$

while the pull-back of the hyperbolic metric is given by

$$ds^2_{(y,\tau)}(v, l) = \|v + ly\|^2 + l^2.$$

This implies that the quotient space

$$(B_r \times \mathbb{R})/_{\langle \tilde{\gamma}_0 \rangle}$$

is actually isometric to the described Riemannian manifold. $\square$

Proposition D.3.12. The interior of a cusp end C of M is diffeomorphic (and not only homeomorphic) to $V \times \mathbb{R}$, where V is a differentiable oriented $(n-1)$-manifold without boundary supporting a Euclidean structure. If M has finite volume then V is compact and C contains a subset C' such that:

$-C \setminus C'$ has compact closure in M;
$-C'$ is isometric to the Riemannian manifold $V \times [0, \infty)$ with the metric

$$ds^2_{(x,t)}(v, l) = e^{-2t} de^2_x(v) + l^2$$

where de^2 is a Euclidean metric on V.

Proof. We refer to the notations of the proof of Lemma D.3.8. For the first assertion we remark that the interior of $\tilde{L}$ is given by

$$\text{int}(\tilde{L}) = \{(y, t) : t > Q(y)\};$$

Γ_1 operates as a group of diffeomorphisms of $\text{int}(\tilde{L})$. Since Q is Γ_1-invariant a quotient function $\overline{Q}$ is well-defined and continuous on $V = \mathbb{R}^{n-1}/\Gamma_1$. The interior of C is the quotient manifold $\text{int}(\tilde{L})/\Gamma_1$, and by the above remarks it is diffeomorphic to

$$\{(x, t) \in V \times \mathbb{R} : t > \overline{Q}(x)\}$$

which is diffeomorphic to $V \times \mathbb{R}$.

Now, assume that $\text{vol}(M) < \infty$; we know that $V = \mathbb{R}^{n-1}/\Gamma_1$ is compact. A fundamental domain for the action of Γ_1 is constructed in the usual way:

$$Y = \{y \in \mathbb{R}^{n-1} : \|y\| \leq \|y - \gamma(0)\| \ \forall \gamma \in \Gamma_1\}.$$

Since V is compact Y is compact too, and then the function Q has maximum r on Y. By the Γ_1-invariance of Q we obtain $Q \leq r$ everywhere, so that

$$\tilde{L} \supseteq \mathbb{R}^{n-1} \times [r, \infty).$$

This set is Γ_1-invariant, and hence its projection C' is isometric to

$$\left(\mathbb{R}^{n-1} \times [r, \infty)\right)/\Gamma_1 \cong V \times [r, \infty),$$

where the metric is given by

$$ds^2_{(x,t)}(v, l) = \frac{1}{t^2} \cdot (dm^2_x(v) + l^2)$$

for a Euclidean metric dm^2 on V. Let us consider now the diffeomorphism

$$f : V \times [0, \infty) \to V \times [r, \infty) \qquad (x, t) \mapsto (x, re^t).$$

The pull-back with respect to f of the above metric has the required form

$$ds^2_{(x,t)}(v, l) = e^{-2t} de^2_x(v) + l^2$$

where de^2 is a positive multiple of dm^2.

For the conclusion we only need to remark that the closure of $C \setminus C'$ is the projection of the compact set

$$\{(y, t) : y \in Y, \ Q(y) \leq t \leq r\}. \qquad \qquad \square$$

Now we concentrate our attention on the three-dimensional case, where the determination of the ends is much more precise (compare [Gro1]):

Theorem D.3.13.   Every end of a complete oriented hyperbolic three-manifold is isometric to one of the following manifolds:

1) $B_r \times S_\rho$, where $r, \rho > 0$, B_r is a closed disc of radius r in $\mathbb{R}^2$, S_ρ is a circle of length ρ and the metric is given by

$$ds^2_{(y,z)}(v,l) = \|v + ly\|^2 + l^2.$$

2) $T_2 \times [0, \infty)$ with the metric

$$ds^2_{(x,t)}(v,l) = e^{-2t} de^2(v) + l^2$$

where de^2 is a Euclidean metric on the torus T_2.

3) $S_\rho \times \mathbb{R} \times [0, \infty)$, where $\rho > 0$, S_ρ is as above and the metric is

$$ds^2_{(z,\tau,t)}(v,w,l) = e^{-2t}(v^2 + w^2) + l^2.$$

If the manifold has finite volume the third case cannot occur.

Proof. We only need to specialize our description of $\tilde{N}$ and $\tilde{L}$, as in the proof of Lemmas D.3.5 and D.3.8.

For $\tilde{N}$, since every element of $SO(2)$ is a rotation of an angle θ, it follows from A.5.8 that each of the closed ellipsoids $(\star)$ is a closed disc centred at the origin, and hence $\tilde{N} \cap (\mathbb{R}^2 \times \{1\})$ is a closed disc centred at the origin. This implies that, for some $\delta > 0$, $\tilde{N}$ is the closed δ-neighborhood of the geodesic line $\{0\} \times \mathbb{R}_+$, and in D.3.11 we checked that the image of such a neighborhood in M is isometric to $B_r \times S_\rho$ with the required metric.

As for $\tilde{L}$ we begin by remarking that a properly discontinuous subgroup of $\mathcal{I}^+(\mathbb{R}^2)$ acting freely on $\mathbb{R}^2$ is generated either by one translation or by two linearly independent translations; in particular the quotient of $\mathbb{R}^2$ by such a group is diffeomorphic either to $S^1 \times \mathbb{R}$ or to T_2. By D.2.2 the group Γ_1 defining $\tilde{L}$ is generated by one or two isometries of the form

$$(y, t) \mapsto (y + b, t).$$

Using A.5.8 it is easily deduced from this that $\tilde{L}$ has the form $\mathbb{R}^2 \times [r_0, \infty)$, and the proof is completed by the same argument presented in D.3.12.

If M has finite volume Proposition D.3.12 implies that the third case cannot occur.   □

Figures D.5 to D.10 illustrate the construction of a tube end and of a cusp based on a torus.

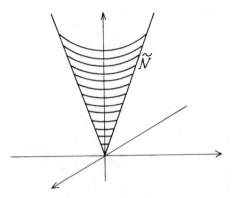

Fig. D.5. Ends of three-dimensional hyperbolic manifolds: tube case. The neighborhood $\tilde{N}$ of the vertical line (axis of a hyperbolic element of the group) projecting to the thin part

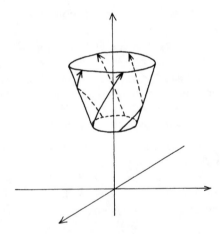

Fig. D.6. Ends of three-dimensional hyperbolic manifolds: tube case. A fundamental domain for the action of Γ_1 on $\tilde{N}$ and the action of γ_0 on its boundary

Fig. D.7. Ends of three-dimensional hyperbolic manifolds: tube case. The resulting tube end: a solid torus

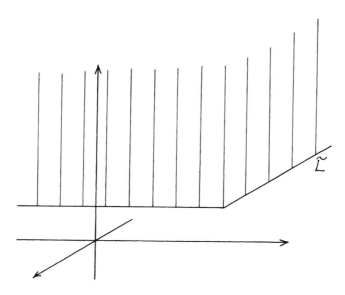

Fig. D.8. Ends of three-dimensional hyperbolic manifolds: case of cusp with compact base. The horoball $\tilde{L}$ centred at ∞ (fixed point of a parabolic element of the group) projecting to the thin part

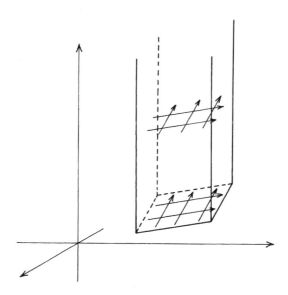

Fig. D.9. Ends of three-dimensional hyperbolic manifolds: case of cusp with compact base. A fundamental domain for the action of Γ_1 on $\tilde{L}$ and action of the generators on its boundary

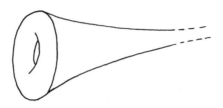

Fig. D.10. Ends of three-dimensional hyperbolic manifolds: case of cusp with compact base. Symbolic representation of the resulting cusp based on a torus

The next result collects some qualitative information about the topology of finite-volume hyperbolic manifolds (the proof is easily deduced from D.2.6, D.3.3, D.3.11, D.3.12 and D.3.13).

Corollary D.3.14. If $n \geq 2$ and M is a connected, oriented, complete hyperbolic finite-volume n-manifold then M is diffeomorphic to the interior of a suitable compact manifold N with (possibly empty) boundary. Each connected component of ∂N (if non-empty) is diffeomorphic to a compact $(n-1)$-manifold supporting a flat structure. In particular if $n = 3$ such components are flat tori.

We briefly recall two of Bieberbach's results which provide interesting informations about the boundary components of a finite-volume hyperbolic manifold also for $n \geq 4$. For a proof and other related results we refer to [Char].

Theorem D.3.15. (1) Each compact flat n-manifold is covered by a flat n-dimensional torus (*i.e.* a flat manifold diffeomorphic to $\mathbb{R}^{n}/\mathbb{Z}^{n}$).
(2) Up to affine equivalence (whence, in particular, up to diffemorphism) there exist only finitely many n-dimensional compact flat manifolds.

Now we confine ourselves to the case $n = 3$ and we prove some topological consequences of the existence of a constant curvature geometric structure (in particular for the hyperbolic case). We recall first a few definitions.

In the following M will denote an arbitrary connected 3-manifold, possibly with boundary. We shall say M is <u>irreducible</u> if each 2-sphere embedded in M is the boundary of a 3-ball embedded in M. If M is compact and $\partial M \neq \phi$ we shall say a subgroup G of $\Pi_1(M)$ is <u>peripheral</u> if there exists a component T of ∂M such that G is conjugate to $i_*(\Pi_1(T))$ (where $i : T \hookrightarrow M$ is the inclusion). M is called <u>atoroidal</u> if all the subgroups of $\Pi_1(M)$ isomorphic to $\mathbb{Z} \oplus \mathbb{Z}$ are peripheral. A surface S embedded in M is called <u>incompressible</u> if the homomorphism $i_* : \Pi_1(S) \to \Pi_1(M)$ (where, as above, $i : S \hookrightarrow M$ is the inclusion) is one-to-one; remark that this is not the standard definition, but it will suffice for our purposes. We shall say M is prime if each connected sum $M_1 \# M_2$ diffeomorphic to M must have one of the summands diffeomorphic to the sphere S^3.

Remark D.3.16. It follows quite easily from the definition that irreducible manifolds are prime.

Proposition D.3.17. If M is endowed with a hyperbolic, elliptic or flat structure then M is irreducible.

Proof. It follows from the Schönflies theorem that S^3 and $\mathbb{R}^3$ are irreducible manifolds. The universal covering of M is diffeomorphic either to S^3 or to $\mathbb{R}^3$, and for the conclusion we only have to recall that S^2 is simply connected, so that we can lift it to the universal covering.   □

Proposition D.3.18. Let M be endowed with a complete hyperbolic structure. Then:

(1) if M is not compact and has finite volume (which implies that it is the interior of a compact manifold $\overline{M}$ whose boundary consists of tori) then these tori are incompressible;

(2) if M has finite volume then it is atoroidal;

(3) if M is diffeomorphic to the interior of a compact manifold whose boundary consists of tori, then M has finite volume.

Proof. (1) Let us remark first that $\Pi_1(\overline{M}) = \Pi_1(M)$ (hint: retract cusps on the boundary tori). If a boundary torus T is obtained from a subgroup of $\Pi_1(M)$ generated by two independent parabolic elements γ_1 and γ_2, then γ_1 and γ_2 correspond to the canonical generators of $\Pi_1(T)$, and hence i_* is one-to-one.

(2) Assume $G < \Pi_1(M) < \mathcal{I}^+(\mathbb{H}^3)$, $G \cong \mathbb{Z} \oplus \mathbb{Z}$. Since G is abelian, the same argument presented for D.2.2 implies G is generated by two parabolic isometries with the same fixed point, and hence, up to conjugation, G is the fundamental group of the corresponding boundary torus.

(3) Assume M has infinite volume. Then M contains a subset diffeomorphic to $S^1 \times \mathbb{R} \times [0, \infty)$, glued to the remainder of M along $S^1 \times \mathbb{R} \times \{0\}$. It follows that there is no way to add tori in order to make M compact, and this is absurd.   □

Chapter E.
The Space of Hyperbolic Manifolds and the Volume Function

In the whole of this chapter we shall always suppose manifolds are connected and oriented. It follows from the Gauss-Bonnet formula B.3.3 (for $n = 2$) and from the Gromov-Thurston theorem C.4.2 (for $n \geq 3$) that the volume of a hyperbolic manifold is a topological invariant. Moreover B.3.3 implies that such an invariant is (topologically) complete for $n = 2$ in the compact case, and it may be proved that in the finite-volume case it becomes complete together with the number of cusp ends ("punctures"). Hence the problem of studying the volume function arises quite naturally: this is the aim of the present chapter.

We start by fixing some notations: for any $n \geq 2$ we shall denote by $\mathcal{H}_n$ the "set" of all n-dimensional complete hyperbolic manifolds up to isometry, and by $\mathcal{F}_n$ the subset of $\mathcal{H}_n$ consisting of the manifolds having finite volume. (Of course we should not call an element of $\mathcal{H}_n$ a "manifold", but we shall often do if there is no risk of misunderstanding.) The volume function vol : $\mathcal{H}_n \to (0, \infty]$ is naturally defined. The main problems we are going to consider are the following:

– what is the structure of the range of the volume function as a topological and as an ordered subspace of $\mathbb{R}$?

– for $c > 0$, what is $\text{vol}^{-1}([0, c])$ made of ?

The key trick for facing such questions is the introduction of a topological structure on the set $\mathcal{H}_n$ with respect to which the volume function has several interesting properties (in particular, it is continuous). The first section of the chapter is devoted to the definition of this topology.

For $n \geq 3$ many properties of the volume fuction will be quite easily deduced from the characterization of convergence for the defined topology and from the two main tools developed in the previous chapters: the rigidity theorem and Margulis' lemma. Our exposition is based on [Th1, ch. 8,9] and (mostly) on [Ca-Ep-Gr, ch. 3].

We can anticipate some qualitative results we shall prove: the cases $n = 2$ and $n \geq 4$ will present many analogies (though the reasons are different), while the case $n = 3$ appears as a complete exception. For instance we shall prove that:

(1) for $n = 2$ and $n \geq 4$ the range of the volume function is a closed discrete subset of $\mathbb{R}_+$, while for $n = 3$ it has limit points;

(2) given $c > 0$, for $n \geq 4$ the set $\mathrm{vol}^{-1}([0, c])$ is finite, and it is finite up to homeomorphism for $n = 2$, while for $n = 3$ if c is large enough it is not finite (and it keeps being infinite up to homeomorphism too).

These facts, in a vague but already quite significant way, show that 3-dimensional hyperbolic geometry plays a special role in this theory, as it summarizes the best features of rigidity and flexibility: on one hand if a 3-manifold has a finite-volume hyperbolic structure then this structure is unique, and on the other hand one has enough flexibility to produce plenty of examples; even more is true: "almost all" manifolds having the same topological description as given in D.3.14 (in particular, compact manifolds) can be endowed with a hyperbolic structure.

E.1 The Chabauty and the Geometric Topology

We begin with a very general definition, first introduced in [Chab]. In the sequel X will be a fixed arbitrary topological space and $\mathcal{C}(X)$ will denote the family of all closed subsets of X (including ϕ). For compact $K \subset X$ and open $U \subset X$ we set

$$\mathcal{U}_1(K) = \{C \in \mathcal{C}(X) : C \cap K = \phi\}$$
$$\mathcal{U}_2(U) = \{C \in \mathcal{C}(X) : C \cap U \neq \phi\}.$$

We define the <u>Chabauty topology</u> on $\mathcal{C}(X)$ as the topology for which

$$\{\mathcal{U}_1(K) : K \text{ compact}\} \bigcup \{\mathcal{U}_2(U) : U \text{ open}\}$$

is a pre-basis. In this section whenever considering a family of closed subsets of a topological space we shall always understand that it is endowed with the relative Chabauty topology.

Lemma E.1.1. (1) $\mathcal{C}(X)$ is a compact space;
(2) if X is a Hausdorff locally compact topological space with a countable basis of open sets, then $\mathcal{C}(X)$ has a countable basis too and moreover it is metrizable.

Proof. (1) According to the Tychonov theorem it suffices to show that a covering of the form

$$\{\mathcal{U}_1(K_i)\}_{i \in I} \bigcup \{\mathcal{U}_2(U_j)\}_{j \in J}$$

has a finite subcovering. Let us define U as the union of all the U_j's and C as the complement of U in X. Since C is closed and $C \notin \mathcal{U}_2(U_j)$ $\forall j$ there exists i such that $C \in \mathcal{U}_1(K_i)$. (Remark that possibly $C = \phi$, and in this case we are just claiming that $I \neq \phi$.) Since the U_j's cover K_i we can find $U_{j_1}, ..., U_{j_p}$ covering K_i too. Given $L \in \mathcal{C}(X)$ if $L \notin \mathcal{U}_1(K_i)$ we have necessarily $L \in \mathcal{U}_2(U_{j_q})$ for some q, and the proof is complete.

(2) Let $C, L \in \mathcal{C}(X)$, $C \neq L$. We can assume by simplicity that there exists $x \in C \setminus L$. Then there exists an open neighborhood U of x whose closure is compact and does not meet L. It follows that

$$C \in \mathcal{U}_2(U) \qquad L \in \mathcal{U}_1(\overline{U}) \qquad \mathcal{U}_1(\overline{U}) \cap \mathcal{U}_2(U) = \phi$$

and hence $\mathcal{C}(X)$ is a Hausdorff space.

Let us prove now that $\mathcal{C}(X)$ has a countable basis (which, together with compactness, implies metrizability). Let $\{U_n\}$ be a basis of open subsets of X such that $\overline{U}_n$ is compact $\forall n$. We claim that $\{\mathcal{U}_1(\overline{U}_n)\} \cup \{\mathcal{U}_2(U_n)\}$ is a pre-basis for the topology of $\mathcal{C}(X)$ (which implies that there exists a countable basis).

Given a compact set K and $L \in \mathcal{U}_1(K)$ (i.e. $L \cap K = \phi$), there exists a covering $\{U_{n_1}, ..., U_{n_p}\}$ of K such that $\overline{U}_{n_i} \cap L = \phi$ for all i; then

$$L \in \bigcap_{i=1}^{p} \mathcal{U}_1(\overline{U}_{n_i}) \subseteq \mathcal{U}_1(K).$$

Given an open set U and $L \in \mathcal{U}_2(U)$, choose $x \in L \cap U$ and find i such that $x \in U_i \subseteq U$; then

$$L \in \mathcal{U}_2(U_i) \subseteq \mathcal{U}_2(U)$$

so that our claim is proved and the proof is over.   □

Maybe the definition of the Chabauty topology is somewhat abstract, but the following result proves that the notion of convergence it entails is quite natural.

Proposition E.1.2. Let X be a locally compact metrizable space. A sequence $\{C_n\}$ in $\mathcal{C}(X)$ converges to $C \in \mathcal{C}(X)$ if and only if the following two conditions are fulfilled:

(1) if $x \in X$ is such that there exist a subsequence $\{C_{n_i}\}$ of $\{C_n\}$ and $x_i \in C_{n_i}$ with $x_i \longrightarrow x$ (in X), then $x \in C$;

(2) given $x \in C$ there exist $x_n \in C_n$ $\forall n$ such that $x_n \longrightarrow x$ in X.

Proof. Assume $C_n \longrightarrow C$ and let us prove (1) and (2).

(1) By contradiction, suppose $x \notin C$ and let K be a compact neighborhood of x such that $K \cap C = \phi$, that is $C \in \mathcal{U}_1(K)$. Then for $i \gg 0$ we have $x_i \in K$ and also $C_{n_i} \in \mathcal{U}_1(K)$, which is absurd.

(2) For $m \in \mathbb{N}$ let U_m be the open ball of radius $1/m$ centred at x. Since $C \in \mathcal{U}_2(U_m)$ there exists an integer h_m such that $C_n \in \mathcal{U}_2(U_m)$ for $n \geq h_m$. Of course we can assume $h_{m+1} > h_m$; then we choose:

$$x_1 \in C_1, ..., x_{h_1-1} \in C_{h_1-1}$$
$$x_{h_1} \in C_{h_1} \cap U_1, ..., x_{h_2-1} \in C_{h_2-1} \cap U_1$$
$$. . .$$
$$x_{h_j} \in C_{h_j} \cap U_j, ..., x_{h_{j+1}-1} \in C_{h_{j+1}-1} \cap U_j$$
$$. . .$$

and the resulting sequence $\{x_n\}$ works.

Let us assume now that (1) and (2) hold. If by contradiction $C_n \not\longrightarrow C$ one of the following cases is verified:

(i) there exists a compact set K such that $C \in \mathcal{U}_1(K)$ and, for infinitely many n's, $C_n \notin \mathcal{U}_1(K)$;

(ii) there exists an open set U such that $C \in \mathcal{U}_2(U)$ and, for infinitely many n's, $C_n \notin \mathcal{U}_2(U)$.

We shall check that both cases are absurd.

(i) There exist a subsequence $\{C_{n_i}\}$ and $x_i \in C_{n_i} \cap K$; since K is compact we can also assume $x_i \longrightarrow x \in K$, but $x \in C$ and $C \cap K = \phi$: absurd.

(ii) For infinitely many indices n, $C_n \cap U = \phi$, so that if $x \in C \cap U$ no sequence $\{x_n\}$ with $x_n \in C_n$ can converge to x: absurd. $\square$

We recall that if X is a compact metric space the <u>Hausdorff</u> <u>distance</u> of two closed subsets K_1, K_2 of X is defined as

$$H(K_1, K_2) = \max_{x_1 \in K_1} d(x_1, K_2) \vee \max_{x_2 \in K_2} d(x_2, K_1).$$

The following result (we shall not make use of) provides another evidence of the naturality of the Chabauty topology. The proof is left as an exercise to the reader.

Proposition E.1.3. If X is a compact metric space then the Hausdorff distance induces the Chabauty topology on $\mathcal{C}(X)$.

Now we come to a more specific situation: we shall consider a fixed Lie group G. We first remark that G is a locally compact Hausdorff metrizable space (for metrizability, remark that any left-invariant Riemannian metric induces the topology of G); moreover we shall suppose that G has a countable basis of open sets.

We shall denote by $\mathcal{S}(G)$ the family of all closed subgroups of G.

Lemma E.1.4. $\mathcal{S}(G)$ is a closed subset of $\mathcal{C}(G)$, and hence it is a compact metrizable topological space with a countable basis of open sets.

Proof. Let $C \in \mathcal{C}(G) \setminus \mathcal{S}(G)$. Then there exist $x, y \in C$ with $x \cdot y^{-1} \notin C$; let K be a compact neighborhood of $x \cdot y^{-1}$ with $K \cap C = \phi$ and let U and V be open neighborhoods of x and y respectively such that $U \cdot V^{-1} \subset K$. The open set

$$\mathcal{U}_1(K) \bigcap \mathcal{U}_2(U) \bigcap \mathcal{U}_2(V)$$

contains C and does not meet $\mathcal{S}(G)$. $\square$

If e denotes the identity of G and U is an open neighborhood of e we set:

$$\mathcal{A}(U) = \left\{ H \in \mathcal{S}(G) : H \bigcap U = \{e\} \right\}$$

$$\mathcal{B}(U) = \left\{ H \in \mathcal{A}(U) : H \text{ is torsion-free} \right\}$$

(we recall that H is <u>torsion-free</u> if it contains no element of finite order but e). Let us remark that for all U the elements of $\mathcal{A}(U)$ are discrete subgroups, and conversely the set $\mathcal{D}(G)$ of all discrete subgroups of G is obtained as the union of all these $\mathcal{A}(U)$'s.

Proposition E.1.5. Let U be an open neighborhood of e; then $\mathcal{A}(U)$ and $\mathcal{B}(U)$ are closed subsets of $\mathcal{S}(G)$ (which implies that they are compact) while $\mathcal{D}(G)$ is open in $\mathcal{S}(G)$.

Proof. Closedness of $\mathcal{A}(U)$ is obvious:

$$\mathcal{A}(U) = \mathcal{S}(G) \setminus \mathcal{U}_2(U \setminus \{e\}).$$

As for closedness of $\mathcal{B}(U)$, assume $\{H_n\}$ is a sequence in $\mathcal{B}(U)$ converging to H (whence $H \in \mathcal{A}(U)$). Assume by contradiction $H \notin \mathcal{B}(U)$, *i.e.* there exists $x \in H \setminus \{e\}$ and $k \in \mathbb{Z}$, $k > 0$, such that $x^k = e$. Proposition E.1.2 implies that there exists a sequence $x_n \in H_n$ such that $x_n \longrightarrow x$, which implies $x_n^k \longrightarrow x^k = e$, and hence for $n \gg 0$ we have $x_n^k \in U$. Since $H_n \in \mathcal{B}(U)$ we must have $x_n = e$ for $n \gg 0$ and hence $x = e$. Contradiction.

Now we turn to the openness of $\mathcal{D}(G)$: let H be a discrete subgroup of G, and let V be an open neighborhood of e in G such that $H \cap V = \{e\}$; moreover we assume the closure of V is compact and V contains no proper subgroup of G. (The existence of a neighborhood of e not containing any proper subgroup of G is easily established: if Z is a neighborhood of 0 in $T_e G$ such that the restriction to Z of the exponential mapping is one-to-one, it suffices to consider the exponential of $(1/2)Z$.)

We choose now U to be an open neighborhood of $\{e\}$ such that $U = U^{-1}$ and $U \cdot U \subset V$; if we set $K = \overline{V} \setminus U$ we have $H \in \mathcal{U}_1(K)$. We shall prove that $\mathcal{U}_1(K) \cap \mathcal{S}(G) \subset \mathcal{D}(G)$, which implies the conclusion.

Let $L \in \mathcal{U}_1(K) \cap \mathcal{S}(G)$: we claim that $L \cap U = \{e\}$ (which, of course, implies the conclusion). By contradiction, let $x \in L \cap U \setminus \{e\}$.

We can prove that $x^k \in U \,\forall k \in \mathbb{Z}$: assume by contradiction that k is the least positive integer such that $x^k \notin U$; then $x^k \in V$ (x and x^{k-1} belong to U, whence $x^k \in U \cdot U \subseteq V$). Then $x^k \in \overline{V} \setminus U = K$, and this is absurd since $x^k \in L$ and $L \cap K = \phi$.

It follows in particular that $x^k \in V \,\forall k \in \mathbb{Z}$, and hence V contains the cyclic group generated by x, which contradicts our choice of V. Our claim is proved and hence the proof is complete.   □

We can consider now the case we are really interested in: the Lie group $\mathcal{I}^+(\mathbb{H}^n)$. For $H \in \mathcal{D}(\mathcal{I}^+(\mathbb{H}^n))$ we can consider the quotient set $\mathbb{H}^n/H$ (which can be viewed as a "singular" hyperbolic n-manifold: for a true manifold we must require that H contains no elliptic element). Remark that we can define the volume of the quotient set $\mathbb{H}^n/H$ also in the singular case, as the volume of any fundamental domain for H (compare C.1.3): we shall denote this (possibly infinite) number by $\mu(H)$.

Proposition E.1.6. Let $\{H_n\}$ be a sequence in $\mathcal{D}(\mathcal{I}^+(\mathbb{H}^n))$ and assume it converges in $\mathcal{S}(\mathcal{I}^+(\mathbb{H}^n))$ to a subgroup H which is discrete too: then

$$\mu(H) \leq \liminf_{n \to \infty} \mu(H_n).$$

Proof. Let $D \neq \phi$ be a subset of $\mathbb{H}^n$ such that $D \cap g(D) = \phi$ for all $g \in H \setminus \{\mathrm{id}\}$ (we shall express this property by saying that D is H-irreducible) and let K be a compact subset of D. We assert that K is H_n-irreducible for $n \gg 0$: since K can be chosen to have volume arbitrarily close to $\mu(H)$ the proposition will follow from this assertion.

Let us set

$$C = \{T \in \mathcal{I}^+(\mathbb{H}^n) : K \cap T(K) \neq \phi \}.$$

Of course C is a compact subset of $\mathcal{I}^+(\mathbb{H}^n)$. We fix now an open neighborhood V of id in $\mathcal{I}^+(\mathbb{H}^n)$ such that $H \cap V = \{\mathrm{id}\}$. $C \setminus V$ is compact and $H \in \mathcal{U}_1(C \setminus V)$, whence $H_n \in \mathcal{U}_1(C \setminus V)$ for $n \gg 0$. The same argument as presented in the third part of the proof of E.1.5 allows one to assume (up to a suitable change of V) that $H_n \cap V = \{\mathrm{id}\}$ for $n \gg 0$. Then for $n \gg 0$ an element $g \in H_n$ has two possibilities: either $g = \mathrm{id}$ or $g(K) \cap K = \phi$, and then our assertion is proved. □

If U is an open neighborhood of id in $\mathcal{I}^+(\mathbb{H}^n)$ and $c > 0$ we set now

$$\mathcal{A}(U,c) = \{H \in \mathcal{A}(U) : \mu(H) \leq c\}$$

$$\mathcal{B}(U,c) = \{H \in \mathcal{B}(U) : \mu(H) \leq c\}$$

The following important result is immediately deduced from E.1.6 and E.1.5:

Corollary E.1.7. $\mathcal{A}(U,c)$ and $\mathcal{B}(U,c)$ are compact sets.

We are now going to use the Chabauty topology on $\mathcal{C}(\mathcal{I}^+(\mathbb{H}^n))$ for the definition of a topology on $\mathcal{H}_n$. We begin by recalling the following elementary fact we have already used several times:

Lemma E.1.8. A discrete subgroup Γ of $\mathcal{I}^+(\mathbb{H}^n)$ operates freely on $\mathbb{H}^n$ if and only if it is torsion-free.

Proof. It was already remarked in D.2.5 that only elliptic isometries have finite order. Conversely if an elliptic isometry is not cyclic then it is easily verified (for instance, in the disc model) that the orbit of some point is not discrete. □

The way to introduce the topology on $\mathcal{H}_n$ is to define a bijection with the space $\mathcal{D}_*(\mathcal{I}^+(\mathbb{H}^n))$ of all discrete torsion-free subgroups of $\mathcal{I}^+(\mathbb{H}^n)$; it is impossible to do this with $\mathcal{H}_n$ (the holonomy is well-defined only up to conjugation), so we need to define other objects. We set:

$$\mathcal{H}_n^* = \{(M,x) : M \text{ hyperbolic}, x \in M\}\big/_\sim$$

(where $(M,x) \sim (N,y)$ if there exists an isometry $f : M \to N$ with $f(x) = y$) and

$$\mathcal{H}_n^{**} = \{(M,x,v) : M \text{ hyp.}, x \in M, v \text{ pos. orthonorm. basis of } T_x M\}\big/_\sim$$

(where $(M,x,v) \sim (N,y,w)$ if there exists an isometry $f : M \to N$ with $f(x) = y$ and $d_x f(v) = w$, i.e. $d_x f(v_i) = w_i$ for $i = 1, ..., n$; 'hyp.' stands for 'hyperbolic', 'pos.' for 'positive' and 'orthonorm.' for 'orthonormal'). We consider the obvious "forgetful" mappings

$$\mathcal{H}_n^{**} \xrightarrow{\phi_2} \mathcal{H}_n^* \xrightarrow{\phi_1} \mathcal{H}_n.$$

Let us fix for the whole section a point z_0 of $\mathbf{H}^n$ and a positive orthonormal basis $t_0 = (a_1, ..., a_n)$ of $T_{z_0}\mathbf{H}^n$. The reason for introducing the space $\mathcal{H}_n^{**}$ is shown in the next result. We recall that we defined $\mathcal{D}_*(\mathcal{I}^+(\mathbf{H}^n))$ to be the set of all discrete torsion-free subgroups of $\mathcal{I}^+(\mathbf{H}^n)$.

Proposition E.1.9. The mapping

$$\Phi : \mathcal{D}_*(\mathcal{I}^+(\mathbf{H}^n)) \ni \Gamma \mapsto \left[\left(\mathbf{H}^n/\Gamma, \pi(z_0), d_{z_0}\pi(t_0)\right)\right] \in \mathcal{H}_n^{**}$$

(where π is the natural projection $\mathbf{H}^n \to \mathbf{H}^n/\Gamma$) is a bijection.

Proof. Surjectivity follows immediately from B.1.7.

Assume Γ_1 and Γ_2 give rise to equivalent triples, *i.e.* there exists a surjective isometry

$$\phi : \mathbf{H}^n/\Gamma_1 \to \mathbf{H}^n/\Gamma_2$$

such that

$$\phi(\pi_1(z_0)) = \pi_2(z_0) \qquad d_{z_0}(\phi \circ \pi_1)(t_0) = d_{z_0}\pi_2(t_0).$$

The mapping $\phi \circ \pi_1 : \mathbf{H}^n \to \mathbf{H}^n/\Gamma_2$ can be lifted to a mapping $k : \mathbf{H}^n \to \mathbf{H}^n$ such that the following diagram is commutative:

$$
\begin{array}{ccc}
\mathbf{H}^n & \xrightarrow{k} & \mathbf{H}^n \\
\downarrow{\pi_1} & \searrow & \downarrow{\pi_2} \\
\mathbf{H}^n/\Gamma_1 & \xrightarrow{\phi} & \mathbf{H}^n/\Gamma_2
\end{array}
$$

The above properties of ϕ imply that k is a local isometry, $k(z_0) = z_0$ and $d_{z_0}k = \mathrm{id}$ so that by A.2.1 k must be the identity.

The natural isomorphism $\phi_* : \Pi_1\left(\mathbf{H}^n/\Gamma_1, \pi_1(z_0)\right) \to \Pi_1\left(\mathbf{H}^n/\Gamma_2, \pi_2(z_0)\right)$ corresponds to $k_* : \Gamma_1 \to \Gamma_2$, and since k is the identity, k_* is the identity too, and then $\Gamma_1 = \Gamma_2$. □

We define now on $\mathcal{H}_n^{**}$ the geometric topology by the requirement that the above Φ is a homeomorphism. The geometric topologies on $\mathcal{H}_n^*$ and $\mathcal{H}_n$ are defined as the finest topologies with respect to which the forgetful mappings ϕ_1 and ϕ_2 are continuous.

Let us remark that the volume function is well-defined (by composition with ϕ_2 and ϕ_1) on $\mathcal{H}_n^{**}$ and $\mathcal{H}_n^*$ (we shall denote these functions by vol^{**} and vol^*, or simply by vol when no confusion arises). Hence the spaces $\mathcal{F}_n^{**}$ and $\mathcal{F}_n^*$ are naturally defined; $\mathcal{F}_n^{**}$, $\mathcal{F}_n^*$ and $\mathcal{F}_n$ will be endowed with the relative geometric topology.

We shall see now the first interesting corollaries of the properties of the Chabauty topology. We shall need another concept: given a point x in a hyperbolic manifold M we shall call <u>injectivity radius</u> of M at x the supremum $r(x, M)$ of all positive numbers r such that the r-ball centred at x in M is isometric to the r-ball in $\mathbb{H}^n$. Remark that for $[(M, x)] \in \mathcal{H}_n^*$ and $[(M, x, v)] \in \mathcal{H}_n^{**}$ the number $r(x, M)$ is well-defined. For $\varepsilon > 0$ and $c > 0$ we set now:

$$\mathcal{H}_n^{**}(\varepsilon) = \left\{ [(M, x, v)] \in \mathcal{H}_n^{**} : r(x, M) \geq \varepsilon \right\}$$
$$\mathcal{H}_n^*(\varepsilon) = \left\{ [(M, x)] \in \mathcal{H}_n^* : r(x, M) \geq \varepsilon \right\}$$
$$\mathcal{F}_n^{**}(\varepsilon, c) = \left\{ [(M, x, v)] \in \mathcal{H}_n^{**}(\varepsilon) : \mathrm{vol}(M) \leq c \right\}$$
$$\mathcal{F}_n^*(\varepsilon, c) = \left\{ [(M, x)] \in \mathcal{H}_n^*(\varepsilon) : \mathrm{vol}(M) \leq c \right\}$$
$$\mathcal{F}_n(c) = \left\{ [M] \in \mathcal{H}_n : \mathrm{vol}(M) \leq c \right\}$$

Theorem E.1.10. For $\varepsilon > 0$ and $c > 0$ the following spaces, endowed with the relative geometric topology, are compact:

$$\mathcal{H}_n^{**}(\varepsilon), \quad \mathcal{H}_n^*(\varepsilon), \quad \mathcal{H}_n, \quad \mathcal{F}_n^{**}(\varepsilon, c), \quad \mathcal{F}_n^*(\varepsilon, c), \quad \mathcal{F}_n(c).$$

Moreover the volume function $\mathrm{vol}^{**} : \mathcal{F}_n^{**} \to \mathbb{R}_+$ is lower semi-continuous and the infimum of its range is positive.

Proof. Let $U = \left\{ T \in \mathcal{I}^+(\mathbb{H}^n) : d(z_0, T(z_0)) < \varepsilon \right\}$. It is immediately checked that U is a neighborhood of the identity and (with the notations of E.1.5 and E.1.9) $\mathcal{H}_n^{**}(\varepsilon) = \Phi(\mathcal{B}(U))$. Since $\mathcal{B}(U)$ is compact, $\mathcal{H}_n^{**}(\varepsilon)$ is compact too.

$\mathcal{H}_n^*(\varepsilon)$ is the image under ϕ_2 of $\mathcal{H}_n^{**}(\varepsilon)$, and hence it is compact.

Compactness of $\mathcal{H}_n$ is deduced from the following assertion: *there exists a constant $\varepsilon > 0$ such that $\mathcal{H}_n$ is the image under ϕ_1 of $\mathcal{H}_n^*(\varepsilon)$.* This assertion can be re-phrased as follows: $\exists \varepsilon > 0$ such that the (2ε)-thick part of all M's in $\mathcal{H}_n$ is non-empty. And this is a direct corollary of the study of the thin-thick decomposition of a hyperbolic manifold: it follows from D.3.3 that for $\varepsilon \leq \varepsilon_n$ the ε-ends of a hyperbolic n-manifold have non-empty boundary, and hence the ε-thick part cannot be empty. Our assertion is proved and compactness of $\mathcal{H}_n$ follows.

Compactness of $\mathcal{F}_n^{**}(\varepsilon, c)$ and $\mathcal{F}_n^*(\varepsilon, c)$ follows again from E.1.7 as

$$\mathcal{F}_n^{**}(\varepsilon, c) = \Phi(\mathcal{B}(U, c)).$$

Finally, the very same argument as above proves that for some $\varepsilon > 0$ the set $\mathcal{F}_n(c)$ is the image under ϕ_1 of $\mathcal{F}_n^*(\varepsilon, c)$.

Lower semi-continuity of the volume function follows at once from E.1.6, while the further assertion is deduced from Margulis' lemma again: for any n-manifold M the ε_n-thick part of M is non-empty, so that the volume of M does not exceed the volume of a $(\varepsilon_n/2)$-ball in $\mathbb{H}^n$ (a universal constant). $\quad\square$

We shall discuss now a different characterization of the geometric topology, in which the "geometric" meaning is much more transparent; we shall define explicitly a fundamental system of neighborhoods of each element of $\mathcal{H}_n^{**}$ and then we shall check that the resulting topology coincides with the one defined above.

The definition we give is quite "strong"; it is possible to prove that much weaker geometric definitions actually lead to the same topology; since we are mainly interested in the study of the volume function and in a precise description of convergent sequences in $\mathcal{F}_n$, we do not discuss these weaker definitions and address the reader to [Ca-Ep-Gr].

We begin with (a special case of) a general definition. Let K be a compact subset of $\mathbf{H}^n$ and let f, g be smooth functions in the neighborhood of K with values in $\mathbf{H}^n$. Recall that for $x, y \in \mathbf{H}^n$ we denoted by $P_{y,x} : T_y \mathbf{H}^n \to T_x \mathbf{H}^n$ the parallel transport along the geodesic arc joining y to x; also, we denoted by d_p the p-differential. Given $z \in K$ and $p \geq 1$ the mapping

$$P_{f(z),z} \circ d_p f(z) - P_{g(z),z} \circ d_p g(z)$$

is a symmetric p-linear mapping on the normed space $T_z \mathbf{H}^n$ and then we can consider its norm. We define the distance between f and g on K by

$$D(f,g)_K = \max_{z \in K} d(f(z), g(z)) +$$

$$+ \sum_{p=1}^{\infty} 2^{-p} \cdot \left(1 \wedge \max_{z \in K} \left\| P_{f(z),z} \circ d_p f(z) - P_{g(z),z} \circ d_p g(z) \right\| \right).$$

A sequence $\{f_i\}$ is said to converge to f with respect to the C^∞ topology on K if $D(f_i, f)_K \to 0$. If $U \subset \mathbf{H}^n$ is open a sequence $\{f_i\}$ of smooth functions on U is said to converge to f with respect to the C^∞ topology on U if such a convergence holds on all compact subsets of U.

Given $[(M, x_0, u)] \in \mathcal{H}_n^{**}$ and $r, \varepsilon > 0$ we define $\mathcal{N}([(M, x_0, u], r, \varepsilon)$ as the set of all $[(N, y_0, v)] \in \mathcal{H}_n^{**}$ such that the following holds: if

$$\pi_M : \mathbf{H}^n \to M \qquad \pi_N : \mathbf{H}^n \to N$$

are the canonical projections (cf. the definition of Φ) there exists a smooth function f in the neighborhood of $\overline{B(z_0, r)}$ with values in $\mathbf{H}^n$, such that $f(z_0) = z_0$ and:
(i) for z_1, z_2 in the domain of f then

$$\pi_M(z_1) = \pi_M(z_2) \iff \pi_N(f(z_1)) = \pi_N(f(z_2))$$

(this property is called equivariance);
(ii) $D(f, \mathrm{id})_{\overline{B(z_0, r)}} < \varepsilon$.

We define another geometric topology τ_G on $\mathcal{H}_n^{**}$ (we deliberately use the same term, as we shall check the equivalence with the previous definition) by taking the class of all the subsets of the form $\mathcal{N}([(M, x_0, u], r, \varepsilon)$ as a pre-basis. It is actually possible to show that

$$\left\{ \mathcal{N}([(M, x_0, u], r, \varepsilon) : r, \varepsilon > 0 \right\}$$

is a fundamental system of neighborhoods of $[(M, x_0, u)]$ for τ_G; since we do not need this fact explicitly we leave it as an exercise to the reader. We also leave to the reader the following very easy:

Lemma E.1.11. Let f be a smooth function in a neigborhood of $\overline{B(z_0, r)}$ with $f(z_0) = z_0$ and $D(f, \mathrm{id})_{\overline{B(z_0, r)}} < \varepsilon$; if ε is small enough there exist $r'(r, \varepsilon), \varepsilon'(r, \varepsilon) > 0$ such that f^{-1} is well-defined and smooth on a neighborhood of $\overline{B(z_0, r'(r, \varepsilon))}$ and $D(f^{-1}, \mathrm{id})_{\overline{B(z_0, r'(r, \varepsilon))}} < \varepsilon'(r, \varepsilon)$. Moreover (for fixed r) as $\varepsilon \to 0$ we have

$$r'(r, \varepsilon) \to r \qquad \varepsilon'(r, \varepsilon) \to 0.$$

Remark E.1.12. The reader may object that the geometric meaning of τ_G is not so clear, as we used the coverings and not the manifolds only. However, if we recall that the r-ball at x_0 in M is the projection of $B(z_0, r)$ and we remark that by equivariance both f and (a suitable restriction of) f^{-1} are well-defined when passing to the manifolds, we obtain that

$$[(N, y_0, v)] \in \mathcal{N}\big([(M, x_0, u], r, \varepsilon\big)$$

implies that the r-ball at x_0 in M is quasi-isometric (depending on ε) to the r-ball at y_0 in N, and moreover this "quasi-isometry" (a formal definition of this notion will be given later) maps x_0 to y_0 and u close to v. Our definition requires actually more, as it is definitely not obvious that such a quasi-isometry can be lifted to an equivariant mapping C^∞ close to the identity. However, it is possible to prove that the topology defined by the existence of a quasi-isometry between the r-balls mapping x_0 to y_0 and u close to v, turns out to be equivalent to τ_G. This is an example of the weaker characterizations of τ_G we decided to omit, addressing the reader to [Ca-Ep-Gr].

We shall denote now by τ_C the Chabauty topology on $\mathcal{D}_*(\mathcal{I}^+(\mathbf{H}^n))$. The remainder of the present section is devoted to the proof of the following:

Theorem E.1.13. The natural bijective mapping

$$\Phi : \big(\mathcal{D}_*\big(\mathcal{I}^+(\mathbf{H}^n)\big), \tau_C\big) \to \big(\mathcal{H}_n^{**}, \tau_G\big)$$

is a homeomorphism.

We are going to prove continuity and openness separately, the latter being much simpler.

Proposition E.1.14. Φ is an open mapping.

Proof. We are going to prove that the inverse mapping is continuous. Since each point has a countable fundamental system of neighborhoods it suffices to prove that Φ^{-1} is continuous along sequences.

Consider a sequence $i \mapsto A_i = [(M_i, x_i, u_i)]$ in $\mathcal{H}_n^{**}$ converging with respect to τ_G to $A_0 = [(M_0, x_0, u_0)]$. According to E.1.2 we must show that for the groups $\Gamma_i = \Phi^{-1}(A_i)$ and $\Gamma_0 = \Phi^{-1}(A_0)$ the following holds:

(1) given an increasing sequence of integers $h \mapsto i_h$ and $\gamma_h \in \Gamma_{i_h}$ such that $\{\gamma_h\}$ converges to $\gamma \in \mathcal{I}^+(\mathbf{H}^n)$, we have $\gamma \in \Gamma_0$;

(2) given $\gamma \in \Gamma_0$ there exist $\gamma_i \in \Gamma_i \, \forall i$ such that the sequence $\{\gamma_i\}$ converges to γ.

Let us remark first that the usual topology on $\mathcal{I}^+(\mathbf{H}^n)$ is equivalent to the topology induced by $D(.,.)_{\overline{B}}$ for an arbitrary closed ball $\overline{B}$ in $\mathbf{H}^n$. We turn now to the proof of (1) and (2).

(1) Up to extraction of a subsequence from $\{A_i\}$ we can assume $\gamma_i \in \Gamma_i$ are such that $\gamma_i \to \gamma \in \mathcal{I}^+(\mathbf{H}^n)$. Let

$$\pi_0 : \mathbf{H}^n \to {\mathbf{H}^n}/_{\Gamma_0} = M_0 \qquad \pi_i : \mathbf{H}^n \to {\mathbf{H}^n}/_{\Gamma_i} = M_i$$

be the canonical projections. Let $\tilde{\alpha}_i$ be the geodesic arc joining z_0 to $\gamma_i(z_0)$; since $\gamma_i(z_0) \to \gamma(z_0)$ we have that the length of $\tilde{\alpha}_i$ is bounded by some constant $r/3$. Now we have, by definition of τ_G, that there exists a sequence of smooth functions in the neighborhood of $\overline{B(z_0,r)}$ such that $f_i(z_0) = z_0$ and:

(i)
$$\pi_0(z_1) = \pi_0(z_2) \iff \pi_i(f_i(z_1)) = \pi_i(f_i(z_2))$$

(ii)
$$D(f_i, \mathrm{id})_{\overline{B(z_0,r)}} \to 0.$$

By E.1.11 we have that for $i \gg 0$ the function f^{-1} is well-defined on $B(z_0, 2r/3)$, so that we can set $\tilde{\beta}_i = f^{-1}(\tilde{\alpha}_i)$; by the equivariance property we have that $\beta_i = \pi_0(\tilde{\beta}_i)$ is a loop at x_0, and moreover if $\psi_i \in \Gamma_0$ corresponds to this loop we have

$$f_i \circ \psi_i = \gamma_i \circ f_i$$

(whenever this relation makes sense: at least on a little ball at z_0). Since $f_i \to \mathrm{id}$ and $\gamma_i \to \gamma$ we get $\psi_i \to \gamma$. Since Γ_0 is closed we get $\gamma \in \Gamma_0$.

(2) Let $r > 2d(z_0, \gamma(z_0))$, and let the above notations be fixed. If $\tilde{\beta}$ is the geodesic arc joining z_0 to γ_0 we set $\tilde{\alpha}_i = f_i(\tilde{\beta})$, $\alpha_i = \pi_i(\tilde{\alpha}_i)$. By equivariance α_i is a loop at x_i and if $\gamma_i \in \Gamma_i$ corresponds to it we have

$$f_i \circ \gamma = \gamma_i \circ f_i$$

(as above, this relation holds at least on some little ball centred at z_0). Since $f_i \to \mathrm{id}$ we obtain that $\gamma_i \to \gamma$ and the proof is over. $\qquad \Box$

Proposition E.1.15. Φ is a continuous mapping.

Proof. Let $\{\Gamma_i\}$ be a sequence of discrete torsion-free subgroups of $\mathcal{I}^+(\mathbf{H}^n)$ converging (with respect to the Chabauty topology τ_C) to a subgroup Γ_0 with the same properties.

We fix the usual notations as above; in particular π_i denotes the projection $\mathbf{H}^n \to {\mathbf{H}^n}/_{\Gamma_i}$. We must prove that given $r > 0$ there exists a sequence $\{f_i\}$ of smooth functions in the neighborhood of $\overline{B(z_0,r)}$ such that $f_i(z_0) = z_0$ and:

(i)
$$\pi_0(z_1) = \pi_0(z_2) \iff \pi_i(f_i(z_1)) = \pi_i(f_i(z_2))$$

(ii)
$$D(f_i, \mathrm{id})\overline{_{B(z_0,r)}} \to 0.$$

We start with a technical step. For any subgroup G of $\mathcal{I}^+(\mathbf{H}^n)$ and $\rho > 0$ we define
$$G(\rho) = \{ g \in G : g(B(z_0, \rho)) \cap B(z_0, \rho) \neq \phi \}.$$

By discreteness we have
$$\Gamma_0(r) = \{ \gamma_1^{(r)}, ..., \gamma_{k_r}^{(r)} \}.$$

The characterization of the convergence for τ_C given in E.1.2 implies that $\forall i$ we can find
$$\gamma_1^{(r)}(i), ..., \gamma_{k_r}^{(r)}(i) \in \Gamma_i$$

in such a way that $\forall \alpha = 1, ..., k_r$ the sequence $\{\gamma_\alpha^{(r)}(i)\}$ converges to $\gamma_\alpha^{(r)}$. (Remark that this implies that for $i \gg 0$ these $\gamma_\alpha(r)(i)$ are different from each other.)

Lemma E.1.16. Up to making r a little bigger there exists $\sigma_1 > 0$ such that
$$\Gamma_0(r + \sigma) = \{ \gamma_1^{(r)}, ..., \gamma_{k_r}^{(r)} \} \quad \text{for } |\sigma| \leq \sigma_1$$

and for $i \gg 0$
$$\Gamma_i(r + \sigma) = \{ \gamma_1^{(r)}(i), ..., \gamma_{k_r}^{(r)}(i) \} \quad \text{for } |\sigma| \leq \sigma_1.$$

Proof. Up to making r a little bigger we can assume no point in the Γ_0-orbit of z_0 has distance precisely $2r$ from z_0, so that we immediately have that for some $\sigma_0 > 0$
$$\Gamma_0(r + \sigma) = \{ \gamma_1^{(r)}, ..., \gamma_{k_r}^{(r)} \} \quad \text{for } |\sigma| \leq \sigma_0.$$

We shall prove the second fact with $\sigma_1 = \sigma_0/2$. Let us remark that it suffices to prove that for $i \gg 0$
$$\Gamma_i(r - \sigma_1) = \Gamma_i(r + \sigma_1) = \{ \gamma_1^{(r)}(i), ..., \gamma_{k_r}^{(r)}(i) \}.$$

We prove this relation with the plus sign, the other case being identical. We fix $\alpha \in \{1, ..., k_r\}$ and claim that for $i \gg 0$ we have $\gamma_\alpha^{(r)}(i) \in \Gamma_i(r + \sigma_1)$. We know there exists $z_\alpha \in B(z_0, r + \sigma_1)$ such that
$$\gamma_\alpha^{(r)}(z_\alpha) \in B(z_0, r + \sigma_1).$$

Since $\gamma_\alpha^{(r)}(i)(z_\alpha) \to \gamma_\alpha^{(r)}(z_\alpha)$ and $B(z_0, r + \sigma_1)$ is open, for $i \gg 0$ we have $\gamma_\alpha^{(r)}(i)(z_\alpha) \in B(z_0, r + \sigma_1)$. Our claim is proved and then
$$\Gamma_i(r + \sigma_1) \supseteq \{ \gamma_1^{(r)}(i), ..., \gamma_{k_r}^{(r)}(i) \}$$

for $i \gg 0$. Assume it is not true that the opposite inclusion holds for $i \gg 0$. Then (up to considering a subsequence) there exists

$$\gamma_i \in \Gamma_i(r + \sigma_1) \setminus \{\gamma_1^{(r)}(i), ..., \gamma_{k_r}^{(r)}(i)\}.$$

Let us consider now the set

$$K = \{g \in \mathcal{I}^+(\mathbf{H}^n) : g\big(\overline{B(z_0, r + \sigma_1)}\big) \cap \overline{B(z_0, r + \sigma_1)} \neq \phi \};$$

K is compact and $\gamma_i \in K$, so that we can assume $\gamma_i \to \gamma \in K$. By the characterization of the Chabauty topology we have $\gamma \in \Gamma_0$. Moreover by our choice of $\sigma_1 = \sigma_0/2$ it is $\gamma = \gamma_\alpha^{(r)}$ for some α. Then we are led to the following situation: we have $\gamma_i, \gamma_i' \in \Gamma_i$ such that $\gamma_i \neq \gamma_i'$ and $\{\gamma_i\}$, $\{\gamma_i'\}$ have the same limit. We can check this is absurd.

It was proved in E.1.5 that for any Lie group G the set $\mathcal{D}(G)$ of all discrete subgroups of G is open with respect to the Chabauty topology in the space of all subgroups. In more detail, using the current notations, we have that there exists a neighborhood U of the identity in $\mathcal{I}^+(\mathbf{H}^n)$ such that for $i \gg 0$ we have $\Gamma_i \cap U = \{\mathrm{id}\}$. Now, for $i \gg 0$ we have $\gamma_i' \cdot \gamma_i^{-1} \in \Gamma_i \cap U$, and then $\gamma_i' = \gamma_i$ which is absurd. $\qquad\square$

Remark E.1.17. The importance of the above lemma resides in the following fact: if we have a smooth mapping f_i in the neighborhood of $\overline{B(z_0, r)}$ close enough to the identity, then, in order to check equivariance, we only need to consider the action of $\gamma_1^{(r)}, ..., \gamma_{k_r}^{(r)}$ and $\gamma_1^{(r)}(i), ..., \gamma_{k_r}^{(r)}(i)$, not the action of the whole groups Γ_0 and Γ_i. (In fact we can assume both the domain and the image of f_i are contained in $B(z_0, r + \sigma_1)$.)

Lemma E.1.18. There exists $\delta > 0$ such that $\forall z \in \overline{B(z_0, r + \sigma_1)}$ the projections π_0 and π_i (for $i \gg 0$) are one-to-one on a neighborhood of $\overline{B(z, \delta)}$.

Proof. Let us assume $\gamma_1^{(r)} = \gamma_1^{(r)}(i) = \mathrm{id}$. The biggest $\tau > 0$ such that $\pi_0\big|_{B(z, \tau)}$ is one-to-one is given by

$$\tau(z) = \min_{\alpha = 2, ..., k_r} d\big(z, \gamma_\alpha^{(r)}(z)\big).$$

By the above lemma we also have that for $i \gg 0$ the biggest $\tau > 0$ such that $\pi_i\big|_{B(z, \tau)}$ is one-to-one is given by

$$\tau_i(z) = \min_{\alpha = 2, ..., k_r} d\big(z, \gamma_\alpha^{(r)}(i)(z)\big).$$

τ is a continuous positive function on $\overline{B(z_0, r + \sigma_1)}$, and $\tau_i \to \tau$ uniformly on this ball, so that it suffices to choose $\delta > 0$ such that

$$2\delta \leq \tau \qquad 2\delta \leq \tau_i \text{ for } i \gg 0. \qquad\square$$

We are now ready for the conclusion of the proof. We recall our aim is to find for i large enough an equivariant smooth mapping in the neighborhood of $\overline{B(z_0, r)}$ which is arbitrarily close to the identity with respect to the C^∞ topology on this ball.

We start by covering $\overline{B(z_0, r)}$ by a finite number of δ-balls $W_1, ..., W_l$ having the centre in $\overline{B(z_0, r)}$ (and moreover we choose W_1 centred at z_0). Of course we can take $\delta > 0$ small enough that

$$\bigcup_{j=1}^{l} \bigcup_{\alpha=1}^{k_r} \gamma_\alpha^{(r)}(\overline{W}_j) \subset B(z_0, r + \sigma_1).$$

For $j = 1, ..., l$ we consider a (strictly decreasing) tower of concentric balls

$$W_j = W_j^1 \supset W_j^2 \supset ... \supset W_j^l$$

and moreover we assume that $W_1^l, ..., W_l^l$ is still a covering of $\overline{B(z_0, r)}$.

We are going to prove now the following recursive assertion (whose final step concludes the proof of E.1.15):

for $t = 1, ..., l$ there exists, if i is sufficiently large, a mapping f_i^t on

$$\bigcup_{j=1}^{t} \bigcup_{\alpha=1}^{k_r} \gamma_\alpha^{(r)}(\overline{W}_j^t)$$

such that $f_i^t(z_0) = z_0$, the equivariance property is verified and the function is arbitrarily close to the identity.

Before giving the explicit construction, we have some remarks:

– closeness to the identity is not precisely the same notion we were condidering before, as the domain of f_i^t is not open; however, this domain is a union of balls, and it is easily checked that the notion is well-defined in this case too;

– the above recursive assertion is somewhat ambigous: we should say that given any $\varepsilon > 0$ all the steps can be carried out in such a way that the final result is ε-close to id; the formalization of this idea as a true recursive assertion requires the introduction of too many tedious details, and we believe the idea is better understood in the present formulation;

– since the mappings we are going to construct are close to the identity on a subset of $B(z_0, r + \sigma_1)$, and by the very construction they will be equivariant with respect to the action of $\gamma_1^{(r)}, ..., \gamma_{k_r}^{(r)}$ and $\gamma_1^{(r)}(i), ..., \gamma_{k_r}^{(r)}(i)$, Remark E.1.17 applies and then no further verification of equivariance is needed;

– the key point for the construction is that by the choice of δ as in E.1.18 for $\alpha \neq \beta$ we have that $\gamma_\alpha^{(r)}(\overline{W}_j)$ and $\gamma_\beta^{(r)}(\overline{W}_j)$ are pairwise disjoint (and then of course the same holds for $\gamma_\alpha^{(r)}(\overline{W}_j^t)$ and $\gamma_\beta^{(r)}(\overline{W}_j^t)$ for $t = 1, ..., l$).

We finally proceed to the construction; as above let $\gamma_1^{(r)} = \gamma_1^{(r)}(i) = \text{id}$.

Basic step. We only need to set

$$f_i^1\big|_{\overline{W}_1^1} = \text{id} \qquad f_i^1\big|_{\gamma_\alpha^{(r)}(\overline{W}_1^1)} = \gamma_\alpha^{(r)}(i) \circ \gamma_\alpha^{(r)-1}.$$

Since $\gamma_\alpha^{(r)}(i) \to \gamma_\alpha^{(r)}$ the resulting function is arbitrarily close to the identity for $i \gg 0$.

Recursive step. Assume f_i^t is defined. We define $f_i^{t+1} = f_i^t$ on

$$\bigcup_{j=1}^{t} \bigcup_{\alpha=1}^{k_r} \gamma_\alpha^{(r)}(\overline{W}_j^{t+1}).$$

Of course it suffices to define f_i^{t+1} on $\overline{W}_{t+1}^{t+1}$ in such a way that it agrees with f_i^t on the intersection with

$$\bigcup_{j=1}^{t} \bigcup_{\alpha=1}^{k_r} \gamma_\alpha^{(r)}(\overline{W}_j^{t+1}).$$

In fact in order to respect equivariance we are forced to have

$$f_i^{t+1}\Big|_{\gamma_\alpha^{(r)}(\overline{W}_{t+1}^{t+1})} = \gamma_\alpha^{(r)}(i) \circ \left(f_i^{t+1}\Big|_{\overline{W}_{t+1}^{t+1}}\right) \circ \gamma_\alpha^{(r)-1}$$

and moreover if $f_i^{t+1}\big|_{\overline{W}_{t+1}^{t+1}}$ is close to the identity then the global f_i^{t+1} is close to the identity too.

The definition of f_i^{t+1} is obtained in the following way; if

$$\overline{W}_{t+1}^{t+1} \subset \bigcup_{j=1}^{t} \bigcup_{\alpha=1}^{k_r} \gamma_\alpha^{(r)}(\overline{W}_j^t)$$

we just define f_i^{t+1} to be the restriction of f_i^t; otherwise we define $f_i^{t+1} = \mathrm{id}$ on

$$\overline{W}_{t+1}^{t+1} \setminus \left(\bigcup_{j=1}^{t} \bigcup_{\alpha=1}^{k_r} \gamma_\alpha^{(r)}(\overline{W}_j^t)\right)$$

and we only need to connect (in a smooth and "close-to-the-identity" way) this definition to the restriction of f_i^t to

$$\overline{W}_{t+1}^{t+1} \cap \left(\bigcup_{j=1}^{t} \bigcup_{\alpha=1}^{k_r} \gamma_\alpha^{(r)}(\overline{W}_j^{t+1})\right).$$

Of course it is possible to do this as suggested in Fig. E.1.

Our recursive assertion, and hence E.1.15, are proved. $\qquad \square$

The equivalence of τ_C and τ_G is finally established. We go back now to Remark E.1.12 and formalize it a little more in view of the description of convergence we shall give in the next section. If X and Y are metric spaces and $k \geq 1$ we shall call a mapping $f : X \to Y$ a <u>k-quasi-isometry</u> if

$$1/k \cdot d_X(a,b) \leq d_Y(f(a), f(b)) \leq k \cdot d_X(a,b) \quad \forall a, b \in X.$$

Remark E.1.19. Let $\{[(M_i, x_i, u_i)]\}$ converge to $[(M_0, x_0, u_0)]$ with respect to τ_G. For $r > 0$ consider a sequence $\{f_i\}$ with the usual properties. Then f_i induces a smooth mapping ϕ_i on a neighborhood of $\overline{B}(x_0, r)$ with values in M_i, such that $\phi_i(x_0) = x_i$ and ϕ_i is a k_i-quasi-isometry, with $k_i \to 1$ as $i \to \infty$. In the next section we shall often be using these quasi-isometries, keeping in mind

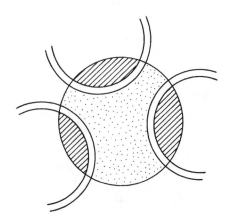

Fig. E.1. Construction of the equivariant extension. The function f_i^{l+1} is given by f_i^l on the shadowed region and by id on the dotted region; since we are dealing with balls in $\mathbb{H}^n$ (and hence with Euclidean balls, if we choose the disc model), there exist effective methods for connecting them

that they are actually induced by smooth equivariant mappings C^∞-close to the identity.

Remark E.1.20. It is easily checked as an application of E.1.16 that the injectivity radius is a continuous function on $\mathcal{H}_n^{**}$. It also follows that if for $\rho > 0$ we define

$$g_\rho\big([(M, x_0, u)]\big) = \min_{x \in B(x_0, \rho)} r(x, M)$$

then the function g_ρ is continuous too. (We shall not need this facts, so we leave them to the reader.) We just mention the fact that the proof of the equivalence with τ_C of the apparently weaker topologies one can define is based on the continuity of these functions g_ρ (see [Ca-Ep-Gr] again).

E.2 Convergence in the Geometric Topology: Opening Cusps.
The Case of Dimension at least Three

We shall discuss now from a topological and geometric viewpoint the notion of convergence with respect to the topology defined in the previous section.

From now on we will always assume $n \geq 3$, as the case $n = 2$ requires different tools; we will consider this case in E.3.3. Moreover we shall be always considering finite-volume manifolds; in the entire section all subsets of $\mathcal{H}_n^{**}$, $\mathcal{H}_n^*$ and $\mathcal{H}_n$ will be endowed with the relative Chabauty (or geometric) topology.

We shall say a sequence

$$\big\{[(M_i, x_i, u_i)]\big\} \subset \mathcal{F}_n^{**}$$

convergent to $\big\{[(M_0, x_0, u_0)]\big\}$ in $\mathcal{F}_n^{**}$ is trivial if $[M_i] = [M_0]$ as an element of $\mathcal{F}_n$ for $i \gg 0$. The very same definiton works for convergent sequences in $\mathcal{F}_n^*$, while in $\mathcal{F}_n$ only constant sequences are trivial.

Remark E.2.1. Let $[(M_0, x_0, u_0)] \in \mathcal{F}_n^{**}$; if y is close enough to x_0 (in such a way that only one length-minimizing geodesic arc joins y to x_0) and if u is an orthonormal basis of $T_y M_0$, then we can consider the orthonormal basis $P_{y,x_0}(u)$ of $T_{x_0}M_0$ obtained by taking the parallel transport of u along the shortest geodesic arc joining y to x_0. Then a trivial sequence in $\mathcal{F}_n^{**}$ convergent to $[(M_0, x_0, u_0)]$ has the following form:

$$\big\{[(M_0, x_i, u_i)]\big\} \quad \text{where } x_i \to x_0 \text{ in } M \text{ and } P_{x_i,x_0}(u_i) \to u_0.$$

Lemma E.2.2. Let $\big\{[(M_i, x_i)]\big\} \subset \mathcal{F}_n^*$ be a sequence converging to a point $[(M_0, x_0)] \in \mathcal{F}_n^*$. Then for $\varepsilon > 0$ small enough we have $x_i \in M_{i[\varepsilon,\infty)}$ for $i \gg 0$. Let $\varepsilon > 0$ be fixed; up to elimination of a finite number of initial terms of the sequence, there exist:

- $\{\sigma_i\} \subset \mathbb{R}$ with $\sigma_i > 0$ and $\sigma_i \to 0$;
- $\{k_i\} \subset \mathbb{R}$ with $k_i > 1$ and $k_i \to 1$;
- $\forall i$ a k_i-quasi-isometry

$$\phi_i : M_{0[\varepsilon,\infty)} \to M_i$$

with the following properties:
- ϕ_i is the restriction of a smooth embedding on a neighborhood of $M_{0[\varepsilon,\infty)}$;
- $\phi_i(x_0) = x_i$;
- $\phi_i\big(M_{0[\varepsilon,\infty)}\big)$ is contained in the interior of $M_{i[\varepsilon-\sigma_i,\infty)}$;
- $\phi_i\big(\partial M_{0[\varepsilon,\infty)}\big)$ does not meet an open neighborhood of $M_{i[\varepsilon+\sigma_i,\infty)}$.

(According to E.1.19, this ϕ_i is actually induced by an equivariant mapping f_i being C^∞-close to the identity.)

Proof. Since the topology of $\mathcal{F}_n^*$ is induced by the geometric topology of $\mathcal{F}_n^{**}$ we can find orthonormal bases u_i of $T_{x_i}M_i$ such that

$$[(M_i, x_i, u_i)] \longrightarrow [(M_0, x_0, u_0)];$$

then we fix as usual the canonical projection π_i of $\mathbf{H}^n$ onto M_i.

Lemma E.1.18 implies in particular the first assertion, as we have that for $i \gg 0$ the projection π_i is one-to-one on a neighborhood of $\overline{B(z_0, \delta)}$, and hence x_i belongs to the δ-thick part of M_i.

As for the second fact, let $r > 0$ be such that $B(x_0, r) \supset M_{0[\varepsilon,\infty)}$ (recall that the volume being finite, thick parts are compact). For $i \gg 0$ we can find an equivariant mapping f_i on a neighborhood of $\overline{B(z_0, r + 4\varepsilon)}$ being arbitrarily C^∞ close to the identity; we know it induces a k_i-quasi-isometry ϕ_i (with k_i arbitrarily close to 1) on $B(x_0, r + 4\varepsilon)$, i.e. on a neighborhood of $M_{0[\varepsilon,\infty)}$. Now, given $z \in \overline{B(z_0, r)}$ we have

$$\pi_0(z) \in M_{0[\varepsilon,\infty)} \Leftrightarrow \min\left\{d(z',z) : z' \neq z, \pi_0(z') = \pi_0(z)\right\} \geq \varepsilon$$
$$\pi_0(z) \in \partial M_{0[\varepsilon,\infty)} \Leftrightarrow \min\left\{d(z',z) : z' \neq z, \pi_0(z') = \pi_0(z)\right\} = \varepsilon.$$

Since f_i is close to the identity we can find small positive numbers α_i, β_i such that

$$\pi_0(z) \in M_{0[\varepsilon,\infty)} \Rightarrow$$
$$\Rightarrow \min\left\{d(f_i(z'), f_i(z)) : z' \neq z, \pi_0(z') = \pi_0(z)\right\} \geq \varepsilon - \alpha_i$$

$$\pi_0(z) \in \partial M_{0[\varepsilon,\infty)} \Rightarrow$$
$$\Rightarrow \min\left\{d(f_i(z'), f_i(z)) : z' \neq z, \pi_0(z') = \pi_0(z)\right\} \leq \varepsilon + \beta_i.$$

Of course we can assume that $\varepsilon + \beta_i \leq 2\varepsilon$ and that the image of f_i covers $B(z_0, r + 2\varepsilon)$; then the equivariance property yields

$$\pi_0(z) \in M_{0[\varepsilon,\infty)} \Rightarrow$$
$$\Rightarrow \min\left\{d(w, f_i(z)) : w \neq f_i(z), \pi_i(w) = \pi_i(f_i(z))\right\} \geq \varepsilon - \alpha_i$$

$$\pi_0(z) \in \partial M_{0[\varepsilon,\infty)} \Rightarrow$$
$$\Rightarrow \min\left\{d(w, f_i(z)) : w \neq f_i(z), \pi_i(w) = \pi_i(f_i(z))\right\} \leq \varepsilon + \beta_i$$

(remark that in these minima we do not need to consider the w's outside $B(z_0, r + 2\varepsilon)$, so that it suffices to consider w in the image of f_i).

These relations mean that

$$\phi_i\big(M_{0[\varepsilon,\infty)}\big) \subset M_{i[\varepsilon-\alpha_i,\infty)}$$
$$\phi_i\big(\partial M_{0[\varepsilon,\infty)}\big) \bigcap M_{i[\varepsilon+\beta_i,\infty)} = \phi$$

and then we only need to take σ_i a little bigger than both α_i and β_i. □

We recall that in Sect. D.3 we defined an end of a hyperbolic n-manifold as a the closure of a connected component of the interior of its ε-thin part, provided $\varepsilon \leq \varepsilon_n$; moreover we called such an end a cusp if it is not compact. As we announced in Chapt. D, in the sequel we will often abbreviate 'closure of a component of the interior' to 'component': the reader can easily check that our constructions and results are coherent with this convention. Of course if A is an element of one of the spaces $\mathcal{F}_n^{**}$, $\mathcal{F}_n^*$ or $\mathcal{F}_n$, the number of its cusp ends is well-defined (and finite).

The proof of the following proposition develops a machinery we shall often make use of in the remainder of the section.

Proposition E.2.3. Let $\{A_i\}$ be a sequence converging to A_0 in one of the spaces $\mathcal{F}_n^{**}$, $\mathcal{F}_n^*$ or $\mathcal{F}_n$; then for $i \gg 0$ the number of cusp ends of A_i does not exceed the number of cusp ends of A_0.

If for $i \gg 0$ the number of cusp ends of A_i equals the number of cusp ends of A_0 then $\{A_i\}$ is a trivial sequence. In particular compact manifolds are isolated points with respect to the geometric topology of $\mathcal{F}_n$.

Proof. Of course it suffices to refer to $\mathcal{F}_n^*$; we fix representatives (M_0, x_0) and (M_i, x_i) of A_0 and A_i $\forall i$.

Let $\varepsilon < \varepsilon_n$ be small enough so that the ε-thin part of M_0 consists of cusps only (we can do this if we take ε less than the length of all axial geodesics of the tube ends). We shall also assume that the boundary components of $M_{0[\varepsilon,\infty)}$ have distance at least $2c\varepsilon$ from each other (we shall discuss later the choice of the constant c; remark anyway that we only need to take ε small enough).

For $i \gg 0$ consider the k_i-quasi-isometry of Lemma E.2.2

$$\phi_i : M_{0[\varepsilon,\infty)} \to M_i$$

(with the properties of that lemma).

Consider i big enough so that $\varepsilon + \sigma_i \leq \varepsilon_n$ and $x_i \in M_{i[\varepsilon+\sigma_i,\infty)}$.

Let us remark that it follows from our classification of ends that under our assumption $n \geq 3$ for $\delta \leq \varepsilon_n$ the δ-thick part of a hyperbolic n-manifold is connected (hint: the boundaries of the ends are connected).

We start by showing that

$$\phi_i\big(M_{0[\varepsilon,\infty)}\big) \supset M_{i[\varepsilon+\sigma_i,\infty)};$$

assume the converse: for $y \in M_{i[\varepsilon+\sigma_i,\infty)} \setminus \phi_i\big(M_{0[\varepsilon,\infty)}\big)$ let α be an arc in $M_{i[\varepsilon+\sigma_i,\infty)}$ joining y to

$$x_i = \phi_i\big(x_0\big) \in \phi_i\big(M_{0[\varepsilon,\infty)}\big).$$

Then α should meet somewhere

$$\partial\phi_i\big(M_{0[\varepsilon,\infty)}\big) = \phi_i\big(\partial M_{0[\varepsilon,\infty)}\big)$$

and it cannot since

$$\phi_i\big(\partial M_{0[\varepsilon,\infty)}\big) \bigcap M_{i[\varepsilon+\sigma_i,\infty)} = \phi .$$

Now, let us fix the closure P of a connected component of

$$M_{i[\varepsilon-\sigma_i,\infty)} \setminus M_{i[\varepsilon+\sigma_i,\infty)}$$

and let us consider the following three different cases:
(1) P is part of a cusp component of $M_{i(0,\varepsilon+\sigma_i]}$;
(2) P is part of a tube component of $M_{i(0,\varepsilon+\sigma_i]}$;
(3) P is an entire tube component of $M_{i(0,\varepsilon+\sigma_i]}$.
(It is quite obvious that these are the only possible cases; remark as well that case (3) occurs if and only if the length of the axial geodesic of the tube has length not less than $\varepsilon - \sigma_i$.)

We discuss the relative situation.

(1) P is homeomorphic to $V \times [0,1]$ (and it shall be identified with it) where V is a compact manifold without boundary and $V \times \{0\}$, $V \times \{1\}$ correspond respectively to a boundary component of $M_{i[\varepsilon+\sigma_i,\infty)}$ and one of $M_{i[\varepsilon-\sigma_i,\infty)}$. We shall prove now, using the properties of ϕ_i, that the subset $P \cap \phi_i\big(M_{0[\varepsilon,\infty)}\big)$ corresponds to a connected submanifold with boundary which

is a neighborhood of $V \times \{0\}$, does not meet a neighborhood of $V \times \{1\}$ and has only one boundary component other than $V \times \{0\}$: as for connectedness, since $\phi_i(M_{0[\varepsilon,\infty)})$ is connected, if $y_1, y_2 \in P \cap \phi_i(M_{0[\varepsilon,\infty)})$ we can find arcs α_1, α_2 in $\phi_i(M_{0[\varepsilon,\infty)})$ joining them to

$$x_i \in M_{i[\varepsilon+\sigma_i,\infty)} \subset \phi_i(M_{0[\varepsilon,\infty)});$$

since x_i does not belong to the interior of P there are subarcs β_1, β_2 in P of α_1, α_2 joining y_1, y_2 to the boundary component of P corresponding to $V \times \{0\}$; since this boundary component is contained in $\phi_i(M_{0[\varepsilon,\infty)})$ and it is connected, the conclusion follows at once. Existence of a boundary component of $\phi_i(M_{0[\varepsilon,\infty)})$ in the interior of P is easily proved (an arc joining $V \times \{0\}$ to $V \times \{1\}$ must meet $\phi_i(\partial M_{0[\varepsilon,\infty)})$) and uniqueness follows from the fact that as soon as $\sigma_i \leq \varepsilon/2$ the diameter of a connected component of

$$M_{i[\varepsilon-\sigma_i,\infty)} \setminus M_{i[\varepsilon+\sigma_i,\infty)}$$

is bounded by $c\varepsilon$, where c is a universal constant depending only on n (this is deduced from our study of the ends); hence our initial assumption that the boundary components of $M_{0[\varepsilon,\infty)}$ have distance at least $2c\varepsilon$ from each other implies that as soon as $k_i \leq 2$ only one can be sent by ϕ_i in P. The other properties of $P \cap \phi_i(M_{0[\varepsilon,\infty)})$ are obvious.

Now, using the fact that the involved quasi-isometries come from equivariant mappings C^∞-close to the identity (as in E.1.19 and E.2.2) and standard arguments of differential topology, one can identify, up to an isotopy of $V \times [0,1]$, the set $P \cap \phi_i(M_{0[\varepsilon,\infty)})$ with $V \times [0,t]$ for some $t \in (0,1)$; if V is the torus (which corresponds to the case $n = 3$) we have the symbolic description represented in Fig. E.2.

(2) Since P is the closure of a tube minus a smaller tube (with the same axial geodesic) P is homeomorphic to

$$S^{n-2} \times S^1 \times [0,1] \qquad (\text{where } S^{n-2} \times S^1 = \partial(\overline{D^{n-1} \times S^1}).)$$

Then $S^{n-2} \times S^1$ plays the same role as V in case (1), and the very same conclusion holds: $P \cap \phi_i(M_{0[\varepsilon,\infty)})$ is obtained (from a topological viewpoint) by removing from the big tube a smaller one; in particular it is homeomorphic to $S^{n-2} \times S^1 \times [0,1]$.

(3) We have now two different possibilities: either the whole of P is contained in $\phi_i(M_{0[\varepsilon,\infty)})$ (in wich case we have nothing more to say) or it is not. If it is not once again we have that P contains precisely one component of $\phi_i(\partial M_{0[\varepsilon,\infty)})$ and $P \cap \phi_i(M_{0[\varepsilon,\infty)})$ is a closed neighborhood of the boundary of the tube, homeomorphic to $S^{n-2} \times S^1 \times [0,1]$.

We are now ready for the conclusion of the proof. The first assertion follows from the following remarks:

(a) the components of $\partial M_{0[\varepsilon,\infty)}$ correspond bijectively to the cusps of M_0;

(b) the cusps of M_i correspond bijectively to the components of

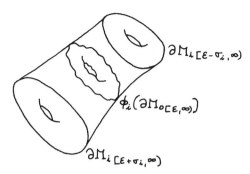

Fig. E.2. If P is part of a cusp based on a torus, and hence homeomorphic to the product of a torus and an interval, then the intersection with P of the image of the thick part under ϕ_i corresponds to the product of the torus with a shorter interval. (The symbol ∂N stands in the figure for "the component of ∂N in question")

$$M_{i[\varepsilon-\sigma_i,\infty)} \setminus M_{i[\varepsilon+\sigma_i,\infty)}$$

being pieces of cusp;

(c) by (1), to each component of $M_{i[\varepsilon-\sigma_i,\infty)} \setminus M_{i[\varepsilon+\sigma_i,\infty)}$ being a piece of cusp there corresponds the component of $\partial M_{0[\varepsilon,\infty)}$ sent into it by ϕ_i.

As for the second assertion, assume M_i has the same number of cusps as M_0; then by the same remarks (a), (b) and (c) we have that case (2) cannot occur and in case (3) we must always have that the whole of P is contained in $\phi_i\big(M_{0[\varepsilon,\infty)}\big)$. It is readily deduced from this that the $(\varepsilon - \sigma_i)$-thin part of M_i contains no tube component and the corresponding thick part $M_{i[\varepsilon-\sigma_i,\infty)}$ is homeomorphic to $\phi_i\big(M_{0[\varepsilon,\infty)}\big)$ and then to $M_{0[\varepsilon,\infty)}$. It follows that both M_0 and M_i are obtained by glueing to the same compact manifold with boundary a cusp on each boundary component; then they are homeomorphic manifolds and the rigidity Theorem C.5.4 implies that the sequence is trivial.

Since a compact manifold has no cusp end, it must be isolated in $\mathcal{F}_n$. □

The following consequence of the above constructions, describing in a very precise way the notion of convergence associated to the geometric topology, can be heuristically expressed in the following way: the only non-trivial convergent sequences in $\mathcal{F}_n$ are characterized, under the topological viewpoint, by the fact that a certain number of tubes open into cusps along the sequence, *i.e.* the axial geodesics of a certain number of tubes are cut out in order to give a cusp based on the boundary torus of the tube. (On the other hand we can say that a non-trivial convergent sequence is obtained by closing some of the cusps of the limit manifold.) Let us remark that it is definitely not obvious at the present time that non-trivial convergent sequences do exist, and in fact they do not for $n \geq 4$ (as we shall prove in E.3 as a corollary of the description of convergence); however, we shall see later that for $n = 3$ they actually exist, and this is the main speciality of the three-dimensional case.

Theorem E.2.4. Let $\{A_i\}$ be a sequence in $\mathcal{F}_n^{**}$, $\mathcal{F}_n^*$ or $\mathcal{F}_n$ convergent to an element A_0 of the same space, let A_0 have k cusp ends, and let $\varepsilon > 0$ be small enough; then, up to elimination of a finite number of initial terms of the sequence, the ε-thick part of A_i is homeomorphic to the ε-thick part of A_0, so that A_0 and A_i have the same number of ε-ends; then $\{A_i\}$ is partitioned into (possibly finite) subsequences

$$\{A_i^{(h)}\} \qquad h = 0, ..., k$$

where $A_i^{(h)}$ has h cusps and $k - h$ tubes in the ε-thin part; moreover the length of each of the $k - h$ axial geodesics of the tube ε-ends of $A_i^{(h)}$ goes to 0 as i goes to infinity.

Proof. As above we refer to $\mathcal{F}_n^*$ and we fix representatives for A_0 and A_i $\forall i$. Let ε be small enough that the constructions of E.2.2 and E.2.3 can be performed, and remark that they can be performed also for $\delta = \varepsilon/2$. We shall never say explicitly "we can do this for $i \gg 0$", but we will need to assume this several times.

We shall prove the proposition with ε replaced by $3\varepsilon/4$.

If we repeat twice the argument of the proof of E.2.2 we obtain that there exist a function $\phi_i : M_{0[\delta,\infty)} \to M_i$ and sequences of positive numbers $\{\tau_i\}$ and $\{\sigma_i\}$ convergent to 0 such that

$$\phi_i\big(M_{0[\varepsilon,\infty)}\big) \subset M_{i[\varepsilon-\sigma_i,\infty)} \qquad \phi_i\big(\partial M_{0[\varepsilon,\infty)}\big) \cap M_{i[\varepsilon+\sigma_i,\infty)} = \phi$$

$$\phi_i\big(M_{0[\delta,\infty)}\big) \subset M_{i[\delta-\tau_i,\infty)} \qquad \phi_i\big(\partial M_{0[\delta,\infty)}\big) \cap M_{i[\delta+\tau_i,\infty)} = \phi .$$

(We first construct ϕ_i for $M_{0[\delta,\infty)}$ and then we check that its restriction works for $M_{0[\varepsilon,\infty)}$.)

As in E.2.3 such conditions imply that

$$M_{i[\varepsilon+\sigma_i,\infty)} \subset \phi_i\big(M_{0[\varepsilon,\infty)}\big) \subset M_{i[\varepsilon-\sigma_i,\infty)}$$

$$M_{i[\delta+\tau_i,\infty)} \subset \phi_i\big(M_{0[\delta,\infty)}\big) \subset M_{i[\delta-\tau_i,\infty)}.$$

Since $M_{0[3\varepsilon/4,\infty)} \cong M_{0[\delta,\infty)}$ and ϕ_i is a homeomorphism the first assertion will be deduced from the following homeomorphism:

$$\phi_i\big(M_{0[\delta,\infty)}\big) \cong M_{i[3\varepsilon/4,\infty)};$$

this is what we are going to prove now.

We also assume $\delta + \tau_i < 3\varepsilon/4 < \varepsilon - \sigma_i$, so that

$$M_{i[\delta+\tau_i,\infty)} \supset M_{i[3\varepsilon/4,\infty)} \supset M_{i[\varepsilon-\sigma_i,\infty)}$$

and hence in particular

$$\phi_i\big(M_{0[\delta,\infty)}\big) \supset M_{i[\varepsilon-\sigma_i,\infty)};$$

then it suffices to show that

$$\phi_i\big(M_{0[\delta,\infty)}\big) \setminus M_{i[\varepsilon-\sigma_i,\infty)} \cong M_{i[3\varepsilon/4,\infty)} \setminus M_{i[\varepsilon-\sigma_i,\infty)}$$

and that they glue to $M_{i[\varepsilon-\sigma_i,\infty)}$ in the same way. Of course ϕ_i sends each component of $\partial M_{0[\delta,\infty)}$ into some component of $M_{i[\delta-\tau_i,\infty)} \setminus M_{i[\varepsilon-\sigma_i,\infty)}$; conversely we shall prove that all such components contain the image of precisely one component of $\partial M_{0[\delta,\infty)}$ and discuss the relative situation. Remark that the components of $M_{i[\delta-\tau_i,\infty)} \setminus M_{i[\varepsilon-\sigma_i,\infty)}$ correspond bijectively to the components of $\partial M_{i[\varepsilon-\sigma_i,\infty)}$.

So, fix a component of $\partial M_{i[\varepsilon-\sigma_i,\infty)}$; there are two cases:

(A) this component is the basis of a cusp; then point (1) in the proof of E.2.3 allows us to describe what happens in a precise way; the situation in the cusp is represented in Fig. E.3.

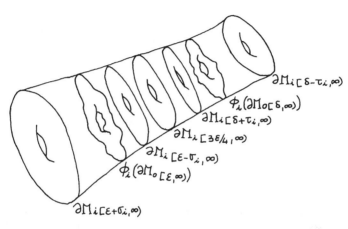

Fig. E.3. Case (A): the variuos abjects occurring inside a piece of cusp. The same convention is used as in Fig. E.2 about the symbol ∂

(B) This component is the boundary of a tube: then it is the boundary of a component P of $M_{i[\varepsilon-\sigma_i,\infty)} \setminus M_{i[\varepsilon+\sigma_i,\infty)}$ which is a proper subset of a tube; then by point (2) of the proof of E.2.3 ϕ_i must send a component B_0 of $\partial M_{0[\varepsilon,\infty)}$ into P; let B_1 be the component of $\partial M_{0[\delta,\infty)}$ on the same cusp, and let C be the piece of cusp delimited by B_0 and B_1; since C is connected ϕ_i sends it in a component of $M_{i[\delta-\tau_i,\infty)} \setminus M_{i[\varepsilon+\sigma_i,\infty)}$, and this component is necessarily the one containing P. Remark that the reason for all this complicated argument is that now we do not need to know whether the component of $M_{i[\delta-\tau_i,\infty)} \setminus M_{i[\varepsilon+\sigma_i,\infty)}$ is an entire tube or a piece, because in any case we have inside it the image of a component of $\partial M_{0[\delta,\infty)}$. The two possible cases are symbolically depicted in Fig. E.4.

It is now obvious that

$$\phi_i\big(M_{0[\delta,\infty)}\big) \setminus M_{i[\varepsilon-\sigma_i,\infty)} \cong M_{i[3\varepsilon/4,\infty)} \setminus M_{i[\varepsilon-\sigma_i,\infty)}$$

and that they glue to $M_{i[\varepsilon-\sigma_i,\infty)}$ in the same way, and then

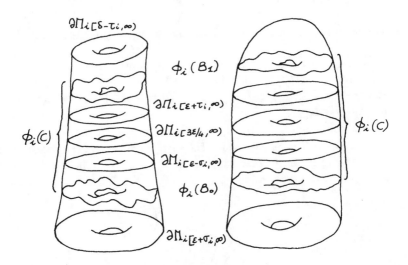

Fig. E.4. Case (B): the left hand side of the figure refers to the case of a piece of tube and the right hand side to an entire tube. The construction allows to prove that the objects occurring inside are the same in both cases. The same convention is used as in Fig. E.2 about the symbol ∂

$$M_{0[\delta,\infty)} \cong M_{i[3\varepsilon/4,\infty)}$$

which implies the first assertion.

The fact that the sequence is partitioned according to the number of cusps is obvious, and we are left to prove that the length of the axial geodesic of a tube opening into a cusp goes to 0. This is easy too: if in the above argument instead of $\delta = \varepsilon/2$ we choose a very small δ, the above construction works for $i \gg 0$, and the tubes in M_i corresponding to cusps in M_0 appear in the $(\delta + \tau_i)$-thin part of M_i, which implies that their axial geodesic has length at most $\delta + \tau_i$, i.e. it is very small. $\square$

We deduce now from the above description of convergence in the geometric topology some very important facts going back to the study of the volume function.

Proposition E.2.5. The functions vol, vol* and vol** are continuous.

Proof. Of course it suffices to prove that vol* is continuous.

We have already proved that it is lower semi-continuous, so we prove upper semi-continuity. Let $\{A_i\}$ converge to A_0 in $\mathcal{F}_n^*$. We use the same notations as in the proof of E.2.4 (we recall that $\delta = \varepsilon/2$ and that M_0 has k cusps); then M_i is covered by $\phi_i\big(M_{0[\varepsilon/2,\infty)}\big)$ and the k components of $M_{i(0,3\varepsilon/4]}$. Given $\lambda > 0$, the maximal volume for a λ-end of a finite-volume hyperbolic n-manifold is a constant $c(\lambda)$ going to 0 as λ goes to 0. Then if we consider the limit as $i \to \infty$ (whence $k_i \to 1$) and simultaneously as $\varepsilon \to 0$ we have (using the fact that ϕ_i is smooth):

$$\text{vol}(M_i) \leq \; k_i^n \; \cdot \; \text{vol}\big(M_{0[\varepsilon/2,\infty)}\big) \; + \; k \cdot c(3\varepsilon/4)$$

$$\downarrow \qquad\qquad \downarrow \qquad\qquad\qquad \downarrow$$

$$1 \qquad\qquad \text{vol}(M_0) \qquad\qquad\qquad 0$$

and hence the maximal limit of $\text{vol}(M_i)$ as $i \to \infty$ does not exceed $\text{vol}(M_0)$, *i.e.* vol^* is upper semi-continuous.   □

Remark E.2.6. Proposition E.2.5 is false in the 2-dimensional case. The point, (which goes back to E.2.2) is that the thick part may be disconnected, so that we cannot be sure that it is definitely the same up to homeomorphism. An explicit counterexample can be constructed as suggested by the following heuristic discussion (see Figg. E.5 and E.6). Let us consider a sequence of hyperbolic structures on the compact surface of genus two, and assume that with respect to these structures the length of a fixed non-trivial loop goes to 0; if we always take the same basepoint the sequence converges to the punctured torus.

Fig. E.5. A convergent sequence of finite-volume hyperbolic surfaces with basepoint

Fig. E.6. The limit surface: a punctured torus

Since the volume of the region "disappearing along the sequence" is bounded away from 0, the volume of the limit manifold is strictly less than the limit of the volumes.

We introduce now two new symbols (using those we defined in the preceding section): for $\varepsilon > 0$ we set:

$$\mathcal{F}_n^{**}(\varepsilon) = \mathcal{H}_n^{**}(\varepsilon) \cap \mathcal{F}_n^{**} \qquad \mathcal{F}_n^{*}(\varepsilon) = \mathcal{H}_n^{*}(\varepsilon) \cap \mathcal{F}_n^{*}.$$

Proposition E.2.7. *The function* vol *and, for* $\varepsilon > 0$, *the restrictions of* vol^* *and* vol^{**} *to* $\mathcal{F}_n^{*}(\varepsilon)$ *and* $\mathcal{F}_n^{**}(\varepsilon)$ *respectively, are proper functions.*

Proof. It suffices to prove that the counter-image of an interval $[a, b]$ is compact. Continuity of the volume function implies that it is closed, and then we only need to recall that by E.1.10 the following spaces are compact:

$$\mathcal{F}_n^{**}(\varepsilon, b) = \left\{ A \in \mathcal{F}_n^{**}(\varepsilon) : \mathrm{vol}^{**}(A) \leq b \right\}$$
$$\mathcal{F}_n^{*}(\varepsilon, b) = \left\{ A \in \mathcal{F}_n^{*}(\varepsilon) : \mathrm{vol}^{*}(A) \leq b \right\}$$
$$\mathcal{F}_n(b) = \left\{ M \in \mathcal{F}_n : \mathrm{vol}(M) \leq b \right\}.$$

$\square$

The following result is deduced from E.2.7, and hence it depends on the underlying condition $n \geq 3$, but it holds also for $n = 2$, as we shall see below.

Corollary E.2.8. $\mathrm{vol}(\mathcal{F}_n)$ is a closed subset of $\mathbb{R}_+$ and it has a positive minimum, so that there exists a hyperbolic n-manifold with minimal volume.

Proof. Closedness follows from E.2.7. Moreover we know (by E.1.10) that $\mathrm{vol}(\mathcal{F}_n)$ has a positive infimum, and the proof is over. $\square$

E.3 The Case of Dimension Different from Three: Conclusions and Examples

The following proposition proving that in dimension bigger than three the geometric topology on $\mathcal{F}_n$ is discrete is based on the description of convergence given in E.2.4.

Proposition E.3.1. If $n \geq 4$ all convergent sequences in $\mathcal{F}_n^{**}$, $\mathcal{F}_n^{*}$ and $\mathcal{F}_n$ are trivial.

Proof. Let $\{A_i\}$ converge to A_0 and let $\{M_i\}$, M_0 be the associated manifolds. Then by E.2.4

$$M_{i[\varepsilon,\infty)} \cong M_{0[\varepsilon,\infty)}$$

for $i \gg 0$. We remark that condition $n \geq 4$ implies that the fundamental group of a hyperbolic n-manifold coincides with the fundamental group of its thick part: in fact cusps retract on their bases and, since the inclusion

$$S^{n-2} \times S^1 = \partial\left(\overline{D^{n-1}} \times S^1\right) \hookrightarrow \overline{D^{n-1}} \times S^1$$

induces an isomorphism at the Π_1-level, the Van Kampen theorem implies that tubes can be cut out without affecting the fundamental group.

Then $\Pi_1(M_i) \cong \Pi_1(M_0)$ and the sharp formulation C.5.4 of the rigidity theorem implies that M_i and M_0 are isometric. $\square$

As a consequence of E.2.8 and E.3.1 we obtain now Wang's theorem (first proved in [Wa] in a more general setting):

Theorem E.3.2. For $n \geq 4$, $c > 0$, the set $\mathcal{F}_n(c) = \left\{ [M] \in \mathcal{F}_n : \mathrm{vol}(M) \leq c \right\}$ is finite. In particular the image of the volume function is a closed discrete subset of $\mathbb{R}_+$.

Proof. $\mathcal{F}_n(c) = \text{vol}^{-1}([0, c])$ is discrete and compact. $\qquad\square$

This result implies, from a heuristic viewpoint, that for $n \geq 4$ complete finite-volume hyperbolic n-manifolds are extremely rare; the question naturally arises whether such manifolds exist or not (remark that at present time we have given examples of finite-volume hyperbolic manifolds only in the two-dimensional case). We will discuss in the remainder of the section a general method producing examples of compact hyperbolic manifolds in all dimensions, but we first settle the study of the volume function in the two-dimensional case (see [Gro1]).

Theorem E.3.3. 1) $\text{vol}(\mathcal{F}_2)$ is a closed discrete subset of $\mathbb{R}_+$;
2) for $c > 0$ the set $\mathcal{F}_2(c) = \{[M] \in \mathcal{F}_2 : \text{vol}(M) \leq c\}$ contains a finite number of topological types, and it is infinite whenever it is non-empty and it is not a point.

Proof. We shall give a complete proof in case $\mathcal{F}_2$ is replaced by the set

$$\mathcal{C}_2 = \{[M] \in \mathcal{F}_2 : M \text{ is compact}\},$$

and we shall outline the proof in the general case.

By the Gauss-Bonnet formula, as we proved in B.3.3, for $[M] \in \mathcal{C}_2$,

$$\text{vol}(M) = -2\pi\chi(M).$$

This implies that the range of the volume function is closed and discrete; moreover the Euler-Poincaré characteristic determines the homeomorphism class of a compact oriented surface, and then the set $\{[M] \in \mathcal{C}_2 : \text{vol}(M) \leq c\}$ contains a finite number of different topological types for any $c > 0$. Since every compact surface supporting a hyperbolic structure supports uncountably many such structures (see B.4.24), such a set is infinite whenever it is not empty.

As for the general case, we begin by remarking that every cusp end of a hyperbolic surface M is diffeomorphic to $S^1 \times [0, \infty)$. If we compactify every end of this type by adding a point at infinity, we construct a compact surface M' such that M is obtained from M' by removing a finite number of points; *i.e.* we can realize M as a punctured compact surface. For such a surface the Euler-Poincaré characteristic is well-defined, and it can be shown that the above formula connecting $\text{vol}(M)$ to $\chi(M)$ holds in this case too. Moreover a necessary and sufficent condition for a punctured compact orientable surface M to support a hyperbolic structure is that $\chi(M) \leq -1$. This structure is unique if $\chi(M) = -1$ (*i.e.* M is a sphere with three punctures) while there exist infinitely many non-isometric structures whenever $\chi(M) \leq -2$. These remarks, together with the fact that there exist only finitely many orientable possibly punctured surfaces having a fixed Euler-Poincaré characteristic, imply the theorem. $\qquad\square$

We describe now a way to construct compact hyperbolic n-manifolds for $n \geq 3$; this method is classical and was widely generalized in [Bor1]. The manifolds we are going to obtain are non-orientable, but for an example of compact oriented hyperbolic n-manifold it suffices to take the two-fold covering of them. Such construction of compact hyperbolic manifolds in all dimensions n has been for a long time the unique effective method known for the case $n \geq 4$ (see [Gr-PS] for examples of different nature); for the case $n = 3$ we shall see in the next sections much more powerful methods. We address the reader to [Bor1] for a generalization of the construction and for some details we are going to omit in the proofs; we remark that for the general case the theory of algebraic groups is needed, while for the particular examples we are going to treat this is not necessary. For another description of the construction we address the reader to [Su2].

For $u > 0$ let us consider in $\mathbb{R}^{n+1}$ the non-degenerate quadratic form of signature $(n, 1)$

$$\phi_u : x \mapsto x_1^2 + \ldots + x_n^2 - u \cdot x_{n+1}^2.$$

We denote by $\mathbb{I}^n(u)$ the upper fold of the hyperboloid defined by the equation $\phi_u = -1$ and by $G(u)$ the subgroup of $\mathrm{Gl}(n + 1, \mathbb{R})$ of the isometries with respect to ϕ_u keeping $\mathbb{I}^n(u)$ invariant. The linear mapping

$$\Delta_u : x \mapsto \left(x_1, \ldots, x_n, \sqrt{u} \cdot x_{n+1} \right)$$

is a natural bijection of $\mathbb{I}^n(u)$ onto $\mathbb{I}^n$ inducing an identification between $G(u)$ and $\mathcal{I}(\mathbb{I}^n)$. Hence for an example of a compact hyperbolic n-manifold it suffices to find a discrete torsion-free subgroup H of $G(u)$ such that

$$\mathbb{I}^n(u)\big/_H$$

is compact.

In order to obtain such an H some hypotheses are needed about u; instead of listing them at once we shall discuss them in detail one at a time and number them by Roman numerals.

I. We assume that u is an algebraic integer, *i.e.* that it is algebraic over $\mathbb{Q}$ and its minimal polynomial with leading coefficient 1 belongs to $\mathbb{Z}[x]$. We shall denote this polynomial by $q(x)$ and its degree by d. Let $I(q(x))$ be the ideal generated by $q(x)$ in $\mathbb{Q}[x]$. Then the quotient ring

$$\mathbb{Q}[x]\big/_{I(q(x))}$$

is canonically identified with the subfield $\mathbb{Q}(u)$ of $\mathbb{R}$ in the following way: the class modulo $I(q(x))$ of any polynomial can be represented in a unique way as

$$\sum_{s=0}^{d-1} c_s \cdot x^s$$

and then we associate to the class of the polynomial the real number

$$\sum_{s=0}^{d-1} c_s \cdot u^s.$$

In particular u corresponds to the class of the polynomial x. Let $\mathbb{Z}[u]$ be the image in $\mathbb{Q}(u)$ of $\mathbb{Z}[x]$; $\mathbb{Z}[u]$ is a subring of $\mathbb{Q}(u)$ and by the assumption that $q(x) \in \mathbb{Z}[x]$ it follows that it consists of all the numbers of the form

$$\sum_{s=0}^{d-1} c_s \cdot u^s$$

with integer coefficients $c_0, \ldots, c_{d-1}$.

The group $\mathrm{Gl}(n+1, \mathbb{Q}(u))$ is a subgroup of $\mathrm{Gl}(n+1, \mathbb{R})$; we shall refer to the canonical basis of $\mathbb{R}^{n+1}$ and then consider it as a group of matrices.

Let us remark now that if $A \in G(u)$ we have $\det(A) = \pm 1$, and hence if we define the group $\Gamma(u)$ as

$$\Gamma(u) = G(u) \bigcap \mathrm{Gl}(n+1, \mathbb{Z}[u]).$$

then we simply have

$$\Gamma(u) = \big\{ (a_{ij}) \in G(u) \bigcap \mathrm{Gl}(n+1, \mathbb{Q}(u)) : \ a_{ij} \in \mathbb{Z}[u] \ \forall i, j \big\}.$$

Now let p be a prime number, denote by $q_p(x)$ the reduction modulo p of the polynomial $q(x)$ and consider the ring

$$R(p, q(x)) = \frac{\mathbb{Z}_p[x]}{I(q_p(x))}.$$

Then we have a homomorphism of reduction modulo p

$$\pi_p : \mathrm{Gl}(n+1, \mathbb{Z}[u]) \to \mathrm{Gl}(n+1, R(p, q(x)))$$

$$\left(\sum_{s=0}^{d-1} c_{ij}^{(s)} \cdot u^s \right)_{ij} \mapsto \left(\Big[\sum_{s=0}^{d-1} [c_{ij}^{(s)}]_p \cdot x^s \Big]_{q_p(x)} \right)_{ij}.$$

It follows that $Ker(\pi_p)$ is a normal group of finite index in $\mathrm{Gl}(n+1, \mathbb{Z}[u])$ (remark that $\mathrm{Gl}(n+1, R(p, q(x)))$ is a finite group).

Lemma E.3.4. It is possible to find p in such a way that $Ker(\pi_p)$ is torsion-free.

Proof. Let $A \in \mathrm{Gl}(n+1, \mathbb{Z}[u])$. The eigenvalues of A are algebraic numbers over $\mathbb{Q}(u)$ with minimal polynomial of degree at most $n+1$. It follows that they are algebraic numbers over $\mathbb{Q}$ with minimal polynomial of degree at most $d \cdot (n+1)$; moreover if A has finite order its eigenvalues are roots of 1. It follows from these facts that we have only a finite number of possibilities for the eigenvalues of an element A of $\mathrm{Gl}(n+1, \mathbb{Z})$ having finite order. Hence there exists a finite set of polynomials with leading coefficient 1

$$r_1(x), \ldots, r_m(x)$$

of degree $n+1$ and different from $(x-1)^{n+1}$ such that if $A \in \mathrm{Gl}(n+1, \mathbb{Z})$ has finite order and is not the identity then the characteristic polynomial $p_A(x)$

of A is one of the $r_j(x)$'s. Then it suffices to choose the prime number p in such a way that

$$r_j(x) \not\equiv (x-1)^{n+1} \mod p \quad \forall j. \qquad \square$$

We are going now to make two more assumptions on u:

II. 0 is the only solution in $\mathbb{Z}^{n+1}$ of the equation $\phi_u(x) = 0$.

III. All the roots of the minimal polynomial $q(x)$ of u have multiplicity 1 and (u excluded) they are real and negative.

We shall say u is admissible if conditions I, II and III are fulfilled.

Lemma E.3.5. For $n = 3$ the choice $u = 7$ is an admissible one.

Proof. Conditions I and III are obviously fulfilled. As for condition II, assume by contradiction that $x \in \mathbb{Z}^4 \setminus \{0\}$ and $\phi_7(x) = 0$ and remark that we can suppose x_1, x_2, x_3, x_4 are relatively prime (otherwise we divide by the common factor). If we consider the reduction modulo 8 of the identity $\phi_7(x) = 0$ and we recall that in $\mathbb{Z}_8$ the squares are 0,1 and 4, we easily obtain that up to a change of order in x_1, x_2, x_3 the only possible cases are the following:

$$x_4^2 \equiv x_1^2 \equiv x_2^2 \equiv x_3^2 \equiv 0 \ (\mathrm{mod}\ 8) \qquad x_4^2 \equiv x_1^2 \equiv 0, \equiv x_2^2 \equiv x_3^2 \equiv 4 \ (\mathrm{mod}\ 8)$$
$$x_4^2 \equiv x_1^2 \equiv x_2^2 \equiv x_3^2 \equiv 4 \ (\mathrm{mod}\ 8) \qquad x_4^2 \equiv x_1^2 \equiv 4, \equiv x_2^2 \equiv x_3^2 \equiv 0 \ (\mathrm{mod}\ 8)$$

and this is absurd since in all cases all the x_i's are even. $\qquad \square$

For $n \geq 4$ no integer choice of u can be admissible, as all positive integers are the sum of four squares of integers, and then condition II cannot hold. However admissible choices exist for all n's, the easiest one being $u = \sqrt{2}$.

We sketch now the proof that to an admissible u is associated a subgroup of $G(u)$ producing a compact hyperbolic n-manifold.

Proposition E.3.6. If u is admissible then:

(a) the group $\Gamma(u)$ described above is discrete and the quotient space

$$\mathbb{I}^n(u) \big/ \Gamma(u)$$

is compact;

(b) if p is chosen as for Lemma E.3.4 then the group

$$H(u, p) = \Gamma(u) \cap Ker(\pi_p)$$

has finite index in $\Gamma(u)$ and it is torsion-free;

(c) the group of isometries of $\mathbb{I}^n$ corresponding to $H(u, p)$ is discrete and torsion-free and the quotient hyperbolic manifold is compact.

Proof. (c) is immediately deduced from (a) and (b), while (b) follows from (a) and Lemma E.3.4. Then we only have to prove (a).

We begin with the special case $n = 3$ and $u = 7$ and then we sketch the generalization. Discreteness of $\Gamma(u)$ is quite obvious in this case, as $\mathbb{Z}[u] = \mathbb{Z}$ is discrete in $\mathbb{R}$ and then $\mathrm{Gl}(n+1, \mathbb{Z})$ is discrete in $\mathrm{Gl}(n+1, \mathbb{R})$. As for compactness of the quotient space, we first recall some general notions.

A <u>lattice</u> in $\mathbb{R}^m$ is an additive subgroup generated by a basis; the set $\mathcal{L}_m$ of all lattices in $\mathbb{R}^m$ is canonically identified with the coset space

$$\mathrm{Gl}(m, \mathbb{R})\big/_{\mathrm{Gl}(m, \mathbb{Z})}$$

(two bases generate the same lattice if and only if they are conjugate in $\mathrm{Gl}(m, \mathbb{Z})$); then we can endow $\mathcal{L}_m$ with the quotient topology. For $L \in \mathcal{L}_m$ the quotient $\mathbb{R}^m\big/_L$ is a flat torus, and then its volume is defined; then we set

$$v(L) = \mathrm{vol}\left(\mathbb{R}^m\big/_L\right).$$

Let us remark soon that if L is generated by $x_1, ..., x_m$ then $v(L)$ is the volume of any "parallelogram" in $\mathbb{R}^m$ having $0, x_1, ..., x_m$ among its vertices, and then it is given by $|\det(x_1...x_m)|$; it follows that $v(L)$ is the absolute value of the determinant of any matrix representing L in $\mathrm{Gl}(m, \mathbb{R})$.

We also set

$$s(L) = \inf_{x \in L \setminus \{0\}} \|x\|$$

(where $\|.\|$ denotes the euclidean norm in $\mathbb{R}^m$).

A proof much like the one we gave for E.1.7, to be found in [Bor2], establishes the following Hermite-Mahler compactness criterion:

Proposition E.3.7. If $\mathcal{A} \subset \mathcal{L}_m$ is such that for some $c, d > 0$

$$s(L) \geq c \text{ and } v(L) \leq d \ \forall L \in \mathcal{A}$$

then the closure of $\mathcal{A}$ is compact in $\mathcal{L}_m$.

Let us go back to our case $n = 3$ and $u = 7$. Since $G(u)$ operates transitively on $\mathbb{I}^3(u)$, the isotropy group $G(u)_y$ of any point y is compact and the natural bijection

$$G(u)\big/_{G(u)_y} \xrightarrow{\sim} \mathbb{I}^3(u)$$

is a homeomorphism, for the proof of compactness of $\mathbb{I}^3(u)\big/_{\Gamma(u)}$ it is enough to prove that $G(u)\big/_{\Gamma(u)}$ is compact.

Under the above identification of

$$\mathrm{Gl}(4, \mathbb{R})\big/_{\mathrm{Gl}(4, \mathbb{Z})}$$

with $\mathcal{L}_4$, the set $G(u)\big/_{\Gamma(u)}$ corresponds to the orbit $\mathcal{A}$ of the canonical lattice $\mathbb{Z}^4$ under the natural action of $G(u)$ on $\mathcal{L}_4$, i.e. to the set of all lattices generated by the columns of some element of a matrix of $G(u)$. We can use the above criterion to check compactness of $\mathcal{A}$. Of course $\forall L \in \mathcal{A}$ we have $v(L) = 1$ and then we have to prove that there exists $c > 0$ such that $\forall L \in \mathcal{A}$, $s(L) \geq c$.

Property II implies that the neighborhood of the origin

$$U = \left\{ x \in \mathbb{R}^4 : |\phi_u(x)| \le 1/2 \right\}$$

does not meet $\mathbb{Z}^4 \setminus \{0\}$ (remark that ϕ_u has integer values on $\mathbb{Z}^4$). Since U is $G(u)$-invariant and each $L \in \mathcal{A}$ is the image of $\mathbb{Z}^4$ under an element of $G(u)$, it follows that

$$(L \setminus \{0\}) \cap U = \phi \qquad \forall L \in \mathcal{A}$$

and then it suffices to choose c as the radius of a euclidean ball centred at 0 and contained in U.

The argument for the special case $n = 3$, $u = 7$ is now complete. For the general case, compactness of $\mathbb{I}^n(u)/\Gamma(u)$ follows by a suitable generalization of the Hermite-Mahler criterion (to be found in [Bor2] too) for the use of which property II again is needed. Discreteness of $\Gamma(u)$ is not as obvious as for integer u (remark that if u is not rational then $\mathbb{Z}[u]$ is not discrete in $\mathbb{R}$) and property III is introduced for this reason. We make explicit the proof for the case $u = \sqrt{2}$, the general argument being an extension of this one.

We start by the remark that $\mathbb{Z}[\sqrt{2}] = \left\{ h + k\sqrt{2} : h, k \in \mathbb{Z} \right\}$. Under the ring-homomorphism

$$h + k\sqrt{2} \mapsto \left(h + k\sqrt{2}, h - k\sqrt{2} \right)$$

$\mathbb{Z}[\sqrt{2}]$ corresponds to a discrete subring of $\mathbb{R} \times \mathbb{R}$, and this allows us to identify $\Gamma(\sqrt{2})$ with a discrete subgroup of

$$\mathrm{Gl}(n+1, \mathbb{R} \times \mathbb{R}) \cong \mathrm{Gl}(n+1, \mathbb{R}) \times \mathrm{Gl}(n+1, \mathbb{R});$$

since $\Gamma(u)$ consists of isometries for $\phi_{\sqrt{2}}$, the image of $\Gamma(\sqrt{2})$ in the second coordinate consists of isometries for the positive-defined quadratic form

$$x \mapsto x_1^2 + \ldots + x_n^2 + \sqrt{2} \cdot x_{n+1}^2$$

and hence it is compact; then the image in the first coordinate must be discrete, and then $\Gamma(\sqrt{2})$ is discrete. □

E.4 The Three-dimensional Case: Jorgensen's Part of the So-called Jorgensen-Thurston Theory

We start with some topological notions we shall need in the remainder of the chapter. A standard reference for these concepts is [Rol].

If $\overline{M}$ is a compact differentiable oriented three-manifold with (possibly empty) boundary $\partial \overline{M}$ made of pairwise disjoint tori $T_1, \ldots, T_k$, for $i = 1, \ldots, k$ we fix generators l_i, m_i of $\mathrm{II}_1(T_i) \cong \mathbb{Z} \oplus \mathbb{Z}$, and we assume these generators are represented by simple smooth closed curves (denoted by l_i and m_i too), meeting transversally in one point and such that $T_i \setminus (l_i \cup m_i)$ is diffeomorphic to an open disc (of course l_i and m_i are the longitude and the meridian of T_i, but we give this abstract definition as T_i is not the standard torus, it is a subset of $\overline{M}$). If $B = S^1 \times \overline{D^2}$ is the standard solid torus, we fix a meridian $m = \{z_0\} \times \partial \overline{D^2} \subset \partial B$. If for some i we are given an orientation-preserving diffeomorphim $\phi_i : \partial B \to T_i$, we can consider the manifold

$$B \coprod_{\phi_i} M$$

obtained by glueing B to $\overline{M}$ along T_i via ϕ_i.

Proposition E.4.1. The manifolds obtained in this way depend, up to diffeomorphism, only on a pair of integers (p_i, q_i) with $(p_i, q_i) = 1$, and to any such a pair there corresponds a manifold obtained in this way. The pair (p_i, q_i) corresponds to the manifold $B \coprod_{\phi_i} \overline{M}$ if $\phi_{i*}(m) = p_i \cdot l_i + q_i \cdot m_i$.

(We recall that $(p_i, q_i) = 1$ means that either p_i and q_i are both non-zero and relatively prime or one of them is 0 and the other one is ± 1. Remark as well that this condition is equivalent to the existence of a matrix in $\mathrm{Sl}(2, \mathbb{Z})$ whose first column is ${}^t(p_i \, q_i)$ and hence to the existence of an orientation-preserving diffeomorphism $\phi_i : \partial B \to T_i$ such that $\phi_{i*}(m) = p_i \cdot l_i + q_i \cdot m_i$. The proof of the uniqueness (up to diffeomorphism) of the manifold corresponding to a given pair of integers follows from the fact that a diffeomorpism of the torus onto itself which maps a meridian to a meridian extends to a diffeomorphism of the solid torus onto itself. Remark that it is false in general that if two pairs of integers are different then the corresponding manifolds are non-diffeomorphic: for instance $(-p_i, -q_i)$ of course gives the same manifold as (p_i, q_i); see [Rol] for other examples.)

Remark E.4.2. Since the loops on the boundary of the solid torus being trivial in the solid torus are precisely the multiples of the meridian, we have the following heuristic explication of the above theorem: if we decide to fill in T_i with a solid torus, in order to know what manifold we get we only need to know what loops of T_i become trivial in the added torus.

We shall say a three-manifold without boundary is obtained from $\overline{M}$ by topological Dehn surgery with coefficients $d_1, ..., d_k$ belonging to the set

$$\mathcal{P}_* = \big\{ (p_i, q_i) \in \mathbb{Z} \times \mathbb{Z} : (p_i, q_i) = 1 \big\} \cup \{\infty\}$$

if it is diffeomorphic to the manifold obtained from $\overline{M}$ in the following way:
− if $d_i = (p_i, q_i)$ a copy of B is glued along T_i via an orientation-preserving diffeomorphism ϕ_i with $\phi_{i*}(m) = p_i \cdot l_i + q_i \cdot m_i$;
− if $d_i = \infty$ the torus T_i is cut out.

We shall denote such a manifold by $\overline{M}_{d_1...d_k}$ (as usual we are only concerned with the diffeomorphism class of the manifold). Recall that, according to the usual notation, we call M the interior of $\overline{M}$, and then with the above symbols we have $M = \overline{M}_{\infty...\infty}$.

Remark E.4.3. If M has no boundary component the only manifold we can obtain from M by Dehn surgery is M itself.

The following result, proved independently by Rochlin, Wallace and Lickorish, plays a central role in the topology of three-manifolds. For a proof we refer again to [Rol]. We recall that a <u>knot</u> in S^3 is a subset of S^3 diffeomorphic to S^1 and a <u>link</u> is the union of a finite number of pairwise disjoint knots.

Theorem E.4.4. Given a compact oriented three-manifold N there exists a link L in S^3, union of the knots $L_1, ..., L_k$, such that N is obtained by Dehn surgery (with suitable coefficients) from the compact oriented three-manifold with boundary consisting of tori

$$S^3 \setminus \left(\bigcup_{i=1,...,k} B_i \right)$$

where the B_i's are pairwise disjoint open tubular neighborhoods of the knots L_i. (We shall also say briefly that N is obtained by <u>Dehn surgery along</u> L.)

We turn now to the situation we are really interested in. From now on we shall always (improperly) refer to the elements of $\mathcal{F}_3$ as manifolds, and identify them with one of their representatives. We recall that

$$\mathcal{F}_3(c) = \left\{ M \in \mathcal{F}_3 : \text{vol}(M) \leq c \right\}.$$

Let $M \in \mathcal{F}_3$; we know that M is diffeomorphic to the interior of a compact manifold $\overline{M}$ with boundary consisting of as many tori as the cusps of M. We shall say a manifold $N \in \mathcal{F}_3$ is obtained from M by (weak) <u>hyperbolic Dehn surgery</u> if it is diffeomorphic to $\overline{M}_{d_1,...,d_k}$ for suitable coefficients $d_i \in \mathcal{P}_*$ and, under such a diffeomorphism, a solid torus glued to $\overline{M}$ corresponds to a tubular neighborhood of a simple closed geodesic in N. We shall say the surgery is <u>strong</u> if, for some $\varepsilon > 0$, N has exactly $h = \#\{j : d_j \neq \infty\}$ simple closed geodesics of length less than ε, and the h solid tori added to $\overline{M}$ are tubular neighborhoods of such h geodesics. The reader should keep in mind the precise definition we give of hyperbolic surgery as this is not the only one used in the literature.

We set now $\mathcal{A}_0 = \mathcal{F}_3$ and we build recursively a (decreasing) sequence $\{\mathcal{A}_n\}_{n \geq 0}$ by defining $\mathcal{A}_{n+1}$ as the set of all limit points of $\mathcal{A}_n$ in $\mathcal{F}_3$ (with respect, as usual, to the geometric topology). We shall say a manifold $M \in \mathcal{F}_3$ has <u>order</u> n, and write $\text{ord}(M) = n$, if

$$M \in \mathcal{A}_n \quad \text{and} \quad M \notin \mathcal{A}_{n+1}.$$

For $c > 0$ we define the order of c as

$$\text{ord}(c) = \sup \left\{ \text{ord}(M) : \text{vol}(M) \leq c \right\}$$

and we introduce another number

$$n(c) = \#\left\{ M \in \mathcal{F}_3 : \text{vol}(M) \leq c, \ \text{ord}(M) = \text{ord}(c) \right\}.$$

Lemma E.4.5. If $M \in \mathcal{A}_n$ then M has at least n cusps.

Proof. We prove this by induction on n, the basic step being obvious.

Let $M \in \mathcal{A}_n$ and choose a sequence $\{M_i\} \subset \mathcal{A}_{n-1}$ such that $M_i \to M$ and $M_i \neq M_j$ for $i \neq j$. Assume by contradiction that M has at most $n-1$ cusps; by the induction hypothesis M_i has at least $n-1$ cusps, and then Proposition E.2.3 implies that (for $i \gg 0$) both M_i and M have precisely $n-1$ cusps, which implies that the sequence is trivial, and this is absurd.   □

Lemma E.4.6. $\forall c > 0$, ord(c) is finite.

Proof. Assume the converse. Then we can find a sequence of manifolds $\{M_i\}$ such that vol$(M_i) \leq c$ $\forall i$ and ord$(M_{i+1}) >$ ord(M_i). Properness of the volume function implies that we can assume that the sequence converges to some $M \in \mathcal{F}_3(c)$; since given any n for $i \gg 0$ the sequence $\{M_i\}$ is a non-constant sequence in $\mathcal{A}_n$, we obtain that $M \in \mathcal{A}_n$ $\forall n$, and then, by the above lemma, M should have infinitely many cusps, and this is impossible.   □

Lemma E.4.7. $\forall c > 0$, $n(c)$ is finite.

Proof. Assume by contradiction that $n(c)$ is infinite; then there exists a sequence $\{M_i\}$ in $\mathcal{F}_3$ such that

$$\text{vol}(M_i) \leq c, \ \text{ord}(M_i) = \text{ord}(c) \ \forall i \qquad M_i \neq M_j \text{ for } i \neq j.$$

Once again we can assume that this sequence converges to $M \in \mathcal{F}_3$ with vol$(M) \leq c$. This implies that $M \in \mathcal{A}_{\text{ord}(c)+1}$, and this contradicts the definition of ord(c).   □

The next result contains the first facts about the study of $\mathcal{F}_3$ and the volume function in the three-dimensional case.

Theorem E.4.8. (1) $\forall c > 0$ there exists a finite subset $\{M_1^{(c)}, ..., M_{p(c)}^{(c)}\}$ of $\mathcal{F}_3(c)$ such that any other element of $\mathcal{F}_3(c)$ is obtained by strong hyperbolic Dehn surgery from one of the $M_j^{(c)}$'s;
(2) $\forall c > 0$ there exists a link L in S^3 such that every element of $\mathcal{F}_3(c)$ is obtained by topological Dehn surgery along L;
(3) $\forall c > 0$, there exists a finite subset $\{N_1^{(c)}, ..., N_{q(c)}^{(c)}\}$ of $\mathcal{F}_3(c)$ such that, for $\varepsilon > 0$ small enough, $\forall N \in \mathcal{F}_3(c)$, the ε-thick part of N is diffeomorphic to the ε-thick part of one of the $N_j^{(c)}$'s.

Proof. (1) Let $k = $ ord(c) and let $M_1^k, ..., M_{m(k)}^k$ be the manifolds in $\mathcal{F}_3(c)$ which have order k. Theorem E.2.4 (together with its proof) imply that for $i = 1, ..., m(k)$ there exists an open neighborhood U_i^k of M_i^k such that if $M \in U_i^k$ then M_i^k is obtained from M by removing some axial geodesics of tubes in M, i.e. M is obtained by filling some boundary tori of $\overline{M}_i^k$ with solid tori corresponding to the tube components of $M_{[\varepsilon,\infty)}$, i.e. M is obtained by strong hyperbolic Dehn surgery from M_i^p.

Now, if we set

$$\mathcal{S}_1 = \mathcal{F}_3(c) \setminus \big(\bigcup_{i=1,...,m(p)} U_i^p\big) \qquad h = \max\big\{\text{ord}(N) : \ N \in \mathcal{S}_1\big\}$$

we have that $h < k$ and the same proof as the one given for E.4.7 allows one to check that there exist finitely many manifolds $M_1^h, ..., M_{m(h)}^h$ in S_1 having order precisely h (remark that S_1 is closed in $\mathcal{F}_3(c)$, so that it is compact too). Then we can find open neighborhoods of such manifolds, $U_1^h, ..., U_{m(h)}^h$, with the property that if $M \in U_i^h$ then M is obtained from M_i^h by strong hyperbolic Dehn surgery. We set

$$S_2 = \mathcal{F}_3(c) \setminus \Big(\bigcup_{i=1}^{m(k)} U_i^k \Big) \setminus \Big(\bigcup_{i=1}^{m(h)} U_i^h \Big)$$

and we go on with the same method: since at each step we decrease the maximal order, we conclude with a finite number of steps and then we obtain at last a finite number of elements of $\mathcal{F}_3(c)$ allowing us to obtain by strong hyperbolic Dehn surgery all other elements of $\mathcal{F}_3(c)$. Remark that at the last step we get a finite number of compact elements of $\mathcal{F}_3$.

(2) This is readily deduced from (1) and Theorem E.4.4, as we can always perturb a little a finite set of links in such a way that their union is a link too.

(3) The proof is the same as the one given for point (1). We only have to remark that, by E.2.4 and its proof, the open neighborhood U_i^k of M_i^k can be chosen in such a way that for $\varepsilon > 0$ small enough the ε-thick part of any $M \in U_i^k$ is diffeomorphic to the ε-thick part of M_i^k. The total number of manifolds needed being finite we have a finite number of upper bounds on ε, and the conclusion follows quite easily.   □

We conclude the present section with some remarks concerning the bibliographical references about the results we have obtained up to now concerning the volume function; if we confine ourselves to the three-dimensional case, existence of a topology on $\mathcal{F}_3$ with respect to which the volume function is continuous and proper, the geometric meaning of convergence described in Theorems E.2.4 and E.4.8 constitute the part due to Jorgensen of the so-called Jorgensen-Thurston theory about finite-volume hyperbolic three-manifolds. Let us remark that, by now, it is not known to the reader that in $\mathcal{F}_3$ non-trivial sequences do exist and that the image of the volume function on $\mathcal{F}_3$ is not discrete in $\mathbb{R}$ (as for $n = 2$ and $n \geq 4$): this is Thurston's part of the theory, to which we dedicate the remainder of the chapter. However, let us go back to the Jorgensen part and discuss the way it was presented in the previous literature.

The main references are [Th1, ch. 5] and [Gro1] (and [Th2] for a list of statements); [Gro1] is widely expository, while [Th1] contains some details, but the order of the treatise is the opposite to the one we chose, and it contains a difficulty we were not able to overcome. Let us explain this with some accuracy.

Assume it is possible to prove preliminarily the following facts:

(A) for $n \geq 4$, for any $\varepsilon \leq \varepsilon_n$ and $c > 0$ the set of all ε-thick parts of hyperbolic n-manifolds with volume at most c realizes a finite number of fundamental groups (up to isomorphism);

(B) for any $\varepsilon \leq \varepsilon_3$ and $c > 0$ the set of all ε-thick parts of hyperbolic 3-manifolds with volume at most c is finite (up to diffeomorphism).

Then our results for $n \geq 4$ (namely, the fact that there exist finitely many manifolds having volume bounded by a fixed constant) would be deduced from (A) and the same application of the Van Kampen theorem given for the proof of E.3.1 (proving that the fundamental group of a hyperbolic n-manifold is the fundamental group of its thick part) together with the sharp formulation C.5.4 of the rigidity theorem. For $n = 3$ we could define (as in [Th1]) a topology on $\mathcal{F}_3$ by setting as a fundamental system of neighborhoods of a manifold M the set

$$\{ \mathcal{V}(M, \varepsilon, k) : \ 0 < \varepsilon \leq \varepsilon_3, \ k > 1 \}$$

where $N \in \mathcal{V}(M, \varepsilon, k)$ if and only if there exists a diffeomorphism of the ε-thick part of M onto the ε-thick part of N being a k-quasi-isometry. Then the geometric meaning of the convergence is the same as the one described in E.2.4 and it is quite easily checked (independently of (B)) that the volume function turns out to be continuous. Properness of the volume function is proved too, but the proof makes use (an essential use, in our opinion) of (B); it is in the proof of (B) suggested by Thurston in [Th1] (a special case of which would allow to prove (A)) that we found the difficulty.

Let us describe the argument, starting from the simpler version for $n \geq 4$ and then considering the case $n = 3$. We keep the constants $\varepsilon \leq \varepsilon_n$ and $c > 0$ fixed.

(A)′ We go back to the notations used for the proof of Proposition D.2.6; for any $M \in \mathcal{F}_n$ with vol$(M) \leq c$ there exists a finite set $Y_0 \subset M_{[\varepsilon, \infty)}$ (whose cardinality is bounded by a constant depending on c and ε) such that the balls of radius $\varepsilon/2$ centred at the points of Y_0 are isometric to balls in $\mathbf{H}^n$ and cover $M_{[\varepsilon, \infty)}$. These balls are simply connected and the intersection of any two of them is connected and simply connected: then it is possible to show that the fundamental group of the union of such balls is isomorphic to the fundamental group of the finite two-dimensional simplicial complex having $\#Y_0$ vertices, in which two vertices are joined by an edge if and only if the corresponding balls have non-empty intersection, and three vertices are the vertices of a triangle if and only if the corresponding three balls have non-empty intersection. It follows (recall that $\#Y_0$ is bounded by a constant depending on c and ε) that the fundamental group of

$$\bigcup \{ B(y, \varepsilon/2) : \ y \in Y_0 \}$$

falls into a certain finite set of possibilities (depending on c and ε).

(B)′ For $n = 3$ it is possible to strengthen the above argument by the remark that (for a generic choice of Y_0) starting from the simplicial two-simplex described above we can construct (in a canonical way) a triangulation of

$$\bigcup \{ B(y, \varepsilon/2) : \ y \in Y_0 \}$$

which implies that the diffeomorphism class of the union falls into a certain finite set of possibilities (depending on c and ε).

All this is true, but we think it is not possible to prove (as suggested by Thurston) (A) and (B) with the same argument as for (A)' and (B)' respectively. In fact we have proved finiteness results on a set,

$$\bigcup \left\{ B(y, \varepsilon/2) : \ y \in Y_0 \right\},$$

containing $M_{[\varepsilon, \infty)}$ but not coinciding with it, so we can say nothing more.

A method we tried to develop for a direct proof of (A) and (B) using the above construction (we could say, "the only possible method") turned out to be unsuccessful, and it is worth explaining why. The idea was to prove that (with the above notations) for $y \in Y_0$ the intersection $B(y, \varepsilon/2) \cap M_{(0, \varepsilon]}$ "cannot be arbitrarily complicated" (at least, the number of its connected components is bounded by a constant depending on c and ε); if this were possible, (A) and (B) could have been deduced by a variation on the argument proving (A)' and (B)', using the covering of the ε-thick part

$$\left\{ B(y, \varepsilon/2) \cap M_{[\varepsilon, \infty)} : \ y \in Y_0 \right\}.$$

Let us remark that, in a heuristic way, if a method for the proof of (A) and (B) existed and was based on the argument proving (A)' and (B)', then necessarily it must be possible to control the complication of $B(y, \varepsilon/2) \cap M_{(0, \varepsilon]}$ by c and ε.

Unfortunately, this is not possible, and we were not able to fill in the gap by any other argument. The reason is that, for $n = 3$, this fact contradicts the results we shall prove in the next sections about hyperbolic Dehn surgery: for instance, a corollary of E.5.1 is that there exists a sequence $\{M_i\} \subset \mathcal{F}_3(c)$ such that $\forall i$ there exists a tube component T of $M_{i(0, \varepsilon]}$ and a point $y \in M_{i[\varepsilon, \infty)}$ with the property that $T \cap B(y, \varepsilon/2)$ has at least i connected components.

We have pointed out the reason we decided to follow an order of exposition opposite to the one presented by Thurston: in fact the construction of the geometric topology a priori and the discussion of its different characterizations allowed us to prove in a complete and self-contained way the most relevant properties of this topology (and hence of $\mathcal{F}_n$ and of the volume function). Our exposition is coherent with the one sketched in [Gro1]; we also remark that in [Ca-Ep-Gr] the topology is defined a priori too.

E.5 The Three-dimensional Case.
Thurston's Hyperbolic Surgery Theorem:
Statement and Preliminaries

One of the main goals of the present chapter is to show that non-trivial convergent sequences do exist in $\mathcal{F}_3$. We know that if $\{M_i\}$ is such a sequence and M is the limit manifold then, for large i, M_i is obtained from M by strong hyperbolic Dehn surgery, which means (in particular) that there exists in M_i

a finite set of simple closed short geodesics $\{\gamma_1, ..., \gamma_k\}$ and a diffeomorphism (close to an isometry on a compact set)

$$\phi : M \xrightarrow{\sim} M_i \setminus \left(\gamma_1 \bigcup ... \bigcup \gamma_k\right).$$

Using the pull-back with respect to ϕ we obtain on M a non-complete hyperbolic structure, "close" to the original complete one, such that (up to isometry) M_i is obtained as the completion of M with respect to such a structure. This fact suggests the way to construct non-trivial convergent sequences.

Given $M \in \mathcal{F}_3$ with at least one cusp (we know this is a necessary condition for M to be a limit point in $\mathcal{F}_3$) we shall deform a little the complete hyperbolic structure given on M and obtain non-complete hyperbolic structures close to it. Then we shall examine the (abstract) metric completion of M with respect to such structures and prove that if they are suitably chosen such a completion is a hyperbolic manifold obtained by Dehn surgery from M; moreover as the deformed structure converges to the original one, the corresponding manifold converges to M in $\mathcal{F}_3$.

We state first the main result, known as Thurston's hyperbolic surgery theorem. We recall that the set $\mathcal{P}_*$ of the coefficients we can use for a Dehn surgery is given by all the pairs of relatively prime integers together with a point ∞, and hence it can be viewed as a subset of S^2, the one-point compactification of $\mathbb{R}^2$; such a sphere will be endowed with its standard topology. Moreover we recall that a hyperbolic three-manifold M is the interior of a compact manifold $\overline{M}$ whose boundary consists of as many tori as the cusp ends of M.

Theorem E.5.1. Let $M \in \mathcal{F}_3$ have k cusp ends and assume $k \geq 1$. Then in the product of k copies of the sphere S^2 there exists a neighborhood U of $(\infty, ..., \infty)$ with the property that if

$$(d_1, ..., d_k) \in U \bigcap \left(\mathcal{P}_* \times .... \times \mathcal{P}_*\right)$$

then the manifold $\overline{M}_{d_1...d_k}$ is obtained by (weak) hyperbolic Dehn surgery from M and has finite volume. Moreover the set of all hyperbolic manifolds obtained in this way contains non-trivial sequences converging to M in $\mathcal{F}_3$.

The crucial point in the proof of E.5.1 consists in showing that the space of the deformations of the original hyperbolic structure on M is big enough (of course we must first define in a formal way what such a space of deformations is). The line we chose for the proof of this fact (as in [Ne-Za]) is the following:
– we shall first show that all elements of $\mathcal{F}_3$ having at least one cusp end (*i.e.* non-compact elements of $\mathcal{F}_3$) can be obtained by glueing the faces of ideal tetrahedra of $\mathbb{H}^3$ (see below for a precise definition);
– for a fixed presentation of a non-compact element of $\mathcal{F}_3$ as glued tetrahedra, the definition of the space of the deformations of the complete structure, the calculation of its dimension, the study of the metric completion of a deformed structure (and all other notions necessary for the proof of E.5.1) are obtained

very elementarily and explicitly (and, in principle, in an automatically computable way); the only underlying fact necessary for developing this machinery is the rigidity theorem.

We remark that the scheme sketched above is not only very simple, but it also suggests an effective method for the construction of finite-volume hyperbolic three-manifolds; moreover almost all remarkable examples explicitly known are obtained in this way. We also remark that the use of ideal tetrahedra for the study of three-dimensional hyperbolic geometry is one of the peculiar themes of Thurston's work (see for instance [Th3] and recall also the proof of the rigidity theorem); on the other hand the method of constructing hyperbolic manifolds (or discrete groups of isometries of the hyperbolic space) by means of polyhedra with identifications is classical and goes back to Poincaré (see for instance [Mask1] and [Ma-Fo]).

Remark E.5.2. We can re-phrase Theorem E.5.1 by saying that "almost all" manifolds obtained by Dehn surgery from $\overline{M}$ can be endowed with a hyperbolic structure. However, such a statement must be interpreted carefully: for a manifold having only one cusp the surgery coefficients excluded are a finite number, but if the cusps are at least two the theorem excludes infinitely many cases (and we shall see in Sect. E.7 by an example that this fact cannot be avoided).

This section is devoted to the following facts:
– definition of a class T_3 of three-manifolds and proof of its first properties;
– definition of a class $\mathcal{H}(M)$ of (possibly non-complete) hyperbolic structures on an element M of T_3 and realization in $\mathcal{H}(M)$ of the complete structure in case M belongs to $\mathcal{F}_3$ too;
– proof that all three-manifolds satisfying certain topological properties can be realized in T_3;
– determination of a large class of links in S^3 whose complement is constructively realized in T_3.

Let us remark that only the first two steps are actually necessary for the proof of the hyperbolic surgery theorem; the other two were included as they prove that our construction is very general, and moreover they suggest effective methods for producing examples of complete hyperbolic finite-volume three-manifolds.

E.5-i Definition and First Properties of T_3 (Non-compact Three-manifolds with "Triangulation" Without Vertices)

We shall denote by $\tilde{T}_3$ the class of all topological spaces $\tilde{Q}$ obtained according to the following procedure:
(i) a finite number n of copies of the standard 3-simplex (*i.e.* a tetrahedron) is fixed; such copies are denoted by $\Delta_1, ..., \Delta_n$;

(ii) if σ_1 and σ_2 are different 2-faces of some of the tetrahedra (possibly of the same tetrahedron, but not the same face) an object $f_{(\sigma_1,\sigma_2)}$ is given, being either the empty set or a simplicial isomorphism of σ_1 onto σ_2, in such a way that if $f_{(\sigma_1,\sigma_2)} \neq \phi$ then $f_{(\sigma_2,\sigma_1)} \neq \phi$ and $f_{(\sigma_2,\sigma_1)} = f_{(\sigma_1,\sigma_2)}^{-1}$;

(iii) the relation $\sim$ on $\coprod\limits_{i=1}^{n} \Delta_i$ defined by:

$x \sim y$ if and only if either $x = y$ or for some faces σ_1 and σ_2
we have $x \in \sigma_1$, $y \in \sigma_2$, $f_{(\sigma_1,\sigma_2)} \neq \phi$ and $f_{(\sigma_1,\sigma_2)}(x) = y$,

is an equivalence relation, and $\tilde{Q}$ is homeomorphic to the quotient topological space

$$\left(\coprod_{i=1}^{n} \Delta_i \right) \Big/ \sim .$$

When considering an element of $\tilde{T}_3$ we shall always refer to its realization via the above construction; this is an improper choice, as such a realization is certainly not unique, but it allows us to simplify the notation and prove the results we are interested in about the class $\tilde{T}_3$. We can overcome this difficulty by assuming that for each $\tilde{Q} \in \tilde{T}_3$ a certain fixed realization following the above scheme is chosen. Given $\tilde{Q} \in \tilde{T}_3$ we shall always denote by π the projection of the union of the tetrahedra onto $\tilde{Q}$.

Remark E.5.3. Each $\tilde{Q} \in \tilde{T}_3$ is naturally endowed with a finite three-dimensional cell complex structure; the 3-cells are as many as the tetrahedra used in the fixed realization. Cell structures are not unique too, but we shall always refer to the structure associated to the fixed realization of $\tilde{Q}$ as a union of tetrahedra with glued faces.

Recall that 0-cells are also called vertices. Let $\tilde{Q} \in \tilde{T}_3$ and let x_0 be one of its vertices; x_0 has a canonical <u>conic</u> neighborhood denoted by $x_0 \cdot L(x_0)$, where $L(x_0)$ denotes the <u>link</u> of x_0 in $\tilde{Q}$; we omit the general definition of such concepts and show their realization in our concrete case.

Let v be a vertex of a tetrahedron Δ such that $\pi(v) = x_0$, consider the second barycentric subdivision of the edges having vertex v and let w_1, w_2, w_3 be the vertices of such subdivisions nearest to v; denote by $L(v)$ the triangle having vertices w_1, w_2, w_3 and by $U(v)$ the tetrahedron having vertices v, w_1, w_2, w_3.

Then we have:

$$L(x_0) = \pi \left(\bigcup_{v \in \pi^{-1}(x_0)} L(v) \right) \qquad x_0 \cdot L(x_0) = \pi \left(\bigcup_{v \in \pi^{-1}(x_0)} U(v) \right).$$

(The reason for considering the second barycentric subdivision instead of the first one is that two vertices of the same tetrahedron could be projected on the same vertex of $\tilde{Q}$, and then we need to divide twice in order to avoid overlaps.)

We shall denote by T_3 the class of all non-compact oriented topological manifolds M without boundary which are homeomorphic to some element $\tilde{Q}$

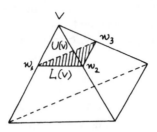

Fig. E.7. A tetrahedron and the link of one of its vertices in the second barycentric subdivision

of $\tilde{T}_3$ deprived of the vertices, where $\tilde{Q}$ is such that for all vertices $x_0 \in \tilde{Q}$ the link $L(x_0)$ in $\tilde{Q}$ is homeomorphic to a torus.

Remark E.5.4. If $M \in T_3$ then M is homeomorphic to the topological manifold obtained by removing from $\tilde{Q}$ the conic neighborhoods of the vertices described above. It follows that M is the interior of an oriented compact manifold $\overline{M}$ ($\tilde{Q}$ deprived of the interior of the conic neighborhoods of the vertices) whose boundary consists of as many tori as the vertices of $\tilde{Q}$.

In the following whenever considering an element M of T_3 we shall always denote by $\overline{M}$ the manifold with boundary described above, and by $\tilde{Q}$ the element of $\tilde{T}_3$ from which M is obtained (by removing the vertices). Remark that $\tilde{Q}$ is obtained from $\overline{M}$ by collapsing the boundary tori to points.

We prove now the first properties of the elements of T_3. We shall keep fixed for a while a manifold $M \in T_3$ obtained by removing the vertices from $\tilde{Q} \in \tilde{T}_3$; for the realization of $\tilde{Q}$ we shall use the same symbols as those used in the definition of $\tilde{T}_3$.

Lemma E.5.5. Given a face σ_1 of one of the tetrahedra, $f_{(\sigma_1,\sigma_2)} \neq \phi$ for precisely one face $\sigma_2 \neq \sigma_1$. Moreover it is possible to fix an orientation on each tetrahedron (inducing one on the faces) such that $f_{(\sigma_1,\sigma_2)}$ is an orientation-reversing isomorphism whenever it is not the empty set.

Proof. The first assertion means that only one face σ_2 is identified with σ_1, and this is a consequence of the fact that if x belongs to the interior of σ_1 then $\tilde{Q}$ is a 3-manifold in the neighborhood of $\pi(x)$ (in particular, some small neighborhood of $\pi(x)$ in $\tilde{Q}$ is homeomorphic to a ball in $\mathbb{R}^3$).

For the second assertion we only need to give each tetrahedron the orientation its interior inherits from M.   □

We shall denote by $\chi(X)$ the Euler-Poincaré characteristic of a space X whenever it is defined.

Lemma E.5.6. If $\tilde{Q}$ contains h_i i-cells (for $i = 0, ..., 3$) we have

$$\chi(\overline{M}) = 0 \qquad \chi(\tilde{Q}) = h_0 \qquad h_1 = h_3.$$

Proof. Consider the manifold $D(\overline{M})$ obtained by "doubling" $\overline{M}$ (*i.e.* obtained by glueing two copies of $\overline{M}$ with opposite orientation along the homologous boundary components, the glueing function being simply the identity). Since $D(\overline{M})$ is a compact oriented 3-manifold without boundary, it is well-known that $\chi(D(\overline{M})) = 0$. On the other hand it is easily checked that

$$\chi(D(\overline{M})) = 2\chi(\overline{M}) - \chi(\partial\overline{M})$$

and moreover $\chi(\partial\overline{M}) = 0$ (as it consists of pairwise disjoint tori) which implies that

$$\chi(\overline{M}) = 0.$$

Since a triangulation of $\tilde{Q}$ is obtained from a triangulation of $\overline{M}$ by replacing the simplices triangulating each boundary component by a vertex, we have

$$\chi(\tilde{Q}) = \chi(\overline{M}) + h_0 - \chi(\partial\overline{M})$$

and hence $\chi(\tilde{Q}) = h_0$. For the last assertion we remark that in $\tilde{Q}$ each 3-cell has four 2-cells in its boundary, and each 2-cell is in the boundary of two 3-cells, which implies that $h_2 = 2h_3$ and hence

$$h_0 = \chi(\tilde{Q}) = h_0 - h_1 + h_2 - h_3 = h_0 - h_1 + h_3,$$

so that $h_1 = h_3$ and the lemma is proved.   □

E.5-ii Hyperbolic Structures on an Element of $\mathcal{T}_3$ and Realization of the Complete Structure

The importance of the construction we are going to describe now of a (possibly empty) set of (generally non-complete) hyperbolic structures on an element of $\mathcal{T}_3$ comes from the fact (we shall see later in this section) that all non-compact three-manifolds M which are the interior of a compact manifold $\overline{M}$ whose boundary consists of tori can be realized as an element of $\mathcal{T}_3$. However only the definition of these structures and the next Theorem E.5.9 we are going to prove are strictly needed for the proof of the hyperbolic surgery theorem.

Consider a fixed element of $\mathcal{T}_3$ (together with its realization) and let $\Delta_1, ..., \Delta_n$ be the tetrahedra used for such a realization of M; we shall refer to the tetrahedron Δ_i deprived of the vertices both as an abstract object and as a subset of M; similarly Δ_i will be also viewed as a subset of $\tilde{Q}$ (the space obtained by glueing the tetrahedra, before removing the vertices). We can realize all the Δ_i's as ideal tetrahedra in $\mathbb{H}^3$, so that the faces are ideal hyperbolic triangles in $\mathbb{H}^3$. Each face lies on a hyperbolic hyperplane, which is isometric to $\mathbb{H}^2$; since in $\mathbb{H}^2$ all ideal triangles are obtained from each other by an isometry and all isometries between hyperbolic hyperplanes in $\mathbb{H}^3$ extend to isometries of $\mathbb{H}^3$ we can realize the glueing isomorphisms between the faces of the tetrahedra as restrictions of isometries of $\mathbb{H}^3$. It follows that we can endow the manifold obtained from M by removing the 1-skeleton, which we shall denote by M^*, with a (non-complete) hyperbolic structure. We shall denote by $\mathcal{H}(M)$ the set of all (not necessarily complete) hyperbolic structures on M which can be obtained as extensions of such a hyperbolic structure on M^*. (Remark that not all hyperbolic structures on M^* extend to M, and the description of those which do is one of the first themes of our discussion in the next section.)

Assume now $M \in \mathcal{T}_3$ belongs to $\mathcal{F}_3$ too, *i.e.* it can be endowed with a finite-volume complete hyperbolic structure; it is a natural question to ask whether such a complete structure is obtained as an element of $\mathcal{H}(M)$ (with respect to the given realization of M in $\mathcal{T}_3$). There exists a natural procedure in order to attempt to do this. The idea is to generalize to the non-compact case the operation of straightening of the singular simplices we have already described in C.4.2. A universal covering $\pi : \mathbb{H}^3 \to M$ associated to the complete structure is fixed (the group of isometries acting on $\mathbb{H}^3$ being denoted by Γ), and $\forall i$ a lifting to $\mathbb{H}^3$ of the interior of the tetrahedron Δ_i (viewed as a simply connected subset of M) is considered; we obtain a subset of $\mathbb{H}^3$ diffeomorphic to the interior of a tetrahedron; we claim that the closure of such a set in $\overline{\mathbb{H}^3}$ contains precisely four points of $\partial \mathbb{H}^3$ corresponding to the vertices of Δ_i; in fact if a sequence $\{x_n\}$ of points of the interior of Δ_i converges to a vertex v of Δ_i in $\overline{M}$ then it is a divergent sequence in M, and then the lifted points constitute a divergent sequence in $\mathbb{H}^3$; then up to extraction of a subsequence such a lifted sequence converges to a certain $x_\infty \in \partial \mathbb{H}^3$; this x_∞ is the fixed point for two parabolic elements of Γ corresponding to the cusp at v; then the lifting of the interior of Δ_i intersected with a neighborhood of v is contained in some horoball at x_∞, which implies that x_∞ is the only limit point above v. Our claim is proved, so that we can replace the lifting of Δ_i by the straight "ideal tetrahedron" in $\mathbb{H}^3$ having these four points as vertices; inverted commas mean that the tetrahedron may be degenerate, which corresponds to four points lying on a hyperplane. If we repeat this procedure for all possible liftings and for all i's we obtain that $\mathbb{H}^3$ is expressed as a Γ-equivariant union of "ideal tetrahedra". If we project to M we obtain that M is covered by "tetrahedra" being isometric to "ideal tetrahedra" in $\mathbb{H}^3$. Unfortunately, this is not in general a realization of M in $\mathcal{T}_3$; in fact, as we said, the tetrahedra may be degenerate; but we have even more: the tetrahedra may overlap, which corresponds to the fact that the lifting of some of the Δ_i's is not "convex", *i.e.* that the ordering of the vertices on the straightened tetrahedron induces the wrong orientation. Figures E.8 and E.10 illustrate this phenomenon: we have generalized the situation to the straightening of a topological triangulation of $\mathbb{H}^n$ in which we possibly have vertices not at infinity. It is easily checked (see Fig. E.9) that examples of the phenomenon of overlapping occur in the two-dimensional case only if we accept vertices not at infinity.

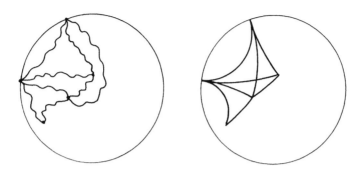

Fig. E.8. A two dimensional example in which the straightening of the triangles produces overlapping

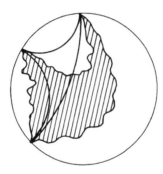

Fig. E.9. Given any three points in the boundary of hyperbolic two-space then any triangle having them as vertices has the same orientation as the geodesic triangle. This implies that an overlapping phenomenon as in Fig. E.8 cannot occur in the two-dimensional case if the vertices are at infinity

We can summarize the two phenomena of degeneracy and overlapping by saying that a realization of an element M of $\mathcal{F}_3$ as an element of $\mathcal{T}_3$ allows us to cover M with ideal tetrahedra, but these tetrahedra have an algebraic volume which is not necessarily positive (the case of volume zero corresponds to a degenerate tetrahedron and the case of negative volume corresponds to a tetrahedron whose vertices induce the wrong orientation). A generalization of the arguments presented for C.4.6 allows to establish:

Lemma E.5.7. The sum of these algebraic volumes is the volume of M.

The above discussion is summarized by the following:

Proposition E.5.8. Let $M \in \mathcal{F}_3$ be given a fixed (up to homeomorphism) realization in $\mathcal{T}_3$. Then the complete structure appears as an element of $\mathcal{H}(M)$

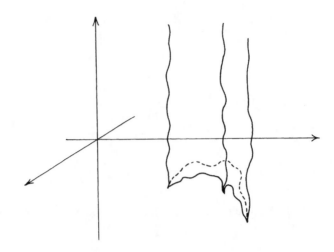

Fig. E.10. An example in the upper three-space model of an ideal tetrahedron whose straightening changes the orientation. Starting from it one can easily construct an example of the overlapping phenomenon for ideal triangulations in the three-dimensional case

(with respect to this realization of M in T_3) if and only if the straightening of all the tetrahedra in (some topological representative of) the realization of M have positive algebraic volume.

The reader will find in [Pe] the description of a simple explicit example of realization in T_3 of an element of $\mathcal{F}_3$ with the property that some of the straightened tetrahedra have volume 0.

We have seen that it is not true in general that if we consider a manifold $M \in \mathcal{F}_3$ and we fix a realization of M in T_3 then the complete structure is obtained as an element of $\mathcal{H}(M)$ with respect to such a realization. On the other hand, we are going to prove now that there exists a suitable realization of M in T_3 with the property that the complete structure does appear as an element of $\mathcal{H}(M)$ with respect to it. This fact plays a central role in our proof of hyperbolic surgery theorem; it was recently obtained by D. B. A. Epstein and R. C. Penner in [Ep-Pe].

Theorem E.5.9. Every non-compact element M of $\mathcal{F}_3$ can be realized as a union of ideal tetrahedra in $\mathbb{H}^3$ with glued faces, *i.e.* it can be realized as an element of T_3 in such a way that the complete hyperbolic structure appears in $\mathcal{H}(M)$.

Proof. The argument in [Ep-Pe] works for all dimensions $n \geq 2$ and not only for $n = 3$, but we confine ourselves to the case we are really interested in. We are going to omit many details and refer to the quoted paper for a complete proof.

As usual we realize M as the quotient space $\mathbb{H}^3/\Gamma$, where Γ is a discrete torsion-free subgroup of $\mathcal{I}^+(\mathbb{H}^3)$.

Let us consider the hyperboloid model $\mathbb{I}^3$ of hyperbolic three-space; we recall it consists (as a set) of the upper fold of the hyperboloid

$$\left\{x \in \mathbb{R}^4 : \langle x|x \rangle_{(3,1)} = -1\right\},$$

where

$$\langle x|y \rangle_{(3,1)} = x_1 \cdot y_1 + x_2 \cdot y_2 + x_3 \cdot y_3 - x_4 \cdot y_4;$$

since the dimension is fixed we set $\langle x|y \rangle = \langle x|y \rangle_{(3,1)}$.

Let us consider in $\mathbb{R}^4$ the cone $L = \left\{x : \langle x|x \rangle = 0\right\}$ and its upper half $L^+ = \left\{x \in L : x_4 > 0\right\}$. For $v \in L^+$ the half-line $r_v = \mathbb{R}_+ \cdot v$ determines a point of $\partial \mathbb{I}^3$ we shall denote by r_v too; moreover v determines the horosphre centred at r_v

$$S_v = \left\{x \in \mathbb{I}^3 : \langle x|v \rangle = -1\right\}$$

being the boundary of the closed horoball

$$B_v = \left\{x \in \mathbb{I}^3 : \langle x|v \rangle \geq -1\right\};$$

moreover all horospheres and closed horoballs are obtained (for a unique v) in this way.

Let us assume now M has $k \geq 1$ cusps; for $\varepsilon \leq \varepsilon_3$ (the third Margulis constant) small enough that there is no tube component in the ε-thin part of M, we number in an arbitrary way the k components of $M_{(0,\varepsilon]}$ and we call i-th cusp the i-th component. For $i = 1,...,k$ let γ_i be a half geodesic line in M asymptotic to the i-th cusp. (If we realize M as a quotient of the hyperbolic space in the half-space model in such a way that the two parabolic isometries corresponding to the i-th cusp have ∞ as a fixed point, then γ_i is obtained as the quotient of a geodesic half-line with endpoint ∞, i.e. a vertical straight half-line; remark as well that the cusp itself is obtained as the projection of a closed horoball centred at ∞.)

For $i = 1,...,k$ we fix a lifting $\tilde{\gamma}_i$ of γ_i to $\mathbb{I}^3$ with respect to the fixed projection $\mathbb{I}^3 \to M = \mathbb{I}^3/\Gamma$; since $\tilde{\gamma}_i$ is a geodesic half-line its endpoint is a point $x_i \in \partial \mathbb{I}^3$; moreover the i-th cusp is the quotient of a horoball centred at x_i; then we can find a unique $v_i \in L^+$ such that $r_{v_i} = x_i$ and the quotient of the closed horoball B_{v_i} is the i-th cusp.

The starting point of the construction is the following important lemma; let us recall that the fixed subgroup Γ of $\mathcal{I}^+(\mathbb{I}^3)$ is canonically identified with a subgroup of the group $O(\mathbb{R}^4, \langle .|. \rangle)$ and hence it operates on $\mathbb{R}^4$; it follows as well that for $\gamma \in \Gamma$ we have $\gamma(L^+) = L^+$.

Lemma E.5.10. For $i = 1,...,k$ the Γ-orbit of v_i is a closed discrete subset of L^+ and of $\mathbb{R}^4$.

Proof. Let $\mathcal{O}$ denote the set of all closed horoballs in $\overline{\mathbb{H}^3}$ and let us endow $\mathcal{O}$ with the Chabauty topology (see Sect. E.1) as a class of closed subsets of

$\overline{\mathbb{H}}^3$; recall that by E.1.3 this topology is induced by the Hausdorff distance between closed subsets of $\overline{\mathbb{H}}^3$ identified with $\overline{D}^3$ and endowed with the natural Euclidean distance. Then it is possible to check that the mapping

$$L^+ \ni v \mapsto B_v \in \mathcal{O}$$

is a homeomorphism.

By the choice of v_i and recalling D.3.4 and D.3.7 we have that if $\gamma \in \Gamma$ and $\gamma(S_{v_i}) \cap S_{v_i} \neq \phi$ then $\gamma(S_{v_i}) = S_{v_i}$. It follows from this that the Γ-orbit of B_{v_i} is closed and discrete in $\mathcal{O}$, and then the Γ-orbit of v_i is closed and discrete in L^+.

For the conclusion that the Γ-orbit of v_i is closed and discrete in $\mathbb{R}^4$ too we remark that the Euclidean diameter of all the horoballs in the Γ-orbit of B_{v_i} is bounded by 2α for some suitable $\alpha < 1$, which implies that $0 \in \mathbb{R}^4$ is not a limit point of the Γ-orbit of v_i.   □

Remark E.5.11. As it was remarked in [Ep-Pe] the above lemma is somewhat surprising, as in the same paper the following facts were established:
– the action of Γ on L^+ is ergodic with respect to the Lebesgue measure, *i.e.* for all non-empty open subset U of L^+ the set $L^+ \setminus \Gamma(U)$ has null measure, which implies quite easily that for almost all $v \in L^+$ the Γ-orbit of v is dense in L^+;
– if the manifold $M = \mathbb{H}^3 / \Gamma$ is compact then for all $v \in L^+$ the Γ-orbit of v is dense in L^+.

Now let C denote the closed convex hull in $\mathbb{R}^4$ of the union of the Γ-orbits of $v_1, ..., v_p$. We first remark that the cone

$$\left\{ x \in \mathbb{R}^4 : \langle x | x \rangle \leq 0, x_4 > 0 \right\}$$

is convex and if we add $\{0\}$ to it, it becomes closed. Since $\Gamma \cdot v_1 \cup ... \cup \Gamma \cdot v_k$ is contained in this cone and it does not have 0 as limit point, it follows that C is contained in the cone. Moreover it is possible to show that C has the following remarkable properties:
(1) $L^+ \cap C = \left\{ \alpha \cdot z : \alpha \geq 1, z \in \Gamma \cdot v_1 \cup ... \cup \Gamma \cdot v_k \right\}$;
(2) if $\partial' C$ denotes the boundary of $C \cap L^+$ (as a subset of the topological space L^+) then for all $v \in L^+$, $r_v \cap \partial' C$ consists of precisely one point (by r_v we are meaning now the entire line $\mathbb{R}_+ \cdot v$);
(3) the boundary ∂C of C in $\mathbb{R}^4$ is decomposed as

$$\partial C = (C \cap L^+) \cup \mathcal{F}$$

where:
(i) $\mathcal{F} = \{F_i\}$ is a countable family of faces of dimension 3;
(ii) all F_i's are the convex hull of a finite number of points belonging to $\Gamma \cdot v_1 \cup ... \cup \Gamma \cdot v_k$;

(iii) for all i's, the restriction of $\langle .|. \rangle$ to the affine hull A_i of F_i is positive-definite, so that it defines a Euclidean structure on A_i, and $A_i \cap L^+$ is a sphere with respect to such a structure;

(iv) $\mathcal{F}$ is locally finite in $\{x \in \mathbb{R}^4 : \langle x|x \rangle < 0, x_4 > 0\}$.

It follows that a locally finite cell complex structure is determined on $\partial C \backslash L^+$; such a structure can be projected to $\mathbb{H}^3$ using the canonical projection to the projective model (see A.1); in fact $\partial C \backslash L^+$ is contained in the open cone $\{x \in \mathbb{R}^4 : \langle x|x \rangle < 0\}$ which has the same projection as $\mathbb{H}^3$. We get a locally finite tesselation of $\mathbb{H}^3$ in which all the cells of dimension at least 1 are geodesic subsets of $\mathbb{H}^3$, and there exists no 0-cell in $\mathbb{H}^3$, $i.e.$ all the vertices are at infinity (in fact all the 0-cells of ∂C are, by construction, in L^+). Remark in particular that 1-cells are entire geodesic lines. Since the construction is Γ-equivariant it follows that all 3-cells of such a tesselation are fundamental domains for the action of Γ. If we fix one of these three-cells P we have that P is a convex geodesic polyhedron in $\mathbb{H}^3$ with all the vertices at infinity, and M is obtained from P by suitable glueings of the faces. Then we only have to consider a decomposition of P into a finite number of geodesic ideal tetrahedra.   □

E.5-iii Elements of $\mathcal{T}_3$ and Standard Spines

We turn now to some further remarks about the class $\mathcal{T}_3$. We remind the reader that the remainder of the section is not necessary for the proof of the hyperbolic surgery Theorem E.5.1.

We are going to show that the obvious necessary conditions for a manifold M to belong to $\mathcal{T}_3$ are sufficient too. The main reference for this fact is [Ma-Fo].

Some notions we shall use in the following come from P.L. geometry and may be found in [Ro-Sa]. Consider a finite polyhedron with a triangulation K; as usual we shall write $|K|$ when referring to the polyhedron as a topological space. Let K_1 be a sub-polyhedron of K; we shall say there exists an elementary collapse of K onto K_1 if there exists a simplex Δ of K and a face F of Δ which is not the face of any other simplex in K such that

$$|K_1| = |K| \backslash \left(\text{Int}(\Delta) \bigcup \text{Int}(F) \right)$$

(where as usual $\text{Int}(A)$ denotes the simplex A deprived of its proper faces). We shall say K collapses on K_1 and write $K \searrow K_1$ if it is possible to obtain K_1 from K by a (finite) sequence of elementary collapses.

We recall that a regular neighborhood of a point x of $|K|$ is obtained for instance by taking the cone with vertex x over the link of x in any triangulation of $|K|$ refining K and containing x as a vertex. We shall say K is quasi-standard if each point of $|K|$ has a regular neighborhood in $|K|$ homeomorphic to one of the three types I, II and II represented in Fig. E.11.

Let us remark that a neighborhood of type II can be viewed as the product of a Y and a closed interval, while a neighborhood of type III is given by the cone over the 1-skeleton of a tetrahedron with the baricentre as vertex.

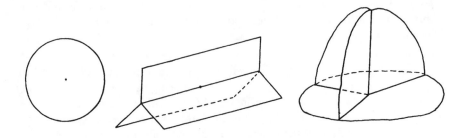

Fig. E.11. The three types of neighborhood, denoted respectively by I, II and III, of a point in a quasi-standard polyhedron

Remark that if K is quasi-standard then K is naturally stratified by closed sub-polyhedra K_0 and K_1 (the former consisting of the points of type III and the latter of the points of types II and III):

$$K_0 \subset K_1 \subset K$$

(it is easily verified that the endpoints of a segment of points of type II cannot be of type I, which proves that K_1 is a sub-polyhedron, while this is obvious for K_0). We shall say K is <u>standard</u> if all the components of $|K_1| \setminus |K_0|$ are homeomorphic to open segments and all the components of $|K| \setminus |K_1|$ are homeomorphic to open discs. If K is standard we shall call the points of type III <u>vertices</u> of K; remark that not all the 0-simplices in K are necessarily vertices.

Let M be a compact 3-manifold with non-empty boundary, and let us fix a triangulation H of M relative to ∂M; a <u>standard spine</u> K in M is a sub-polyhedron of H being standard and such that $H \searrow K$.

The importance of standard spines comes from the next result, being actually the starting point of the theory of standard spines. We state it without proof and address the reader to [Cas].

Theorem E.5.12. (a) Let M_1 and M_2 be triangulated compact three-manifolds with non-empty boundary, and assume they have homeomorphic standard spines: then they are homeomorphic too;

(b) all triangulated compact three-manifolds with non-empty boundary admit standard spines (with respect to a suitable refinement of the original triangulation).

We only want to remark that the proof of existence in point (b) is constructive.

The next result (whose proof we just outline, addressing the reader to [Ma-Fo] and the references quoted therein) establishes the relationship between standard spines and elements of $\mathcal{T}_3$.

Theorem E.5.13. Let $\overline{M}$ be a compact three-manifold with non-empty boundary consisting of tori, and let M denote the interior of $\overline{M}$; then we have that:

(i) to each realization of M as an element of T_3 using n tetrahedra it is possible to associate a standard spine of $\overline{M}$ with n vertices;

(ii) to each standard spine in $\overline{M}$ with n vertices it is possible to associate a realization of M as an element of T_3 using n tetrahedra.

Proof. We first remark that the realizations of M as an element of T_3 correspond bijectively to the realizations of $\tilde{Q}$ (the space obtained from $\overline{M}$ by collapsing the boundary tori to points) as an element of $\tilde{T}_3$ (and the number of tetrahedra involved is the same).

(i) The construction is explicit: inside each tetrahedron appearing in the realization of $\overline{M}$ as an element of $\tilde{T}_3$ we consider the portion of spine represented in Fig. E.12; since this subset of the polyhedron can be realized as a sub-polyhedron of a barycentric subdivision of the tetrahedron, the different portions are nicely glued (we recall that the identifications between the faces of the tetrahedra are simplicial isomorphisms) and then the resulting subset of $\overline{M}$ is a standard spine. Moreover precisely one vertex of the spine belongs to each tetrahedron.

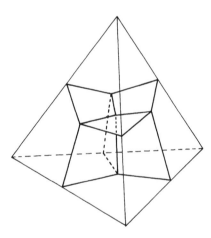

Fig. E.12. Construction of a standard spine starting from a tetrahedra decomposition

(ii) To each vertex of the spine we can associate a tetrahedron (recall that a regular neighborhood of a vertex is precisely the cone over the 1-skeleton of a tetrahedron). The combinatorial structure of the spine (*i.e.* the way the boundaries of the regular neighborhood of the vertices are glued together) induces a way to glue the faces of the tetrahedra thus obtained. The assumption that the spine is standard, together with the possibility of reconstructing $\overline{M}$ starting from the spine, imply that the manifold obtained by glueing the above tetrahedra in the required way is (homeomorphic to) $\tilde{Q}$. $\qquad\square$

The following straight-forward consequence of E.5.12 and E.5.13 concludes the first part of our argument about the size of the class T_3; remark that by D.3.14 the result applies to all non-compact elements of $\mathcal{F}_3$.

Corollary E.5.14. Let M be a three-manifold which is the interior of a compact manifold $\overline{M}$ with non-empty boundary consisting of tori; then M can be realized as an element of T_3 using as many tetrahedra as the vertices of a standard spine for $\overline{M}$.

E.5-iv Some Links Whose Complements Are Realized as Elements of T_3

We are going to describe an explicit realization in T_3 (actually of a very special type) of the complement of a large class of links in S^3; this construction is partially inspired by [Me]. The reader will find a generalization of this construction and some details we are going to omit in [Pe]; this paper also describes an algorithmic method producing the hyperbolicity equations (which we are going to meet in Sect. E.6-i below) associated to the realization in T_3 of the link complement.

Let us fix a link L in S^3 represented by a regular projection G on the horizontal plane $H \subset \mathbb{R}^3 \subset \mathbb{R}^3 \cup \{\infty\} = S^3$; G is a finite graph whose vertices are crossings with the branch passing above the other one being specified by the usual symbology. We will refer to $H \cup \{\infty\}$ as S^2.

In our construction we shall make several hypotheses about G; some of them have technical motivation, and they do not reduce the generality of the construction (that is, one can easily find another link isomorphic to L whose projection satisfies them); on the other hand, some other hypotheses we shall make actually reduce the generality of the construction: we shall number them by Roman numerals.

First of all we assume that the projection cannot be trivially simplified, *i.e.* that each vertex of G meets the closure of four different components of $S^2 \setminus G$ (these components will be called <u>regions</u>) and that there exists no region having only two edges one of which passes below at both the vertices.

Moreover we assume L is not the unknot in its usual projection, *i.e.* that G does have crossings.

We come to the first serious hypothesis;

I. *G is connected.*

Remark E.5.15. It is easily checked that condition I is equivalent to the fact that each region is homeomorphic to the open disc. Moreover it follows from the previous assumption that each crossing meet the closure of four different regions and from assumption I that the closure of each region is homeomorphic to the closed disc.

Remark E.5.16. Assumption I is automatically verified if L is a knot, but not in the general case. Remark that condition I is actually necessary if we want the complement of the link to be irreducible (in particular, if we want

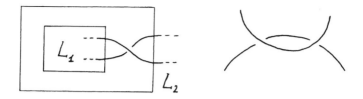

Fig. E.13. These link projections admit trivial simplifications. They are ruled out by our assumptions on G

it to possess a finite-volume hyperbolic structure, see D.3.17). In fact if G were disconnected we could find two non-empty sublinks of L separated by an embedded 2-sphere.

Since G is a graph in S^2 a region D of $S^2 \setminus G$ can be referred to as a polygon; when doing this we will call vertices of D the crossings meeting $\overline{D}$ and edges of D the edges of G meeting $\overline{D}$.

Remark E.5.17. $S^3 \setminus S^2$ has two connected components homeomorphic to the open 3-ball D^3, both having S^2 as boundary. Under the above assumptions G induces a cell decomposition of S^3 (the 0-cells being the crossings, the 1-cells being the edges of G, the 2-cells being the regions and the 3-cells being the components of $S^3 \setminus S^2$).

The first step of the construction consists in associating to G another representation of S^3 as a cell complex, having in particular the following properties:

–(a link isomorphic to) L is a subcomplex of such a complex;

–only two 3-cells are involved.

We first describe heuristically the construction and then we fill in the details. The key idea is to loosen the crossings, *i.e.* to modify G in a small neighborhood of each crossing in the following way: since we know what branch passes above the other one, we raise a little such a branch and lower a little the other one, and then we add a short vertical segment joining the branches.

Then the 0-cells are the endpoints of these short segments, the 1-cells are the short segments and the edges of G slightly modified near the crossings, the 2-cells are the regions slightly modified near the crossings as suggested by Fig. E.15; a look at the same figure showing what is happening near a vertex of G, allows one to conclude that what is left out actually consists of two open 3-balls which we take as 3-cells.

Let us describe the above construction in a more formal way. We fix pairwise disjoint neighborhoods of the crossings in G; the skeletons of dimension 0,1 and 2 of the cell complex we are going to construct differ from those of the cell decomposition of S^3 associated to G only inside these neighborhoods, so we consider a crossing at a time, and work near it.

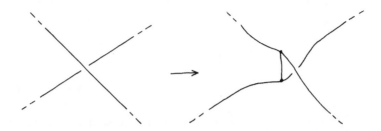

Fig. E.14. How to loosen a crossing and add a vertical segment

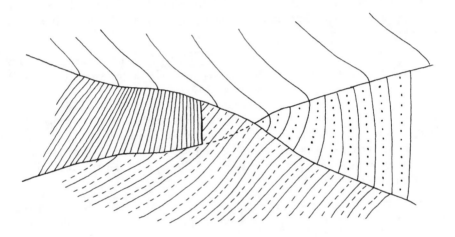

Fig. E.15. Overview of the two-skeleton of the cell complex near a crossing

For a fixed crossing we consider an open square neighborhood Q as in Fig. E.16, and remove it from S^2. We fix an orientation on ∂Q and choose a vertex of Q on the branch passing below the other one (there are two possible choices for the orientation and two for the vertex, but it is immediately verified that our construction is independent of both). Starting from the fixed vertex and following the orientation, we denote the edges of Q by a, b, c, d. We consider now a rectangle R as in Fig. E.17, and glue it to $S^2 \setminus Q$ along a and c giving it half a twist, in such a way that the arc $q_3 \cup q_4$ passes below the arc $q_1 \cup q_2$.

The curves

$$k_1 = q_1 \cup g \cup q_4 \cup b \qquad\qquad k_2 = q_2 \cup g \cup q_3 \cup d$$

are closed loops, so that we can find disjoint sets D_1 and D_2 having closure homeomorphic to a closed disc, not meeting $(S^2 \setminus Q) \cup R$ and having respectively k_1 and k_2 as boundary.

We assume this construction is performed for all crossings and describe the resulting cell decomposition of S^3 (actually we only describe a partition

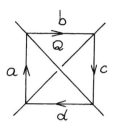

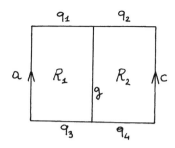

Fig. E.16. The square neighborhood of the vertex we remove from the sphere

Fig. E.17. The rectangle we are going to glue in

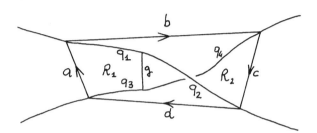

Fig. E.18. The rectangle glued in

into cells, as it is obvious that there exist continuous glueing functions from ∂D^i to the union of the cells of dimension at most $i - 1$):

– the 0-cells are the endpoints of the segments of type g;

– the 1-cells are the segments of type g and the segments obtained by replacing in each segment joining two crossings in G its intersection with the square with the suitable segment of type q_i (at both sides);

– the 2-cells are the discs obtained by modifying the regions of $S^2 \setminus G$ near the crossings, by substituting their intersection with the square with the suitable disc of type R_i or D_i (at all the vertices). Fig. E.19 shows (in the situation considered above) the pieces of 2-cell near a crossing corresponding to two different regions.

– the 3-cells are the connected components of S^3 minus the lower-dimensional cells; we have already remarked that these components are two open 3-balls.

From now on we will denote by L the link isomorphic to the initial one which is a subcomplex of the described realization of S^3 as a cell complex. We consider the quotient space $S^3/_L$, i.e. (by definition) the quotient space of S^3 with respect to the following equivalence relation:

$$x \sim y \iff x = y \text{ or both } x \text{ and } y \text{ belong to the same component of } L.$$

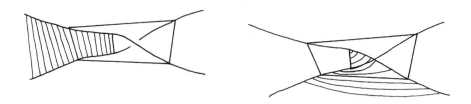

Fig. E.19. Two modified regions

We will denote by π the projection of S^3 onto $S^3/_L$; the cell complex structure on S^3 induces one on $S^3/_L$ described as follows:

– $S^3/_L$ has as many 0-cells as the components of L; moreover $S^3/_L$ deprived of the 0-cells is canonically homeomorphic to $S^3 \setminus L$;

– $S^3/_L$ has as many 1-cells as the crossings in G (and they are the image under π of the segments of type g described above;)

– the 2-cells in $S^3/_L$ correspond to those in the cell decomposition of S^3, and hence to the regions of $S^2 \setminus G$;

– in $S^3/_L$ there are again two 3-cells; as they come from the half-spaces lying above and below the plane H we will denote them by B_+ and B_-.

Hence the number of cells appearing in $S^3/_L$ is easily determined starting from G. What about the glueing functions? The glueing of the 1-cells to the 0-cells is evident, so we describe the higher-dimensional situation.

We start by fixing arbitrarily an orientation on each of the segments of type g in S^3, inducing one on the corresponding 1-cells in $S^3/_L$. As usual if γ and δ are paths we denote by $\gamma \cdot \delta$ the path obtained by following first γ and then δ (remark that we need that the second endpoint of γ equals the first endpoint of δ) and by γ^{-1} the path γ followed backwards. Let us consider a crossing as in Fig. E.16 and let us associate to the orientation of g an orientation on the edges of Q in the (natural) way described in Fig. E.20.

Now we can describe the glueing function of a disc corresponding to a region D; the apparently complicated construction will be soon clarified by an example. Let $g_1, ..., g_k$ be the segments associated to the vertices of D; we write the symbol g_i near the (oriented) edge of square lying inside D and corresponding to the same crossing as g_i; the disc obtained by removing from D the open squares is now bounded by a $2k$-gon, k alternate edges of which have g_* written near them; we collapse the other k edges to points; the boundary of the resulting k-gon is a now loop naturally represented by an expression as $g_1^{\pm 1} \cdot ... \cdot g_k^{\pm 1}$; such an expression also represents a loop in the 1-skeleton of $S^3/_L$, and the glueing function of the of the 2-cell corresponding to D is given by any homeomorphism between $S^1 = \partial D^2$ and such a loop.

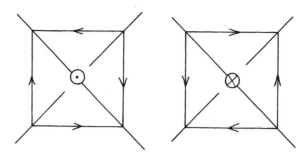

Fig. E.20. How to associate to the orientation of g an orientation of the edges of Q. As usual, $\odot$ means that a vertical vector points upwards and $\otimes$ means that it points downwards

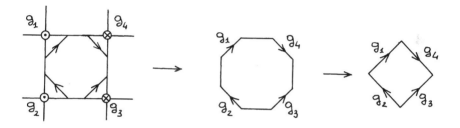

Fig. E.21. How to obtain the glueing function of a two-cell. In this example the resulting expression is $g_1^{-1} \cdot g_2^{-1} \cdot g_3 \cdot g_4^{-1}$

The example given in Fig. E.21 with $k = 4$ clarifies the situation (and proves that the construction is completely natural).

The reason for the use of such a detail in the above construction is that we want to keep track of the fact that all steps necessary for the realization of $S^3 \setminus L$ as an element of $\mathcal{T}_3$ can be performed algorithmically, and this is the case for the above construction.

Before turning to the description of the glueing functions of the 3-cells we make another important hypothesis on the link L and discuss it a little. Since our goal is to represent

$$S^3 \setminus L = \left(S^3 / L \right) \setminus \{0\text{-cells}\}$$

as an element of $\mathcal{T}_3$, and hence S^3 / L as an element of $\tilde{\mathcal{T}}_3$, it is necessary to obtain a realization of S^3 / L as a cell complex in which there exists no 2-cell being a "bigon", *i.e.* a 2-cell glued to two segments having the same endpoints (see Fig. E.22); in fact such a situation cannot occur for an element of $\tilde{\mathcal{T}}_3$.

If we assume that the orientations of the two edges of a bigon (segments of type g) agree, we can identify these segments and get rid of the bigon.

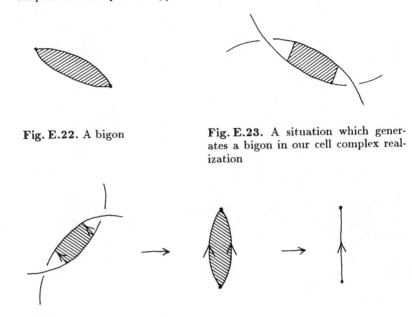

Fig. E.22. A bigon

Fig. E.23. A situation which generates a bigon in our cell complex realization

Fig. E.24. If the edges of a bigon have the same orientation then we can get rid of the bigon and identify them

Remark that in $S^3/_L$ we are contracting a 2-cell to a segment, and we can do this by enlarging the surrounding 2-cells (so that the new cell complex we obtain still represents $S^3/_L$). But if we want to do this for all the bigons we need that:

(a) there exists a region which is not a bigon;

(b) the orientations of the segments of type g can be chosen in such a way that the two edges of each bigon have agreeing orientation;

(c) while performing the eliminations one never arrives to a bigon whose edges are already identified (for otherwise the bigon would give a sphere in $S^3/_L$, and the sphere cannot be switched to a segment); remark that such a case can actually occur as when one eliminates a bigon with edges g_1 and g_2 the other copies of g_1 and g_2 are identified as well.

A case in which (a)-(c) are certainly verified is when the bigons are isolated (*i.e.* different bigons have disjoint closures). The reader will find in [Pe] a detailed discussion of the necessary and sufficient conditions for (a)-(c) to hold. Here we confine ourselves to the simple case and assume:

II. *The bigons are isolated.*

Remark E.5.18. It is easily checked that the orientations of the edges of a bigon agree if and only if different symbols ($\odot$ and $\otimes$) are attached to its two vertices.

So in the representation of $S^3/_L$ as a cell complex we get rid of all the bigons; it follows that the 2-cells in such a representation have at least three edges (segments of type g) in their boundary.

We come now to the description of the glueing functions of B_+ and B_- to the 2-skeleton; in this case too everything we are going to do can be performed algorithmically starting from G. As we saw above each 2-cell D in $S^3/_L$ is represented by an expression like $g_1^{\pm 1} \cdot \ldots \cdot g_k^{\pm 1}$ meaning the loop in the 1-skeleton to which the boundary of D is glued. (Remark that by the above assumption no expression like $g_1 \cdot g_2$ can occur, and if an expression like $g_1 \cdot g_2^{-1}$ occurs we get rid of the corresponding 2-cell and set $g_2 = g_1$.) Then D is represented by a k-gon ($k \geq 3$) with $g_*^{\pm 1}$ written near each edge. Giving the glueing function of B_+ (or B_-) means giving a function from S^2 to the 2-skeleton; if we recall the situation we come from it is easily verified that the function is obtained in the following way: a realization of S^2 is given as the union of the polygons representing the 2-cells of $S^3/_L$, each polygon appearing exactly once and two polygons being nicely adjacent (i.e. along vertices or edges), and the glueing function is the one whose restriction to each polygon is the "identity". Hence it only remains to determine adjacencies along edges (those along vertices follow), and this is very easily done; in a crossing as represented in Fig. E.25 (with associated segment g) the pairs (D_1, D_4) and (D_2, D_3) are adjacent along g in B_+, while the pairs (D_1, D_2) and (D_3, D_4) are adjacent along g in B_-.

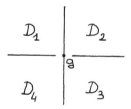

Fig. E.25. Determining adjacencies along the vertical segment g. In this situation we have that the adjacent pairs are on the up-down direction in B_+ and on the left-right direction in B_-

Example E.5.19. In Figg. E.26 to E.31 we explain the construction on the famous example of the <u>eight knot</u>.

From the figures it is quite evident how the complement of the eight knot can be realized as an element of $\mathcal{T}_3$: in fact $S^3/_L$ is obtained by glueing the faces of two tetrahedra.

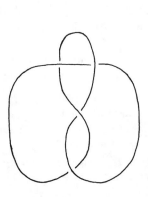

Fig. E.26. The usual projection of the eight knot

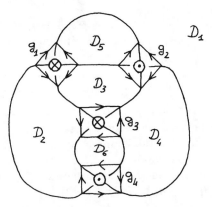

Fig. E.27. Choice of the orientation of the segments of type g

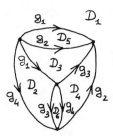

Fig. E.28. Representation of the sphere ∂B_+ before eliminating the bigons

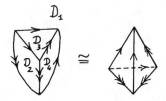

Fig. E.29. Representation of the sphere ∂B_+ after eliminating the bigons

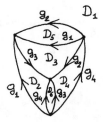

Fig. E.30. Representation of the sphere ∂B_- before eliminating the bigons

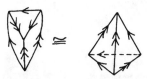

Fig. E.31. Representation of the sphere ∂B_- after eliminating the bigons

Example E.5.20. Figures E.32 and E.33 refer to a projection of the trefoil knot, but not the usual one. The construction is somewhat harder; we shall see why later in E.5.22.

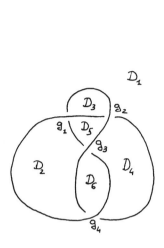

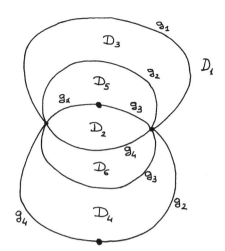

Fig. E.32. A projection of the trefoil knot

Fig. E.33. Representation of the sphere ∂B_+, without eliminating the bigons

Let us repeat in other words what we just saw: the glueing functions of B_+ and B_- are given by a realization of ∂B_+ and ∂B_- (both identified with S^2) as polyhedra in which the 2-faces are the 2-cells of $S^3/_L$. (In this context we use the term polyhedron for a union of nicely glued polygons, *i.e.* we do not require the faces to be triangles.)

The goal we have in mind, as already for assumption II, induces us to make a further hypothesis; we will say a 2-cell is <u>not</u> self-adjacent if its closure is homeomorphic to the closed disc; we will say two cells have <u>minimal adjacency</u> if the intersection of their closures is either empty, or a single vertex or a single closed edge.

III. *In ∂B_+ and ∂B_- no 2-cell is self-adjacent, and any two 2-cells have minimal adjacency.*

Once again the condition we are giving is not the weakest we could give, it is just the most natural one; we address to [Pe] for a detailed discussion of the most general conditions under which the construction works.

Since the construction of the polyhedron structure on ∂B_+ and ∂B_- is algorithmic and starts from G, the above condition III can be easily checked. However, we are going to show there are simpler criteria.

We say G has <u>alternate crossings</u> if for all components N of L, given any orientation on N, each crossing at which N passes below is followed by one at which it passes above, and conversely.

Lemma E.5.21. If condition III holds then G has alternate crossings.

Proof. By contradiction, one of the situations of Fig. E.34 must occur.

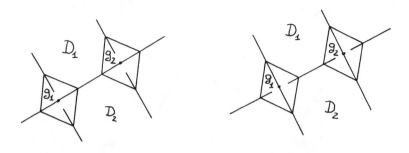

Fig. E.34. If the projection does not have alternate crossings then one of these situations occurs

Then the 2-cells in $S^3/_L$ corresponding to D_1 and D_2 are adjacent along both g_1 and g_2 in ∂B_- and ∂B_+ respectively; remark that by the first assumptions on G, the regions D_1 and D_2 cannot be bigons, so that they are not eliminated in the 2-skeleton of $S^3/_L$.   □

We prove now that for links having projection with alternate crossings (and hence, in particular, for links satisfying condition III) the construction of the glueing functions of B_+ and B_- is much simplified; remark that the projection of the eight knot considered in E.5.19 does have alternate crossings, while the projection of the trefoil knot considered in E.5.20 does not.

Proposition E.5.22. Assume G has alternate crossings; then:
(1) the graphs representing the realizations of S^2 as a polyhedron associated to the glueing functions of B_+ and B_- *before* removing the bigons are canonically identified with G itself;
(2) the graphs representing the realizations of S^2 as a polyhedron associated to the glueing functions of B_+ and B_- *after* removing the bigons are canonically identified with the graph $\tilde{G}$ obtained from G by reducing double joins between vertices to single joins.

Proof. (1) To each edge in G separating two regions D_1 and D_2 there corresponds a segment of type g separating the corresponding 2-cells in $S^3/_L$, and conversely, according to the scheme given in Fig. E.35.
(2) is readily deduced from (1).   □

The above results imply the following easily applicable criterion for condition III.

Corollary E.5.23. Condition III is satisfied if and only if:
(i) G has alternate crossings;
(ii) in $S^2 \setminus \tilde{G}$ no region is self-adjacent and any two regions have minimal adjacency.

Let us summarize what our construction has lead to for a link L given by a projection satisfying I, II and III (see [Pe] for a formal proof):

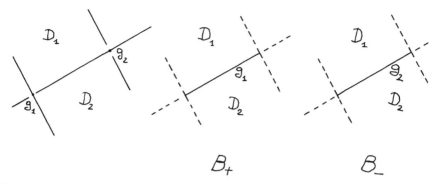

Fig. E.35. What an edge of g with alternate endpoints gives in B_+ and B_- respectively

(1) a finite number of polygons $D_1, D_2, \ldots$ having at least three edges;

(2) two realizations (we will denote by S_+^2 and S_-^2) of S^2 as a polyhedron involving the D_i's (each appearing exactly once) in such a way that:

(3) in both realizations no D_i is self-adjacent and any two D_i's have minimal adjacency;

(4) S^3/L is obtained by considering two copies (B_+ and B_-) of the closed 3-ball bounded by S_+^2 and S_-^2 respectively and glueing each $D_i \subset S_+^2$ to the corresponding $D_i \subset S_-^2$;

(5) $S^3 \setminus L$ is obtained by removing the vertices of the D_i's in such a realization of S^3/L.

We are now ready to conclude our construction: we ony need to extend the polyhedra representations of S_+^2 and S_-^2 to triangulations of B_+ and B_- in such a way that:

(i) no vertex is added;

(ii) the glueings on S_+^2 and S_-^2 are respected.

This is done very easily. First of all one adds inside each D_i having more than three vertices a suitable number of edges in order to triangulate it, and then he copies these edges in S_+^2 and S_-^2 in the obvious way. Property (3) implies that S_+^2 and S_-^2 are now triangulated. The next (and final) step is to extend these triangulations to triangulations of B_+ and B_- by adding edges and faces in the interior. This is done by picking a vertex $v_+ \in S_+^2$ and realizing S_+^2 (a similar construction working for B_-, of course) in such a way that all the triangles not containig v_+ lie on a plane π in $\mathbb{R}^3$, $v_+ \notin \pi$ and the triangles containing v_+ are cones with vertex v_+ and basis in π.

Then one only needs to add some other cones with vertex v_+ based at vertices and edges in π (in Fig. E.36 the two edges to be added are dotted; the seven triangles to be added are not drawn).

This concludes our construction.

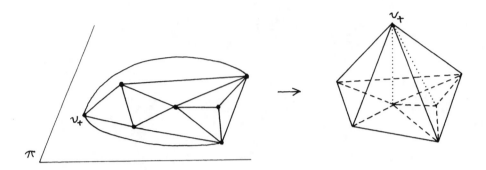

Fig. E.36. Realization of S_+^2 with all the vertices but one in a same plane

Remark E.5.24. We want to emphasize the fact that given a general projection of a link L in S^3, the verification of conditions I, II and III can be performed algorithmically (and very easily). Moreover many links satisfy these conditions, so that a large number of explicit examples of elements of $\mathcal{T}_3$ can be produced.

Remark E.5.25. We have already analysed the case of the eight knot in E.5.19; we can say in heuristic terms that the complement of the eight knot is the simplest element of $\mathcal{T}_3$ we can obtain with the above construction, as at least two tetrahedra are always involved; however, it is not the only element of $\mathcal{T}_3$ obtained by glueing the faces of two tetrahedra: there exists another one (see [Ma-Fo]).

Example E.5.26. Figures E.37 to E.39 describe three famous links to which our construction applies (providing a realization of their complements as elements of $\mathcal{T}_3$).

Fig. E.37. Whitehead's link   Fig. E.38. Borromeo's rings

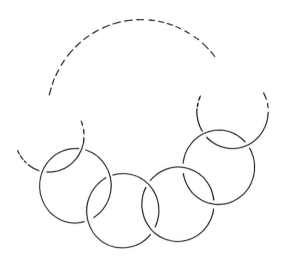

Fig. E.39. The "cyclic" link with $k \geq 3$ components

E.6 Proof of Thurston's Hyperbolic Surgery Theorem

In this section we prove the hyperbolic surgery Theorem E.5.1, following this scheme:

– given an element M of $\mathcal{T}_3$ (obtained by glueing n tetrahedra and removing the k vertices) we prove that $\mathcal{H}(M)$ can be identified with an algebraic subset of the product of n copies of the upper half-plane $\Pi^{2,+}$; we find the explicit equations for $z \in \left(\Pi^{2,+}\right)^n$ to represent an element of $\mathcal{H}(M)$ and to represent a complete hyperbolic structure; we obtain in particular that $\mathcal{H}(M)$ is an algebraic subset of $\left(\Pi^{2,+}\right)^n$;

– using a combinatorial argument we prove that the dimension of $\mathcal{H}(M)$ is at least k everywhere;

– in case $M \in \mathcal{F}_3$ we consider its realization as an element of $\mathcal{T}_3$ with the property that the complete structure appears as a point z^0 of $\mathcal{H}(M)$, we prove that $\mathcal{H}(M)$ has dimension precisely k around z^0 and we study a neighborhood of z^0 in $\mathcal{H}(M)$ (the space of deformations of the structure);

– in the same hypothesis that $M \in \mathcal{F}_3$ we describe the completion of a structure $z \in \mathcal{H}(M)$ close to z^0; in particular we obtain that for some special values of z such a completion is an element of $\mathcal{F}_3$ obtained from M by hyperbolic surgery and deduce from the previous step the conclusion of the proof of E.5.1.

We want to emphasize that the first two steps do not require the assumption *a priori* that M can be endowed with a finite-volume complete hyperbolic structure. Since the resulting equations are easily calculated starting from the combinatorial data of the realization in $\mathcal{T}_3$ the first step provides a powerful method for producing examples of elements of $\mathcal{F}_3$: the problem reduces to looking for the solution of a system of algebraic equations. It is worth remarking that the existence of a solution for these equations is a sufficient condition for an element of $\mathcal{T}_3$ to belong to $\mathcal{F}_3$, but not a necessary one: we saw in E.5-ii that given an element M of $\mathcal{F}_3$ and an arbitrary realization of M in $\mathcal{T}_3$, it is not possible in general to obtain the complete structure as an element of $\mathcal{H}(M)$ (see E.7 for a further discussion of this point).

E.6-i Algebraic Equations of $\mathcal{H}(M)$
(Hyperbolic Structures Supported by $M \in \mathcal{T}_3$)

Let us fix an element $M \in \mathcal{T}_3$ (together with its realization as glued tetrahedra with the vertices removed); we recall that we denoted by $\mathcal{H}(M)$ the set of the hyperbolic structures on M extending a hyperbolic structure on M^* (M deprived of the 1-skeleton) obtained by a realization of the tetrahedra as ideal tetrahedra in $\mathbb{H}^3$. We are going to show that $\mathcal{H}(M)$ can be explicitly described and prove that the question of whether $\mathcal{H}(M)$ contains a complete structure or not can be easily answered. For this reason we consider the parametrization of ideal tetrahedra in $\mathbb{H}^3$ we have already used in Chapt. C. We recall that the isometry class of an ideal tetrahedron in $\mathbb{H}^3$ is determined by the similarity class of a Euclidean triangle, the correspondence being obtained by realizing the tetrahedron in the half-space model with a vertex at infinity and considering the intersection of the tetrahedron with a suitably high horizontal plane (*i.e.* a horosphere centred at infinity). Since we need everything to be oriented we endow ideal tetrahedra with the orientation they inherit from $\mathbb{H}^3$ and consider the relation of existence of an orientation-preserving isometry; it is very easily checked that the quotient set is parametrized by the orientation-preserving similarity classes of Euclidean triangles; of course Euclidean triangles are endowed with the orientation they inherit from $\mathbb{R}^2$.

Now, consider a triangle in the Euclidean plane: we shall describe the way to associate to it a complex number with positive imaginary part (or, more precisely, three such complex numbers which are easily obtained from each other). As usual we shall denote the set of complex numbers with positive imaginary part by $\Pi^{2,+}$. Think of the Euclidean plane as the complex plane, fix a vertex v_0 of the triangle, denote by v_1 the vertex following v_0 with respect to the positive orientation of the boundary of the triangle and let ϕ be the only orientation-preserving similarity of the plane mapping v_0 to 0 and v_1 to 1; then the third vertex of the triangle is mapped by ϕ to a complex number $z \in \Pi^{2,+}$. It is easily checked that the other two choices of the starting vertex produce with the same method the numbers

$$\frac{1}{1-z} \qquad \text{and} \qquad 1 - \frac{1}{z};$$

conversely if $z \in \Pi^{2,+}$ we can associate to it the similarity class of the triangle having vertices 0,1 and z, and it is easily checked that z' produces the same oriented similarity class if and only if

$$z' = \frac{1}{1-z} \quad \text{or} \quad z' = 1 - \frac{1}{z}.$$

We shall call the number z associated to the triangle by a choice of a starting vertex v the <u>modulus</u> of the triangle with respect to v; remark that the inner angle at v of the triangle is the argument of the modulus of the triangle with respect to v (from now on we shall call the argument of a non-zero complex number x the only real number $\vartheta \in [0, 2\pi)$ such that $x = |x|e^{i\vartheta}$, and denote it by $\arg(x)$).

For $z \in \Pi^{2,+}$ we shall set

$$w_1(z) = z, \quad w_2(z) = \frac{1}{1-z}, \quad w_3(z) = 1 - \frac{1}{z}.$$

It is easily checked that for $i = 1, 2, 3$ two numbers $c_i, d_i \in \{-1, 0, 1\}$ are uniquely determined in such a way that

$$w_i(z) = \pm z^{c_i} \cdot (1-z)^{d_i}.$$

Remark that these numbers are uniquely determined as we want these relations to hold for all z's, while for $z = e^{i\pi/3}$ the $w_i(z)$'s coincide (which corresponds to the triangle being equilateral). If a vertex v_1 of the triangle is specified, if z is the modulus of the triangle with respect to v_1 and if v_2, v_3 are the other vertices chosen in such a way that the ordering v_1, v_2, v_3 induces the positive orientation on the triangle then the modulus of the triangle with respect to v_i is $w_i(z)$, and we shall look to these three expressions as different from each other even if they produce the same complex number.

If T is an ideal tetrahedron in $\mathbb{H}^3$ and an edge E of T is fixed we can associate to T a number $z \in \Pi^{2,+}$ (called the <u>modulus</u> of T with respect to E) in the following way: realize T in the half-space model in such a way that one of the endpoints of the preferred edge is at infinity, consider the Euclidean triangle being the intersection of T with a suitably high horizontal plane and let z be the modulus of the triangle with respect to the vertex lying on the preferred edge. The six choices of a preferred edge produce the numbers $w_i(z)$, $i = 1, 2, 3$, each being obtained twice; these numbers may coincide, but, once an edge is fixed, we shall regard them as different as we shall be interested not only in the number but also in its realization as $\pm z^{c_i} \cdot (1-z)^{d_i}$; the point is that when changing z no ambiguity must appear.

Now, consider again $M \in \mathcal{T}_3$; as in the preceding section we shall denote by $\tilde{Q}$ the space obtained as a union of tetrahedra with glued faces, so that M is the space obtained by removing from $\tilde{Q}$ the 0-cells. We shall introduce now several notations we shall need in the remainder of the section; we suggest the reader writes them down in a scheme in order to follow in a proper way

our arguments. Let $T_1, ..., T_n$ be the tetrahedra glued in $\tilde{Q}$ and fix on each
of them a preferred edge. A realization of $T_1, ..., T_n$ as ideal tetrahedra in $\mathbb{H}^3$
is determined by n complex numbers $z_1, ..., z_n$ in $\Pi^{2,+}$, where z_r represents
the modulus of the ideal tetrahedron corresponding to T_r with respect to
the fixed edge. Let such complex numbers be given. We recall that in $\tilde{Q}$ the
1-cells are n too (Proposition E.5.6); we shall denote them by $e_1, ..., e_n$ and
think of them both as subsets of $\tilde{Q}$ and of the ideal tetrahedra; remark in
particular that e_r is the edge of several different tetrahedra: we shall denote
these tetrahedra having e_r as an edge by $T(r)_j$ for $j = 1, ..., x_r$; remark as well
that the same tetrahedron may appear more than once between these $T(r)_j$'s,
as e_r may be obtained by glueing two or more edges of a certain tetrahedron.
The modulus of $T(r)_j$ with respect to e_r will be denoted by $m(r,j)$; remark
that if $\nu(r,j) \in \{1, ..., n\}$ is such that $T(r)_j = T_{\nu(r,j)}$ we have that

$$m(r,j) \in \{w_i(z_{\nu(r,j)})\}_{i=1,2,3};$$

moreover a well-determined $i(r,j) \in \{1, 2, 3\}$ is fixed in such a way that

$$m(r,j) = w_{i(r,j)}(z_{\nu(r,j)}).$$

We remark once again that such $i(r,j)$ is possibly not the only $i \in \{1, 2, 3\}$
such that

$$m(r,j) = w_i(z_{\nu(r,j)})$$

as a number, but $i(r,j)$ is well-determined in any case. Moreover we set for
$r = 1, ..., n$, $j = 1, ..., x_r$ and $i = 1, 2, 3$

$$t_i(r,j) = \begin{cases} 1 & \text{if } i = i(r,j) \\ 0 & \text{otherwise} \end{cases}$$

in such a way that

$$m(r,j) = \prod_{i=1}^{3} w_i(z_{\nu(r,j)})^{t_i(r,j)}.$$

We begin now the description (by means of the parameters $z_1, ..., z_n$) of
the conditions for a hyperbolic structure on M^* to extend to the 1-skeleton
(*i.e.* to define an element of $\mathcal{H}(M)$) and to be complete. For the first problem
(extension of the structure to the 1-skeleton) the answer is that the structure
extends if and only if for all r's the tetrahedra $T(r)_1, ..., T(r)_{x_r}$ glue properly
along e_r; if we realize e_r as a vertical line in the half-space model and the
tetrahedra having e_r as an edge in such a way that the glued faces containing
e_r coincide, such a condition is graphically represented in Fig. E.40 (where
we have assumed $x_r = 5$); as above, remark that these five tetrahedra are not
necessarily different from each other.

We recall that for $j = 1, ..., r$ we have denoted the modulus of $T(r)_j$ with
respect to e_r by $m(r,j)$. If we glue in the required way the horizontal triangles
one after the other as in Fig. E.41, we obtain that the algebraic conditions for
the glueings along e_r to produce a hyperbolic structure which can be extended
on e_r are the following:

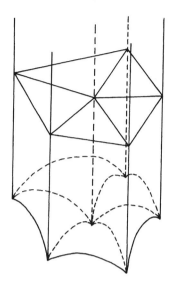

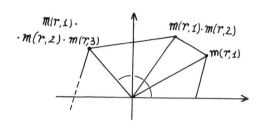

Fig. E.40. Geometric condition that the various tetrahedra fit properly around an edge

Fig. E.41. How to translate this geometric condition into equations involving the moduli

$$\sum_{j=1}^{x_r} \arg(m(r,j)) = 2\pi \qquad \prod_{j=1}^{x_r} m(r,j) = 1.$$

Let us remark that *a priori* the second formula does not imply the first one, it only implies that the sum of the arguments is a (positive) integer multiple of 2π. However we have the following:

Lemma E.6.1. If for $r = 1, ..., n$ we have

$$\prod_{j=1}^{x_r} m(r,j) = 1$$

then we also have

$$\sum_{j=1}^{x_r} \arg(m(r,j)) = 2\pi \quad \text{for } r = 1, ..., n.$$

Proof. Let $N(r) \in \mathbb{N}$ be such that

$$\sum_{j=1}^{x_r} \arg(m(r,j)) = 2\pi \cdot N(r).$$

Then we have

$$2 \sum_{r=1}^{n} \sum_{j=1}^{x_r} \arg(m(r,j)) = 4\pi \sum_{r=1}^{n} N(r).$$

What does the sum on the left hand side mean? The angle relative to each edge of each tetrahedron appears precisely twice in such sum; then the sum is the sum over all tetrahedra of the inner angles of the four triangles shown in Fig. E.42; then each tetrahedron contributes with 4π, so that the total sum must be $4\pi \cdot n$. Each of the $N(r)$'s being a non-negative natural number it follows that $N(r) = 1$ for all r's and the proof is over.

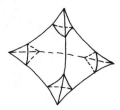

Fig. E.42. The sum of the angles of the four triangles is 4π

□

The above discussion and Lemma E.6.1 are summarized by the following:

Lemma E.6.2. The hyperbolic stucture on M^* determined by the parameters $z_1, ..., z_n$ extends to the 1-skeleton, *i.e.* to an element of $\mathcal{H}(M)$, if and only if

$$\prod_{j=1}^{x_r} m(r,j) = 1 \quad \text{for } r = 1, ..., n.$$

We re-write now such equations in a more convenient way. We have, using the notations introduced above:

$$\prod_{j=1}^{x_r} m(r,j) = \prod_{i=1}^{3} \prod_{j=1}^{x_r} w_i(z_{\nu(r,j)})^{t_i(r,j)} =$$

$$= \prod_{i=1}^{3} \prod_{j=1}^{x_r} \left(\pm z_{\nu(r,j)}{}^{c_i} \cdot (1 - z_{\nu(r,j)})^{d_i} \right)^{t_i(r,j)}$$

so that the equations can be re-written according to the next result.

Proposition E.6.3. For $r, \nu \in \{1, ..., n\}$ there exist integer numbers $\theta_1(r, \nu)$, $\theta_2(r, \nu)$ and $\varepsilon(r) = \pm 1$ (depending only on the combinatorial data of the realization of M and of the initial choice of a preferred edge on each tetrahedron) such that the hyperbolic structure on M^* associated to the parameters $z_1, ..., z_n$ extends to the whole of M if and only if

$$\prod_{\nu=1}^{n} z_{\nu}^{\theta_1(r,\nu)} \cdot (1 - z_{\nu})^{\theta_2(r,\nu)} = \varepsilon(r) \quad \text{for } r = 1, ..., n.$$

By the above discussion such numbers $\theta_l(r, \nu)$ can be easily calculated, and we shall need in the following the explicit way they were obtained.

This way of writing the equations immediately implies the following:

Corollary E.6.4. The subset of $\left(\Pi^{2,+}\right)^n$ corresponding to the parameters giving an element of $\mathcal{H}(M)$ is an algebraic subset.

From now on we shall refer to $\mathcal{H}(M)$ as a subset of $\left(\Pi^{2,+}\right)^n$, the identification being given as described above.

We turn now to the determination of the equations for $z \in \mathcal{H}(M)$ representing the condition that the hyperbolic structure associated to z is complete. The notation and the machinery we are going to develop for this purpose will be used again several times in this section.

We denote by $L_1, ..., L_k$ the links of the removed 0-cells in $\tilde{Q}$; by hypothesis these links are homeomorphic to tori. We shall think of these L_j's as fixed up to isotopy and use suitable concrete representatives of them as subsets of M. We remark first that the decomposition of M in tetrahedra induces a "triangulation" on each of these links; we use inverted commas as it may happen that two triangles have two common vertices without having in common the edge joining such vertices, and moreover an edge can have coincident endpoints, and these phenomena are forbidden according to the usual definition of triangulation: but from now on we shall use this term triangulation in this weaker sense in order to include our case. We recall that the group of orientation-preserving similarities of the real plane coincides, under the canonical identification $\mathbb{R}^2 = \mathbb{C}$, with the group $\mathrm{Aff}(\mathbb{C})$ of affine automorphisms of the complex line $\mathbb{C}$; we call similarity structure on a torus a $(\mathbb{C}, \mathrm{Aff}(\mathbb{C}))$-structure, according to the general definition we gave in Sect. B.1.

We start by remarking that for all $z \in \mathcal{H}(M)$ on all the L_j's a similarity structure depending on z is defined. In fact these links can be viewed as the union of Euclidean triangles and, as we saw above, the condition that $z \in \mathcal{H}(M)$ implies that these triangles are nicely glued to each other along the edges, so that the similarity structure of the triangles globalizes. Remark that by the construction, given $z \in \mathcal{H}(M)$ the triangulation of L_j associated to the decomposition of M in tetrahedra is performed by triangles which are straight with respect to the similarity structure on L_j induced by z. The following result gives a geometric characterization of the elements of $\mathcal{H}(M)$ representing complete structures.

Proposition E.6.5. The hyperbolic structure associated to $z \in \mathcal{H}(M)$ is a complete one if and only if the similarity structures induced on the tori $L_1, ..., L_k$ are actually Euclidean.

Proof. Assume the structure is complete; for $j = 1, ..., k$ we can deprive L_j of two simple simplicial loops in order to make it simply connected and lift it to the universal covering $\mathbb{H}^3$ to a horizontal set, whose closure is a triangulated polygon. The identifications on the edges of such a polygon giving L_j are induced by isometries belonging to a discrete subgroup of $\mathcal{I}^+(\mathbb{H}^3)$ and keeping ∞ fixed, and then it follows from what we saw in Chapt. D that they act horizontally as Euclidean isometries. Then the similarity structure of L_j is actually a Euclidean one.

Assume the structure of L_j is Euclidean; we are going to prove that the subset of M obtained by removing the vertex from the conic neighborhood in $\tilde{Q}$ of the vertex based on L_j is complete: of course this is enough as M is covered by k such sets and a compact set. Let us remark that it follows from the above definition of the similarity structure on L_j that the set we are interested in is the quotient of a set of the form $P \times [t_0, \infty) \subset \mathbb{H}^3$, where P is a compact polyhedron in $\mathbb{R}^2$, under certain identifications on the vertical faces induced by similarities operating horizontally. The similarity structure of L_j being Euclidean these similarities are actually horizontal isometries, so that they are global isometries of the hyperbolic space, and it follows that our set is the quotient of $\mathbb{R}^2 \times [t_0, \infty)$ under the action of a dicrete subgroup of $\mathcal{I}^+(\mathbb{H}^3)$ whose elements keep ∞ fixed, and then it is complete. $\qquad \square$

We are now going to translate the equivalent conditions of this lemma into equations. Let us first recall that a similarity structure on L_j given by $z \in \mathcal{H}(M)$ induces a conjugacy class of homomorphism $\Pi_1(L_j) \to \mathrm{Aff}(\mathbb{C})$ called the holonomy of the structure. We have the following:

Lemma E.6.6. Consider a torus endowed with a similarity structure with holonomy ρ. The following conditions are pairwise equivalent:

(a) the similarity structure is Euclidean;

(b) ρ is injective and its image consists of translations;

(c) the image of ρ consists of translations;

(d) the image of ρ contains a translation different from the identity.

Proof. (a) $\Rightarrow$ (b) is well-known and (b) $\Rightarrow$ (c) is obvious. We sketch the proof of (c) $\Rightarrow$ (a). We cut the torus open along a longitude-meridian pair, lift the set thus obtained to the universal covering of the torus and map such a lift to $\mathbb{C}$ by the developing mapping of the structure to which ρ is associated. We obtain a "quadrilateral" whose interior carries the similarity structure of the torus and whose opposite edges are paired by the images under ρ of the longitude and the meridian. We use inverted commas as the edges of the quadrilateral are not in general straight segments; and moreover *a priori* the quadrilateral might degenerate to an annulus, *i.e.* have a pair of coincident opposite edges, in case the longitude or the meridian have trivial holonomy (remark that only one of them can be the identity, as otherwise we would have embedded the torus in the plane, which is impossible). We can actually check that this case of degeneration does not occur: in fact we are assuming (c), *i.e.* that the image of ρ consists of translations, and in case of degeneration to an annulus we would have the boundary components of the annulus paired by a translation, and this cannot be the case. Now we have that the torus is obtained from the quadrilateral by glueing the opposite edges according to the prescribed pairings (non-trivial translations), and hence it is endowed with a Euclidean structure.

(b) $\Rightarrow$ (d) is obvious and (d) $\Rightarrow$ (c) follows from the fact that an element of $\mathrm{Aff}(\mathbb{C})$ commuting with a non-trivial translation is itself a translation. $\square$

In the setting of the above lemma, let us remark that there are actually cases when ρ maps a non-trivial loop to the identity, so that the quadrilateral of the proof degenerates to an annulus; as an example, consider the similarity structure on the torus obtained from the annulus $\{w \in \mathbb{C} : 1 \leq |w| \leq 2\}$ by glueing the boundary components via the multiplication by 2. This fact implies that condition (d) cannot be replaced by the requirement that the image under ρ of a non-trivial loop is a translation, as this translation might be trivial.

As we stated above, the group of orientation-preserving similarities of $\mathbb{R}^2$ is canonically identified with $\mathrm{Aff}(\mathbb{C})$ and hence with the semi-direct product of groups $\mathbb{C} \coprod \mathbb{C}^*$, a pair (a, b) with $b \neq 0$ operating by

$$\mathbb{C} \ni x \mapsto a + b \cdot x;$$

we shall say b is the <u>dilation component</u> of the similarity (or complex-affine automorphism) associated to (a, b). Since for $(a, b) \in \mathbb{C} \coprod \mathbb{C}^*$ we have $(a, b)^{-1} = (-a/b, 1/b)$, so that

$$(a, b)^{-1} \cdot (a_0, b_0) \cdot (a, b) = (*, b_0),$$

the dilation component of the holonomy of the similarity structure induced by $z \in \mathcal{H}(M)$ on L_j is a well-defined homomorphism (not only a conjugation class)

$$\delta(z)_j : \Pi_1(L_j) \to \mathbb{C}^*.$$

By simplicity of notation we define a mapping (not a homomorphism!) $\delta(z)$ on the disjoint union of $\Pi_1(L_1), ..., \Pi_1(L_k)$ with values in $\mathbb{C}^*$ by

$$\delta(z)\big|_{\Pi_1(L_j)} = \delta(z)_j.$$

Since a similarity is a translation if and only if its dilation component is 1, Lemma E.6.6 (equivalence of (a) and (c)) and E.6.5 imply:

Proposition E.6.7. An element z of $\mathcal{H}(M)$ represents a complete structure if and only if $\delta(z) \equiv 1$.

We fix now for the remainder of the section generators l_j and m_j of the fundamental group of L_j, chosen in such a way that they can be represented by simple oriented loops intersecting at one point only (corresponding to the canonical longitude and meridian of the abstract torus). Moreover, since a triangulation is defined (up to isotopy) on L_j, we shall choose fixed representatives of l_j and m_j being simple simplicial oriented loops with respect to such a triangulation, and denote them by l_j and m_j too. Recall that a choice of $z \in \mathcal{H}(M)$ induces on L_j a similarity structure with respect to which the triangulation consists of straight triangles.

We shall need to compute the value of $\delta(z)$ on l_j and m_j, and this is achieved by the following result. We first introduce some notation: given $z \in \mathcal{H}(M)$, if Δ is a triangle of the triangulation of L_j and v is a vertex of Δ then the modulus of Δ with respect to v is defined as the modulus of the only tetrahedron containing Δ with respect to the edge which contains v; such a number will be denoted by $y(\Delta, v)$. Moreover if γ is a simple oriented simplicial loop on L_j and v is a vertex of γ, the set of all the triangles touching γ in v and lying "on the right" of γ is well-defined and will be denoted by $R(\gamma, v)$; finally, we shall denote by γ_0 the set of the vertices of γ and by γ itself the element of $\Pi_1(L_j)$ represented by the loop γ.

Lemma E.6.8. Let all the above notations be fixed; then

(1)
$$\delta(z)(\gamma) = (-1)^{\#\gamma_0} \cdot \prod_{v \in \gamma_0} \prod_{\Delta \in R(\gamma, v)} y(\Delta, v);$$

(2) if $\delta(z)(\gamma) = 1$ we have that the holonomy of γ is non-trivial if and only if

$$\sum_{v \in \gamma_0} \sum_{\Delta \in R(\gamma, v)} \arg\big(y(\Delta, v)\big) = \#\gamma_0 \cdot \pi;$$

moreover if $\#\gamma_0 \leq 2$ this relation is automatic.

Proof. (1) Let us remark first that given two oriented segments of the plane there exists one and only one orientation-preserving similarity mapping one onto the other and preserving the orientation of the segments. As in Fig. E.43, let us take the two segments in such a way that the second vertex of the first one coincides with the first vertex of the second one, and moreover the angle at the common vertex lying on the right of the segments is less than π. We denote by w the modulus of the triangle having these segments as edges with respect to the common vertex; then it is very easily checked that the dilation component of the similarity which maps the first segment onto the second one is given by $-w$. As a generalization of this fact it is readily proved that in the situation of Fig. E.44 the dilation component of the complex-affine automorphism of the plane mapping s onto s' is given by $-w_1 \cdot \ldots \cdot w_p$.

Let us fix a starting vertex on γ and let us remark that by definition $\delta(z)(\gamma)$ is the dilation component of the holonomy of the homotopy class of γ. This is easily represented in the following way: consider a lifting of γ (as a curve) to the universal covering of L_j and map it to the complex plane via a developing mapping; the resulting curve consists of a finite number of straight segments $s_1, \ldots, s_m$, where $m = \#\gamma_0$ and the segments are ordered and oriented according to the way they appear on γ starting from the fixed point. If we repeat the procedure starting at the second endpoint of s_m we can denote the first segment we obtain by s_1', and then the complex-affine automorphism of the plane representing the holonomy of γ is characterized by the fact that it maps s_1 onto s_1'. Its dilation component is the product of

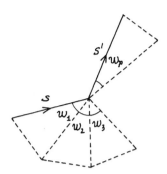

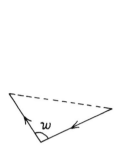

Fig. E.43. Dilation component of the similarity which maps a segment onto another. Case when the segments enclose a triangle

Fig. E.44. Dilation component of the similarity which maps a segment onto another. Case when the segments enclose a finite number of triangles

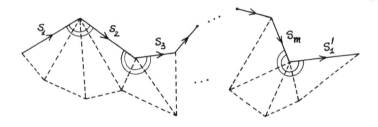

Fig. E.45. Computation of the holonomy of the meridian with respect to the similarity structure

the dilations components of the complex-affine automorphisms of the plane mapping s_1 onto s_2, s_2 onto s_3, ..., s_m onto s_1'; then the conclusion follows by the above remark, as suggested in Fig. E.45.

(2) We have that the holonomy of γ is trivial if and only if in the situation of Fig. E.45 one has that $s_1' = s_1$, and it is not difficult to see that this is the case if and only if the sum of the arguments in question is not $\#\gamma_0 \cdot \pi$ (namely: this sum is $(\#\gamma_0 + 2)\pi$ if the loop $s_1 \cup ... \cup s_m$ wraps counter-clockwise, $(\#\gamma_0 - 2)\pi$ otherwise. The last remark can be easily seen by calculations; it corresponds to the geometric fact that at least three straight segments are necessary to construct a loop.   □

Remark E.6.9. With the notation of the above lemma we have in general that if $\delta(z)(\gamma) = 1$ then the sum of arguments of part (2) is $(\#\gamma_0 + 2p)\pi$ for some $p \in \{-1, 0, 1\}$.

Remark E.6.10. According to Lemma E.6.6 (equivalence of (a) and (d)) the condition on a similarity structure to be Euclidean is equivalent to the existence of a loop whose holonomy is a non-trivial translation. By part (2) of the lemma triviality of the holonomy is determined by the same objects determining the dilation components. This implies that if one knows the objects $y(\Delta, \gamma)$ for a non-trivial loop γ on each of the k tori, then he can represent the completeness condition on $z \in \mathcal{H}(M)$ by k rational equations and k additional equations involving the arguments. This is useful when explicitly constructing hyperbolic structures with this method.

We are not going to use part (2) of the above lemma soon; we shall need it later in E.6-iii when specializing our study to a neighborhood of a complete structure in $\mathcal{H}(M)$.

Lemma E.6.8 (part (1)) immediately implies the following:

Proposition E.6.11. There exist integer constants $\lambda_l(j, \nu)$ and $\mu_l(j, \nu)$ with $l = 1, 2$, $j = 1, ..., k$ and $\nu = 1, ..., n$ (depending only on the combinatorial data of the realization of M in $\mathcal{T}_3$ and on the initial choice of a preferred edge on each tetrahedron) such that

$$\delta(z)(l_j) = \pm \prod_{\nu=1}^{n} z_\nu^{\lambda_1(j,\nu)} \cdot (1 - z_\nu)^{\lambda_2(j,\nu)}$$

$$\delta(z)(m_j) = \pm \prod_{\nu=1}^{n} z_r^{\mu_1(j,\nu)} \cdot (1 - z_\nu)^{\mu_2(j,\nu)}.$$

As above for the numbers $\theta_l(r, \nu)$ we shall make use in the sequel of the explicit way to obtain these $\lambda_l(j, \nu)$ and $\mu_l(j, \nu)$, not only their existence.

The above proposition together with Lemma E.6.6 (equivalence of (a) and (c)) implies the following:

Proposition E.6.12. An element $z \in \mathcal{H}(M)$ represents a complete structure if and only if it satisfies the following system of $2k$ rational equations:

$$\begin{cases} \delta(z)(l_j) = 1 & \forall j = 1, ..., k \\ \delta(z)(m_j) = 1 & \forall j = 1, ..., k. \end{cases}$$

Moreover the coefficients of these algebraic equations are determined by the combinatorial data of the realization of M in $\mathcal{T}_3$.

E.6-ii Dimension of $\mathcal{H}(M)$: General Case

In view of E.6.4, the problem naturally arises of calculating the dimension of $\mathcal{H}(M)$ as an algebraic subset of $\left(\Pi^{2,+}\right)^n$; since it is defined by n equations it may have dimension 0, but this is actually not the case. We recall that we denoted by k the number of boundary components of $\overline{M}$, i.e. the number of 0-cells of $\hat{Q}$; our goal is to prove now that $\mathcal{H}(M)$ has dimension at least k in the neighborhood of all its points; in case $\mathcal{H}(M)$ contains a complete structure we shall see later that this dimension is precisely k. We follow the line of proof presented by W. D. Neumann and D. Zagier in [Ne-Za]. Their argument is quite fascinating as it uses only elementary combinatorial techniques but it turns out to be quite complicated; we shall present fully detailed proofs.

Remark that we are still dealing with an element of $\mathcal{T}_3$ on which no further assumption is made, so that there is no reason a priori for $\mathcal{H}(M)$ to be non-empty. Of course our assertion that $\mathcal{H}(M)$ has dimension at least k in the neighborhood of its points makes no sense if $\mathcal{H}(M)$ is empty, so let us fix a point $z^0 = (z_1^0, ..., z_n^0)$ in $\mathcal{H}(M)$.

For all ν's we can find in the neighborhood of z_ν^0 holomorphic determinations of the functions

$$z_\nu \mapsto \log(z_\nu) \quad \text{and} \quad z_\nu \mapsto \log(1 - z_\nu)$$

so that in the neighborhood of z^0 the equations giving $\mathcal{H}(M)$ we obtained in E.6.3 can be re-written in the more convenient form

$$\sum_{\nu=1}^{n} \left(\theta_1(r, \nu) \cdot \log(z_\nu) + \theta_2(r, \nu) \cdot \log(1 - z_\nu) \right) = c(r) \quad \text{for } r = 1, ..., n$$

(where the $c(r)$'s are suitable constants and the $\theta_l(r, \nu)$'s are those obtained in E.6.3).

The Jacobian matrix of this system of equations in the point z is given by the product of the two matrices Θ and $Z(z)$, where $\Theta = (\Theta_1 \, \Theta_2)$ with

$$\Theta_l = \begin{pmatrix} \theta_l(1,1) & \cdots & \theta_l(1,n) \\ \vdots & & \vdots \\ \theta_l(n,1) & \cdots & \theta_l(n,n) \end{pmatrix}$$

and

$$Z(z) = \begin{pmatrix} 1/z_1 & \cdots & 0 \\ \vdots & \ddots & \vdots \\ 0 & \cdots & 1/z_n \\ -1/(1-z_1) & \cdots & 0 \\ \vdots & \ddots & \vdots \\ 0 & \cdots & -1/(1-z_n) \end{pmatrix}.$$

We are going to prove that Θ has rank at most $n - k$, which implies that the Jacobian matrix of the system has rank at most $n - k$ too and hence in the neighborhood of z^0 the algebraic set $\mathcal{H}(M)$ has dimension at least k.

In order to prove this we need of course to "involve" in some way the ends of M, i.e. the k removed 0-cells of $\tilde{Q}$. For this reason we consider the constants $\lambda_l(j, \nu)$ and $\mu_l(j, \nu)$ we determined in E.6.11 and we set for $l = 1, 2$

$$\Lambda_l = \begin{pmatrix} \lambda_l(1,1) & \cdots & \lambda_l(1,n) \\ \vdots & \ddots & \vdots \\ \lambda_l(k,1) & \cdots & \lambda_l(k,n) \end{pmatrix} \qquad \mathcal{M}_l = \begin{pmatrix} \mu_l(1,1) & \cdots & \mu_l(1,n) \\ \vdots & \ddots & \vdots \\ \mu_l(k,1) & \cdots & \mu_l(k,n) \end{pmatrix}.$$

Moreover we set $\Lambda = (\Lambda_1 \ \Lambda_2)$ and $\mathcal{M} = (\mathcal{M}_1 \ \mathcal{M}_2)$. Let us remark that $\Lambda_1, \Lambda_2,$ $\mathcal{M}_1$ and $\mathcal{M}_2$ are $k \times n$ matrices while Θ_1 and Θ_2 are $n \times n$ matrices, so that Λ and $\mathcal{M}$ are $k \times 2n$ and Θ is $n \times 2n$, and hence the definition

$$U = \begin{pmatrix} \Lambda \\ \mathcal{M} \\ \Theta \end{pmatrix}$$

makes sense and represents a $(2k+n) \times 2n$ matrix. We consider now for natural h the matrix

$$J_{2h} = \begin{pmatrix} 0 & -I_h \\ I_h & 0 \end{pmatrix}$$

where I_h denotes the $h \times h$ identity matrix.

The core of the argument about the dimension of $\mathcal{H}(M)$ resides in the following lemma, whose proof is very long and technical: we suggest the reader at a first reading skips this proof and concentrates on the way this lemma allows one to prove that Θ actually has rank at most $n - k$ (Proposition E.6.14).

Lemma E.6.13.
$$U \cdot J_{2n} \cdot {}^tU = 2 \begin{pmatrix} J_{2k} & 0 \\ 0 & 0 \end{pmatrix}.$$

Proof. The matrix U depends only on the decomposition of M and on the initial choice of a preferred edge on each tetrahedron, but for the proof it is convenient to fix an arbitrary $z \in \left(\Pi^{2,+} \right)^n$ and consider the geometric meaning with respect to z of the constants appearing in U.

According to the explicit definition of U the formulae we have to prove are the following:

(1)
$$\begin{cases} \sum_{\nu=1}^{n} \left(\lambda_1(j,\nu) \cdot \lambda_2(t,\nu) - \lambda_2(j,\nu) \cdot \lambda_1(t,\nu) \right) = 0 & j,t = 1,...,k \\ \sum_{\nu=1}^{n} \left(\mu_1(j,\nu) \cdot \mu_2(t,\nu) - \mu_2(j,\nu) \cdot \mu_1(t,\nu) \right) = 0 & j,t = 1,...,k \\ \sum_{\nu=1}^{n} \left(\lambda_1(j,\nu) \cdot \mu_2(t,\nu) - \lambda_2(j,\nu) \cdot \mu_1(t,\nu) \right) = 2\delta_{jt} & t = 1,...,k \end{cases}$$

(2)
$$\begin{cases} \sum_{\nu=1}^{n} \left(\lambda_1(j,\nu) \cdot \theta_2(r,\nu) - \lambda_2(j,\nu) \cdot \theta_1(r,\nu) \right) = 0 & j = 1,...,k, r = 1,...,n \\ \sum_{\nu=1}^{n} \left(\mu_1(j,\nu) \cdot \theta_2(r,\nu) - \mu_2(j,\nu) \cdot \theta_1(r,\nu) \right) = 0 & j = 1,...,k, r = 1,...,n \end{cases}$$

(3)
$$\sum_{\nu=1}^{n} \left(\theta_1(r,\nu) \cdot \theta_2(s,\nu) - \theta_2(r,\nu) \cdot \theta_1(s,\nu) \right) = 0 \quad r,s = 1,...,n$$

We want to re-write formulae (1) and (2) in a more convenient way. Let us first recall that we chose simple simplicial oriented loops l_j and m_j on L_j

corresponding to the usual longitude and meridian and that the constants $\lambda_l(j,\nu)$ and $\mu_l(j,\nu)$ were obtained from these loops as deduced by Lemma E.6.8 and Proposition E.6.11; moreover the constants $\theta_l(r,\nu)$ were determined in Lemma E.6.2 and Proposition E.6.3.

We recall now a general definition: if N is an oriented triangulated surface and α and β are simple simplicial oriented loops on N we define their algebraic intersection number $I(\alpha,\beta)$ as the sum over all edges of α being not edges of β too, but touching β in a vertex, of a contribution ± 1 according to the rules described in Fig. E.46.

Fig. E.46. Definition of the algebraic intersection of two curves on a surface. The edges of α in the situation of the left hand side give a contribution $+1$, those in the situation of the right hand side give -1. Remark that the two situations can occur simultaneously, so that the corresponding edge of α gives a contribution 0. In the figure we mean that the portion of surface represented has the canonical orientation of the plane

This definition of $I(\alpha,\beta)$ is not the only possible one, but we shall need it in this form. It is well-known that $I(\alpha,\beta)$ depends actually only on the homotopy classes of α and β and moreover it immediately follows from the definition that $I(\alpha,\alpha) = 0$ and $I(\beta,\alpha) = -I(\alpha,\beta)$.

Let us apply this concept to our loops on the tori; we have

$$I(l_j, m_j) = -I(m_j, l_j) = 1$$

and moreover it is natural to set

$$I(l_j, l_t) = I(l_j, m_t) = I(m_j, m_t) = 0$$

for $j \neq t$ (in fact the loops lie on different tori and hence they do not intersect). Then we can re-write (in fact, generalize) formulae (1) and (2) in the following way: let α and β be simple simplicial oriented loops on some of the tori $L_1, ..., L_k$ (not necessarily on the same one); then by E.6.8 for $\nu = 1, ..., n$ there exist integer numbers $a_1(\nu)$, $a_2(\nu)$, $b_1(\nu)$ and $b_2(\nu)$ such that

$$\delta(z)(\alpha) = \pm \prod_{\nu=1}^{n} z_\nu^{a_1(\nu)} \cdot (1 - z_\nu)^{a_2(\nu)}$$

$$\delta(z)(\beta) = \pm \prod_{\nu=1}^{n} z_\nu^{b_1(\nu)} \cdot (1 - z_\nu)^{b_2(\nu)}$$

(where we have denoted by α and β themselves the homotopy classes represented by these loops on the torus they lie on. Our argument will make use of the explicit method to obtain such constants, as suggested by E.6.8). In order to prove (1) and (2) it is then enough to establish the following formulae:

$$(1)' \qquad \sum_{\nu=1}^{n} \big(a_1(\nu) \cdot b_2(\nu) - a_2(\nu) \cdot b_1(\nu) \big) = 2 \cdot I(\alpha, \beta)$$

$$(2)' \qquad \sum_{\nu=1}^{n} \big(a_1(\nu) \cdot \theta_2(r, \nu) - a_2(\nu) \cdot \theta_2(r, \nu) \big) = 0 \quad \text{for } r = 1, ..., n$$

We develop now some technical preliminaries we are going to use for the proof. Let us consider in $\mathbb{Z}^2$ the bi-linear skew-symmetric form defined by

$$(\phi_1, \phi_2) \wedge (\psi_1, \psi_2) = \phi_1 \cdot \psi_2 - \phi_2 \cdot \psi_1.$$

Let us consider the three pairs of integers

$$\eta(1) = (1, 0) \quad \eta(2) = (0, -1) \quad \eta(3) = (-1, 1);$$

it is easily checked that

$$\eta(i) \wedge \eta(i+1) = 1$$

for all i's, where the indices are meant modulo 3 and represented in $\{1, 2, 3\}$. These three pairs are chosen in the following way: if we consider a triangle in the plane represented by $z \in \Pi^{2,+}$ then its moduli relative to the three vertices ordered in a positive way are respectively

$$\pm\tilde{\eta}(i)(z), \ \pm\tilde{\eta}(i+1)(z), \ \pm\tilde{\eta}(i+2)(z)$$

for a suitable choice of i, where the same convention holds on the indices and we have set

$$\tilde{\eta}(i)(z) = z^{\eta(i)_1} \cdot (1-z)^{\eta(i)_2}.$$

Now, let E and F be edges of one of the tetrahedra T_ν (an ideal tetrahedron in $\mathbb{H}^3$ represented by $z_\nu \in \Pi^{2,+}$). Then the moduli of T_ν with respect to E and F are respectively given by $\tilde{\eta}(i)(z_\nu)$ and $\tilde{\eta}(j)(z_\nu)$ for suitable choices of i and j. Let us remark that by construction the left hand sides of formulae (3), (2)' and (1)' are the sum of a certain number of contributions of the type $\eta(i) \wedge \eta(j)$, where $\eta(i)$ and $\eta(j)$ are obtained in this way. We want to express this number in a more convenient way. Let us remark first that if E and F are opposite edges we have $\eta(i) = \eta(j)$ and hence $\eta(i) \wedge \eta(j) = 0$. Conversely, if E and F are adjacent (which means that they have a common endpoint) we introduce a number $\sigma(\nu, E, F)$ in the following way: denote by p_0 the common endpoint, by p_1 the other endpoint of E, by p_2 the other endpoint of F and by p_3 the vertex of T_ν left out; then we set $\sigma(\nu, E, F) = 1$ if the ordering of the vertices p_0, p_1, p_2, p_3 induces on T_ν the positive orientation, and $\sigma(\nu, E, F) = -1$ otherwise. It is easily checked that, in the above case that the moduli of T_ν with respect to E and F are respectively given by $\tilde{\eta}(i)(z_\nu)$ and $\tilde{\eta}(j)(z_\nu)$, we have

$$\eta(i) \wedge \eta(j) = \sigma(\nu, E, F).$$

(Remark that both expressions take values $+1$ or -1 and are skew-symmetric in their arguments.)

It follows that the left hand sides of formulae (3), (2)' and (1)' are the sum of a certain number of contributions of the type $\sigma(\nu, E, F)$, where $\nu \in \{1, ..., n\}$ and E and F are adjacent edges of T_ν. Of course we shall have to prove that many of these contributions delete each other: the following criteria are easily established and will be useful in the sequel:

(i) $\sigma(\nu, F, E) = -\sigma(\nu, E, F)$;

(ii) if $T_{\nu'}$ is adjacent to (i.e. is glued to) T_ν along the face containing E and F then

$$\sigma(\nu', E, F) = -\sigma(\nu, E, F);$$

(iii) if E, F and G are the three edges of a face of T_ν then

$$\sigma(\nu, E, F) = \sigma(\nu, F, G) = \sigma(\nu, G, E).$$

We are going to prove now formulae (3), (2)' and (1)'.

(3) $\quad \displaystyle\sum_{\nu=1}^{n} \big(\theta_1(r, \nu) \cdot \theta_2(s, \nu) - \theta_2(r, \nu) \cdot \theta_1(s, \nu)\big) = 0 \ \text{ for } r, s = 1, ..., n$

Let us recall that for $r = 1, ..., n$ we denoted by e_r the r-th edge in the 1-skeleton of the manifold (e_r is obtained by glueing several edges of the tetrahedra). According to E.6.2 and E.6.3 $\big(\theta_1(r, \nu), \theta_2(r, \nu)\big)$ is the sum over all edges E of T_ν being identified with e_r of the pairs $\eta(i)$, where the modulus of T_ν with respect to E is $\tilde\eta(i)(z_\nu)$. It follows that the left hand side is the sum of all the contributions of the type $\sigma(\nu, E, F)$, where $\nu = 1, ..., n$, E and F are adjacent edges of T_ν being identified with e_r and e_s respectively.

Given such ν, E, F there exists precisely one other index ν' such that $T_{\nu'}$ is glued to T_ν along the face containig E and F; moreover the contribution $\sigma(\nu', E, F)$ appears in the sum too and $\sigma(\nu', E, F) = -\sigma(\nu, E, F)$, which implies that the sum vanishes.

(2)' $\quad \displaystyle\sum_{\nu=1}^{n} \big(a_1(\nu) \cdot \theta_2(r, \nu) - a_2(\nu) \cdot \theta_2(r, \nu)\big) = 0 \ \text{ for } r = 1, ..., n$

Let us start describing the way the pair $\big(a_1(\nu), a_2(\nu)\big)$ is obtained (following Lemma E.6.8). Let L be the torus containing α; then $\big(a_1(\nu), a_2(\nu)\big)$ is the sum of all the pairs $\eta(i)$, where E is an edge of T_ν, the modulus of T_ν with respect to E is $\tilde\eta(i)(z_\nu)$ and E has the following property: an endpoint of E corresponds to the torus L, the image of E in the manifold has a common vertex with α and the triangle correponding to T_ν on L lies on the right of α at such vertex. If we also use the above characterization of the pair $\big(\theta_1(r, \nu), \theta_2(r, \nu)\big)$ we have that the left hand side is the sum of all the contributions $\sigma(\nu, E, F)$ where:

– $\nu = 1, ..., n$;

– E and F are adjacent edges of T_ν;

– an endpoint of E corresponds to the torus L, the image of E in the manifold has a common vertex with α and the triangle correponding to T_ν on L lies on the right of α at such vertex;

– the image of F in the manifold is e_r.

We are going to see that these contributions cancel each other; we shall do this by considering four differerent possibilities for a triple ν, E, F and showing that to each contribution ± 1 there corresponds an opposite contribution due to a triple ν', E', F' being in the same case; the verification that these contributions actually cancel each other in pairs will then be obvious.

Given ν, E, F let us consider as preferred endpoint of E the one corresponding to the torus L. The four cases we are going to consider correspond to the possible answers to the following questions:

(A) *is the vertex common to E and F the preferred endpoint of E?*

(B) *does the face of T_ν containing E and F also contain an edge of α?*

(From now on the proof will make extensive use of pictures; we are going to realize all the tetrahedra in $\mathrm{III}^{3,+}$; as for major evidence, in the situation in exam, we are going to give E the orientation towards its preferred endpoint: this is not so useful now, as only E has a preferred endpoint, but it will be very useful later when proving (1)′. In the present situation we are going to realize always the preferred endpoint of E as ∞.)

(A) *Yes.* (B) *Yes.*

If $\sigma(\nu, E, F) = -1$ the situation is the one represented in Fig. E.47. Then we can consider the edge of α coming before the one lying on the face of T_ν containing E and F. Let $T_{\nu'}$ be the tetrahedron corresponding to the triangle lying on the right of this edge of α; with the choice of $F' = F$ and E' as in Fig. E.48 we obviously have that the triple ν', E', F' contributes to the sum too and $\sigma(\nu', E', F') = +1$.

In case we start from $\sigma(\nu, E, F) = +1$ we choose the edge of α coming after the one on the face of T_ν, and everything else works in a similar way.

(A) *Yes.* (B) *No.*

If $\sigma(\nu, E, F) = -1$ the situation is described in Fig. E.49, in which we have already represented the construction of the triple ν', E', F': we have chosen $E' = E$, $F' = F$ and ν' as the only index for which $T_{\nu'}$ glues to T_ν along the face containing E and F. Of course the construction works also if we start from $\sigma(\nu, E, F) = +1$.

(A) *No.* (B) *Yes.*

If $\sigma(\nu, E, F) = +1$ the situation is represented in Fig. E.50, which contains the construction too. We have chosen $\nu' = \nu$, $F' = F$ and E' as in the picture; use the same construction if $\sigma(\nu, E, F) = -1$.

(A) *No.* (B) *No.*

If $\sigma(\nu, E, F) = +1$ the situation is the same as the one described in Fig. E.49, the only difference being the fact that in the present situation F is the edge

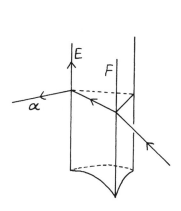

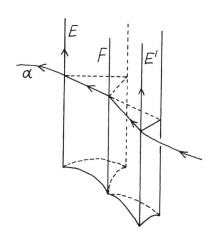

Fig. E.47. Case (A) Yes (B) Yes

Fig. E.48. How to find the canceling triple

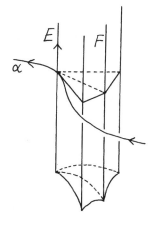

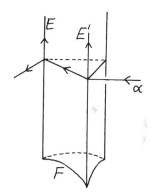

Fig. E.49. Case (A) Yes (B) No

Fig. E.50. Case (A) No (B) Yes

of the face common to the two tetrahedra not having ∞ among their vertices. Then the same choice of ν' and of $E' = E$, $F' = F$ leads to the cancelling triple.

$$(1)' \qquad \sum_{\nu=1}^{n} \big(a_1(\nu) \cdot b_2(\nu) - a_2(\nu) \cdot b_1(\nu)\big) = 2 \cdot I(\alpha, \beta).$$

According to what we saw above the left hand side is the sum of all the contributions $\sigma(\nu, E, F)$ where:

$-\nu = 1, ..., n$;

$- E$ and F are adjacent edges of T_ν;

– one of the endpoints of E corresponds to the torus L_α on which α lies, the image of E in the manifold has a common vertex with α and the triangle corresponding to T_ν on L_α lies on the right of α at such a vertex;

– one of the endpoints of F corresponds to the torus L_β on which β lies, the image of F in the manifold has a common vertex with β and the triangle corresponding to T_ν on L_β lies on the right of β at such a vertex.

As above we shall consider as preferred endpoint of E (F) the vertex corresponding to L_α (L_β) and in the pictures we shall give the edges the orientation towards their preferred vertex.

We are going to prove that many contributions cancel each other, and we shall study what is left out. Given a triple ν, E, F we shall examine several different cases; as above we shall show that the contributions cancel by exhibiting a triple ν', E', F' which is in the same case (with an exception we shall point out). Moreover when we confine ourselves to the description of a cancelling triple for the triple with the condition that either $\sigma(\nu, E, F) = +1$ or $\sigma(\nu, E, F) = -1$ we tacitly mean that the other case is treated in a similar way.

We shall denote by Δ the face of T_ν containing E and F. We start now to divide the cases: we understand that the symbol $C_1.C_2...C_t$ followed by a condition concerning the triple ν, E, F means that condition $C_1.C_2...C_{t-1}$ holds too.

1 The common endpoint of E and F is not the preferred endpoint for both.
(Let us remark that under this condition we have $L_\alpha \neq L_\beta$ and so $I(\alpha, \beta) = 0$; then we have to prove that all the contributions cancel.)

1.1 Δ does not contain edges of α or β.

1.1.1 The common endpoint to E and F is the preferred endpoint of F.
If $\sigma(\nu, E, F) = -1$ the situation is as in Fig. E.51; the choice of $E' = E$, $F' = F$ and ν' such that $T_{\nu'}$ glues to T_ν along Δ produces the required cancelling triple. For the sake of simplicity we do not represent in the picture the new tetrahedron, but the construction is quite evident.

1.1.2 The common endpoint to E and F is the preferred endpoint of E.
The case $\sigma(\nu, E, F) = -1$ corresponds to Fig. E.51, in which the roles of α, β and E, F are interchanged, and then the construction is just the same.

1.1.3 The common endpoint of E and F is not the preferred endpoint of E or of F.
In case $\sigma(\nu, E, F) = +1$ the situation is represented in Fig. E.52; once again we choose $E' = E$, $F' = F$ and ν' as usual.

We are going to assume now for a while that no tetrahedron has on a same face edges of both α and β: we do this in order to avoid too many tedious distinctions and explain the ideas of the constructions, and afterwards we shall drop this assumption and consider the general case.

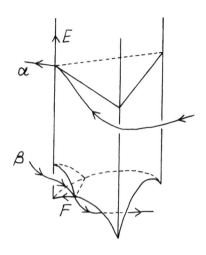

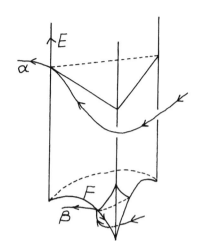

Fig. E.51. Case 1.1.1 (and 1.1.2)   **Fig. E.52.** Case 1.1.3

1.2 Δ contains an edge of α.

1.2.1 The other edge of Δ touched by α is not F.

Let us remark that it follows that the preferred edge of E cannot be common to E and F. Then we are left to consider two cases.

1.2.1.1 The common endpoint to E and F is the preferred endpoint of F.
In case $\sigma(\nu, E, F) = +1$ the situation is as in Fig. E.53; then we choose $\nu' = \nu$, $F' = F$ and E' as in the picture (the third edge of Δ).

1.2.1.2 The common endpoint is not the preferred endpoint of F.
In case $\sigma(\nu, E, F) = +1$ the situation is as in Fig. E.54; as above we choose $\nu' = \nu$, $F' = F$ and E' as in the picture (the third edge of Δ).

1.2.2 The other edge of Δ touched by α is F.

We deduce from this that the preferred endpoint of E belongs to F; then in case $\sigma(\nu, E, F) = -1$ the situation is as in Fig. E.55.

Then we divide again according to the way the other branches of α and β appear while turning around F; in other words we consider the other tetrahedra glueing to T_ν along F and containing the other branch of α and the two branches of β, and examine the relative position of these tetrahedra. The different cases we can meet, which we denote by *1.2.2.1, 1.2.2.2* and *1.2.2.3*, are described in Fig. E.56 (in which the view is "from above"). Remark that by the assumption that no face of the tetrahedra contains an edge of both α and β these three are actually the only possible cases; afterwards we shall discuss the limit cases in which one of the branches we are interested in lies on a same face with another. In Fig. E.56 we also outline the triangle suggesting the tetrahedron we are going to consider; remark that in all cases the triangle lies on the right of α and β (since α must be looked at from above and β from below).

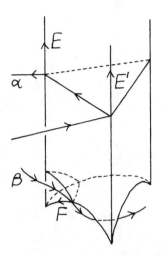

Fig. E.53. Case 1.2.1.1

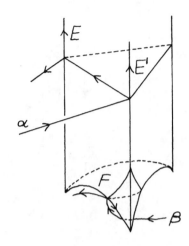

Fig. E.54. Case 1.2.1.2

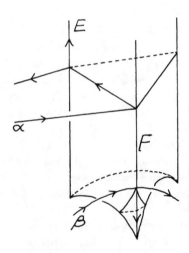

Fig. E.55. Case 1.2.2

The construction is the same for the first two cases, which are represented in Fig. E.57 and Fig. E.58 respectively, where $\sigma(\nu, E, F) = -1$. In both cases we take ν' corresponding to the new tetrahedron depicted, $F' = F$ and E' as in the figure.

The case 1.2.2.3 entails the construction of Fig. E.59: ν' corresponds to the new tetrahedron and E', F' are depicted.

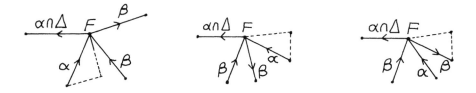

Fig. E.56. The three possible situations one can get when considering how the other branches of α and β appear around F. These cases are numbered respectively as 1.2.2.1, 1.2.2.2 and 1.2.2.3

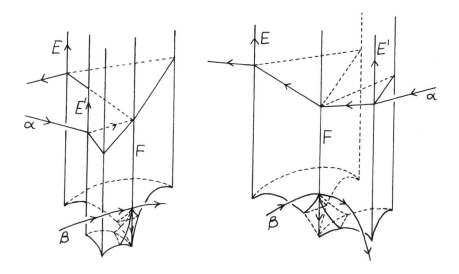

Fig. E.57. Case 1.2.2.1   **Fig. E.58.** Case 1.2.2.2

Let us remark that we have a new phenomenon here, as the new triple does not fall in the same case as the original one: if we denote by Δ' the face of $T_{\nu'}$ containing E' and F', we still have in the first two cases 1.2.2.1 and 1.2.2.2 that Δ' contains an edge of α and the other edge of Δ' touched by α is F', but in the third case 1.2.2.3 Δ' contains an edge of β and the other edge of Δ' touched by β is E'. Of course this is not a problem: if we make an analogous construction (starting from 1.2) in case Δ contains an edge of β it is easily checked that globally the contributions to the sum cancel in pairs.

Let us now get rid of the assumption that no face contains an edge of both α and β by examining directly the contributions of this type. Then, let Δ_0 be a face of the tetrahedron T_{ν_0} containing an edge of both α and β; denote by a, b and c the edges of Δ_0 and consider for instance the situation of Fig. E.60. Then in the sum in question we have contributions of the type

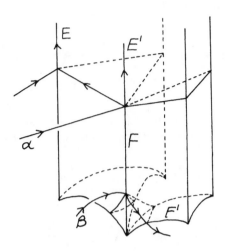

Fig. E.59. Case 1.2.2.3

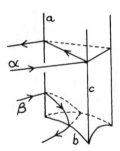

Fig. E.60. The situation we can get when we drop the assumption that no tetrahedron has on a same face edges of both α and β

$$\sigma(\nu_0, E_0, F_0)$$

with $(E_0, F_0) \in \{(a,b), (c,b), (c,a)\}$ two of which (but not the third one) cancel each other. (We remark that this corresponds to the fact that in such a case a construction like the one we made for 1.2.1 can be performed with respect to both α and β.) The contribution left is that of a triple ν_0, E_0, F_0 in which we have that either α touches F_0 or β touches E_0. Then this contribution cancels with the one of ν, E, F where the triple is such that in case 1.2.2.3 we have the limit situation *1.2.2.4* of Fig. E.61.

To such a triple ν, E, F corresponds the same construction we did for 1.2.2.3, given for completeness in Fig. E.62. Remark that indeed the resulting triple ν', E', F' satisfies the properties of the triple ν_0, E_0, F_0 considered above and moreover all such triples can be obtained in this way.

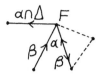

Fig. E.61. The limit situation 1.2.2.4

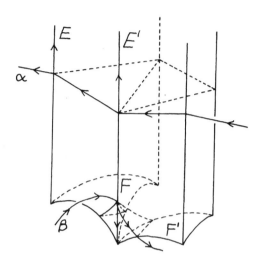

Fig. E.62. Case 1.2.2.4

Fig. E.63. A "doubly" limit situation

We finally remark that two contributions of triples of the type ν_0, E_0, F_0 considered above may cancel each other: this is the case for instance in the "doubly" limit situation considered in Fig. E.63.

It is now proved that if condition 1 is satisfied the sum on the left hand side of $(1)'$ vanishes, so that we can turn to the next case.

2 The common endpoint of E and F is the preferred one for both.
It follows from this that α and β lie on the same torus L; moreover the segment e of the triangle corresponding to T_ν on L and having the endpoints on E and F joins a vertex of α to a vertex of β.

2.1 e is contained neither in α nor in β.

The cancelling triple is chosen with $E' = E$, $F' = F$ and ν' as usual. The situation is as in Fig. E.64.

  2.2 e is contained in both α and β.
The cancelling triple is obtained with $\nu' = \nu$, $E' = F$ and $F' = E$, as represented in Fig. E.65.

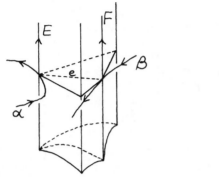
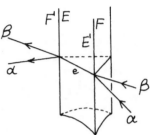

Fig. E.64. Case 2.1   **Fig. E.65.** Case 2.2

It follows from these first two cases that the contributions left correspond to the edges of α being not common to β but touching it, and to the edges of β being not common to α but touching it. The study the resulting contributions is carried out in Figg. E.66 to E.69; in all these figures the right hand side represents the situation on the torus L corresponding to the left hand side.

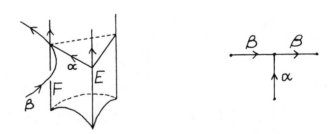

Fig. E.66. First possibility for an edge of α not in common with β but touching it. The resulting contribution is +1

It follows from this study that the edges of α give the same contribution as they give to the sum defining $I(\alpha, \beta)$, while the edges of β give the opposite contribution with respect to the one they give to the sum defining $I(\beta, \alpha)$; then we finally have that the sum on the left hand side of (1)' is indeed

$$I(\alpha, \beta) - I(\beta, \alpha) = 2 \cdot I(\alpha, \beta). \qquad \Box$$

Fig. E.67. Second possibility for an edge of α not in common with β but touching it. The resulting contribution is -1

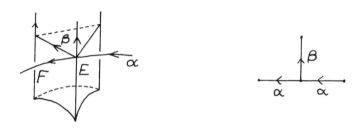

Fig. E.68. First possibility for an edge of β not in common with α but touching it. The resulting contribution is $+1$

Fig. E.69. Second possibility for an edge of β not in common with α but touching it. The resulting contribution is -1

We can deduce now from the above lemma the information we were looking for about the dimension of $\mathcal{H}(M)$ around one of its points.

Proposition E.6.14. Θ has rank at most $n - k$.

Proof. We shall think of $\mathbb{C}^{2n}$ as the space of row-vectors of length $2n$, and consider on $\mathbb{C}^{2n}$ the symplectic bi-linear form

$$\langle x|y \rangle = x \cdot J_{2n} \cdot {}^t y$$

and use the symbol $\perp$ to denote orthogonality with respect to this form. Moreover if A is a matrix with $2n$ columns we shall denote by $[A]$ the subspace of $\mathbb{C}^{2n}$ spanned by the rows of A. Let us re-write in extended form the statement of E.6.13:

$$\begin{pmatrix} \Lambda \\ \mathcal{M} \\ \Theta \end{pmatrix} \cdot J_{2n} \cdot \begin{pmatrix} {}^t\Lambda & {}^t\mathcal{M} & {}^t\Theta \end{pmatrix} = 2 \begin{pmatrix} J_{2k} & 0 \\ 0 & 0 \end{pmatrix}.$$

This formula implies the following:

– the $2k$ row vectors of $\begin{pmatrix} \Lambda \\ \mathcal{M} \end{pmatrix}$ are (up to a constant factor) a symplectic basis of the space they span; in particular they are linearly independent so that

$$\dim \begin{bmatrix} \Lambda \\ \mathcal{M} \end{bmatrix} = 2k;$$

– $[\Theta] \subseteq [U]^\perp$ and hence in particular

$$\dim[\Theta] \leq \dim[U]^\perp;$$

– $[\Theta] \cap \begin{bmatrix} \Lambda \\ \mathcal{M} \end{bmatrix} = \{0\}$ (in fact the restriction of the symplectic form to $\begin{bmatrix} \Lambda \\ \mathcal{M} \end{bmatrix}$ is non-degenerate, while all vectors of $[\Theta]$ are orthogonal to the whole of $[U]$); in particular we have that

$$\dim[U] = \dim \begin{bmatrix} \Lambda \\ \mathcal{M} \end{bmatrix} + \dim[\Theta] = 2k + \dim[\Theta];$$

– since the form is non-degenerate we have

$$\dim[U]^\perp = 2n - \dim[U].$$

These facts easily imply the conclusion:

$$\dim[\Theta] \leq \dim[U]^\perp = 2n - \dim[U] = 2n - 2k - \dim[\Theta]$$

$$\Rightarrow \quad \dim[\Theta] \leq n - k. \qquad \square$$

Since we remarked that $\mathcal{H}(M)$ is locally determined by a system of n equations whose Jacobian matrix at z is $\Theta \cdot Z(z)$ for a suitable $2n \times n$ matrix $Z(z)$ we immediately deduce from the above proposition the following:

Corollary E.6.15. $\mathcal{H}(M)$ has dimension at least k in the neighborhood of any of its points.

E.6-iii The Case M is Complete Hyperbolic: the Space of Deformations

We are now going to specialize our discussion to manifolds supporting a finite-volume complete hyperbolic structure. It has been proved in E.5.9 that given $M \in \mathcal{F}_3$ there exists a realization of M in T_3 with the property that the complete structure appears in $\mathcal{H}(M)$. We fix M and such a realization in T_3, and we denote by z^0 the point of $\mathcal{H}(M)$ corresponding to the complete structure. We shall consider an arbitrarily small neighborhood Def(M) of z^0 in $\mathcal{H}(M)$; such a neighborhood is called the space of _deformations_ of the complete structure.

The following result has central importance in our proof of the hyperbolic surgery theorem; it is worth remarking that its proof makes use of the rigidity theorem; we shall sketch in E.7.7 a result sharpening it.

Proposition E.6.16. The elements of $\mathcal{H}(M)$ corresponding to complete structures are isolated points.

Proof. Denote by $\mathcal{H}_c(M)$ the subset of $\mathcal{H}(M)$ corresponding to complete structures. According to E.6.12 $\mathcal{H}_c(M)$ is an algebraic subset of $\left(\Pi^{2,+}\right)^n$; then if a point of $\mathcal{H}_c(M)$ were not isolated $\mathcal{H}_c(M)$ would have dimension at least 1 in the neighborhood of such a point, so that in particular $\mathcal{H}_c(M)$ would be uncountable: we are going to check that this is absurd.

Let us remark first that by the rigidity theorem all the elements of $\mathcal{H}_c(M)$ induce on M the same hyperbolic structure; then if $\mathcal{H}_c(M)$ is uncountable we have that the complete hyperbolic manifold M has uncountably many different decompositions in ideal tetrahedra. Such decompositions are determined by the edges of the tetrahedra, and these edges are geodesic lines asymptotic to cusps, _i.e._ simple non-closed geodesic lines $\gamma : \mathbb{R} \to M$ such that for any $\varepsilon > 0$ there exists t_0 such that $\gamma([t_0, \infty))$ and $\gamma((-\infty, -t_0])$ are contained in the ε-thin part. We claim that the set of these lines in M is at most countable, which implies the conclusion that $\mathcal{H}_c(M)$ cannot be uncountable.

Fix a presentation of M as $\mathbb{H}^3/\Gamma$, where $\Gamma \cong \Pi_1(M)$ is a discrete group of $\mathcal{I}^+(\mathbb{H}^3)$ acting freely. Since Γ is discrete and $\mathcal{I}^+(\mathbb{H}^3)$ has a countable basis of open sets, Γ is countable and then in particular it contains countably many parabolic elements. Denote by F the set of the fixed points in $\partial \mathbb{H}^3$ of the parabolic elements of Γ. Then F is countable and it easily follows from the description of the cusps (Chapt. D) that a geodesic line in M is asymptotic to cusps if and only if it is the projection of a geodesic line in $\mathbb{H}^3$ having both the endpoints in F; such lines in $\mathbb{H}^3$ are countably many, and the proof is over. $\square$

Lemma E.6.17. For $j = 1,...,k$ the two equations $\delta(z)(l_j) = 1$ and $\delta(z)(m_j) = 1$ are equivalent on Def(M). Hence the two systems of k rational equations

$$\{\delta(z)(l_j) = 1 \qquad \forall j = 1, ..., k$$
$$\{\delta(z)(m_j) = 1 \qquad \forall j = 1, ..., k$$

are equivalent on $\mathrm{Def}(M)$, so z^0 is the only point of $\mathrm{Def}(M)$ satisfying them.

Proof. If $\delta(z)(l_j) = 1$ we have that the sum of the arguments of the moduli appearing in $\delta(z)(l_j)$ is $(\#l_{j_0} + 2p)\pi$ for some $p \in \{-1, 0, 1\}$ (where $\#l_{j_0}$ denotes the number of vertices of l_j). By lemmas E.6.6 (equivalence of (a) and (c)) and E.6.8 (part (2)) we have that when $z = z^0$ this sum is $\#l_{j_0} \cdot \pi$, so it must be $\#l_{j_0} \cdot \pi$ for $z \in \mathrm{Def}(M)$. Again, Lemmas E.6.6 (equivalence of (a) and (d)) and E.6.8 (part (2)) imply that the similarity structure on L_j is Euclidean and hence that $\delta(z)(m_j) = 1$. The converse follows by the same argument.

According to what we have just proved both the systems of equations imply the completeness of $z \in \mathcal{H}(M)$; hence by E.6.16 only $z^0 \in \mathrm{Def}(M)$ can satisfy them. $\qquad\square$

Corollary E.6.18. The image of both the mappings

$$\mathrm{Def}(M) \ni z \mapsto \left(\delta(z)(l_j)\right)_{j=1,...,k} \in \mathbb{C}^k \quad \text{and}$$
$$\mathrm{Def}(M) \ni z \mapsto \left(\delta(z)(m_j)\right)_{j=1,...,k} \in \mathbb{C}^k$$

covers a neighborhood of $(1, ..., 1) \in \mathbb{C}^k$.

Proposition E.6.19. The matrix Θ has rank precisely $n - k$ and $\mathcal{H}(M)$ has dimension k.

Proof. Assume by contradiction that Θ has rank strictly less than $n - k$, which implies that $\mathrm{Def}(M)$ has dimension strictly more than k. By E.6.17 we have that that $\mathrm{Def}(M)$ must have dimension at most k in the neighborhood of z^0, and hence we get a contradiction. $\qquad\square$

Proposition E.6.20. It is actually possible to define $\mathcal{H}(M)$ with precisely $n - k$ equations.

Proof. We know that Θ has rank $n - k$; then we can assume the first $n - k$ rows of Θ are linearly independent; let us recall that the equations defining $\mathcal{H}(M)$ can be written in the form

$$\sum_{\nu=1}^{n} \left(\theta_1(r, \nu) \cdot \log(z_\nu) + \theta_2(r, \nu) \cdot \log(1 - z_\nu)\right) - c(r) = 0 \quad \text{for } r = 1, ..., n.$$

For $s > n - k$ let

$$\theta_l(s, \nu) = \sum_{r=1}^{n-k} \lambda_r^s \cdot \theta_l(r, \nu) \quad \forall l = 1, 2, \nu = 1, ..., n.$$

Consider the following linear combination of the first $n - k$ equations:

$$\sum_{r=1}^{n-k} \lambda_r^s \cdot \left(\sum_{\nu=1}^n \big(\theta_1(r,\nu) \cdot \log(z_\nu) + \theta_2(r,\nu) \cdot \log(1 - z_\nu) \big) - c(r) \right) = 0.$$

We have that it differs from the s-th equation possibly only in the constant term

$$\sum_{r=1}^{n-k} \lambda_r^s \cdot c(r);$$

however, we know that z^0 satisfies simultaneously all the n equations, and then we actually have

$$\sum_{r=1}^{n-k} \lambda_r^s \cdot c(r) = c(s)$$

which implies that for $s > n - k$ we can eliminate the s-th equation without altering the set $\mathcal{H}(M)$.   □

We go back now to the above Corollary E.6.18 and carry out the main technical step of the proof. Since we have $\delta(z^0)(l_j) = \delta(z^0)(m_j) = 1 \ \forall j$ we can consider the usual holomorphic determination of the logarithm function around 1 and define the functions of $z \in \text{Def}(M)$

$$u_j(z) = \log \big(\delta(z)(l_j) \big) \qquad j = 1, ..., k$$

$$v_j(z) = \log \big(\delta(z)(m_j) \big) \qquad j = 1, ..., k.$$

Remark that such functions are restrictions to $\text{Def}(M)$ of global holomorphic functions on $\left(\Pi^{2,+} \right)^n$. Corollary E.6.18 immediately yields:

Corollary E.6.21. The image of both the mappings

$$\text{Def}(M) \ni z \mapsto \big(u_j(z) \big)_{j=1,...,k} \in \mathbb{C}^k \quad \text{and}$$

$$\text{Def}(M) \ni z \mapsto \big(v_j(z) \big)_{j=1,...,k} \in \mathbb{C}^k$$

covers a neighborhood of $0 \in \mathbb{C}^k$.

We are going to prove now a result concerning the way u and v are related to each other.

Lemma E.6.22. For $j = 1, ..., k$ there exists an analytic function τ_j such that

$$v_j(z) = u_j(z) \cdot \tau_j(z)$$

and moreover $\tau_j(0) \notin \mathbb{R}$.

Proof. u_j and v_j are holomorphic functions of z and moreover by definition

$$u_j(z) = 0 \ \Leftrightarrow \ \delta(z)(l_j) = 1$$
$$v_j(z) = 0 \ \Leftrightarrow \ \delta(z)(m_j) = 1$$

and then by E.6.17 we have that $u_j(z) = 0 \ \Leftrightarrow \ v_j(z) = 0$, so that we easily have

$$v_j(z) = u_j(z) \cdot \tau_j(z)$$

for some holomorphic function τ_j defined whenever $u_j(z) \neq 0$; we are going to prove that the limit as z converges to z^0 of τ_j exists and is not real (in particular, it is not 0) which implies the conclusion at once.

For $z \in \text{Def}(M)$ such that $u_j(z) \neq 0$ set by simplicity $a(z) = \delta(z)(l_j)$ and $b(z) = \delta(z)(m_j)$; remark that by definition we have $u_j(z) = \log(a(z))$ and $v_j(z) = \log(b(z))$. If we cut L_j open along l_j and m_j we can realize the resulting "quadrilateral" in $\mathbb{C}$ preserving the similarity structure (as above inverted commas are used as the edges of the quadrilateral are not in general straight segments). We can arrange the quadrilateral in such a way that its vertices and edges are very close to those of a parallelogram giving the Euclidean structure of L_j corresponding to $z = z^0$.

Then we have that $a(z)$ and $b(z)$ represent the dilation components of the complex-affine automorphisms of the plane identifying the opposite edges of this quadrilateral. Let us denote by $p_1(z), ..., p_4(z)$ the vertices of the quadrilateral arranged in a counter-clockwise order. Then

$$\lim_{z \to z^0} \frac{p_4(z) - p_1(z)}{p_2(z) - p_1(z)} = \frac{p_4(z^0) - p_1(z^0)}{p_2(z^0) - p_1(z^0)}$$

and this is a non-real complex number τ_j as $p_1(z^0), ..., p_4(z^0)$ are the vertices of a parallelogram. (This number is actually a modulus of the Euclidean structure induced y z^0 on L_j.) Now we assume (for fixed $z \neq z^0$) that there exist $\alpha, \beta \in \mathbb{C}$ such that

$$p_2(z) = b(z)p_1(z) + \beta$$
$$p_4(z) = a(z)p_1(z) + \alpha$$
$$p_3(z) = a(z)p_2(z) + \alpha = b(z)p_4(z) + \beta$$

(we may possibly need to interchange the roles of $a(z)$ and $b(z)$: we will discuss what this change implies). Since the number

$$\frac{p_4(z) - p_1(z)}{p_2(z) - p_1(z)}$$

is of course invariant under similarities we can assume without loss of generality that $p_1(z) = 1$ and $p_2(z) = a(z)$. Hence $\beta = 0$ and using the two expressions for $p_3(z)$ (and the fact that $b(z) \neq 1$) it is easily checked that $\alpha = 0$. So

$$\frac{p_4(z) - p_1(z)}{p_2(z) - p_1(z)} = \frac{a(z) - 1}{b(z) - 1}.$$

Hence we have

$$\mathbb{R} \not\ni \tau_j = \lim_{z \to z^0} \frac{a(z) - 1}{b(z) - 1} = \lim_{z \to z^0} \frac{\log a(z)}{\log b(z)} = \lim_{z \to z^0} \frac{v(z)}{u(z)}.$$

Remark that an interchange of the roles of $a(z)$ and $b(z)$ would lead to

$$\lim_{z \to z^0} \frac{v(z)}{u(z)} = 1/\tau_j \notin \mathbb{R}$$

so the conclusion holds in any case.

$\square$

According to the above result, for $z \in \mathrm{Def}(M)$ if $u_j(z) \neq 0$ there exists a unique pair of real numbers $p_j(z), q_j(z) \in \mathbb{R}^2$ satisfying the relation

$$p_j(z) \cdot u_j(z) + q_j(z) \cdot v_j(z) = 2\pi i;$$

the reason for considering this equation is not so clear now, but we shall discuss its importance when examining the completion of the structures of $\mathrm{Def}(M)$ and see that to a co-prime integer solution $(p_j(z), q_j(z))$ there corresponds a hyperbolic surgery of coefficients $(p_j(z), q_j(z))$ at the j-th end. For $u_j(z) = 0$ we define the solution of the above equation to be ∞. As usual we identify $\mathbb{R}^2 \cup \{\infty\}$ with the sphere S^2 and consider on it the usual topology. The main technical result from which the proof of the hyperbolic surgery theorem will be deduced is the following:

Proposition E.6.23. The image of the mapping

$$\mathrm{Def}(M) \ni z \mapsto \left((p_j(z), q_j(z))\right)_{j=1,\ldots,k} \in \left(S^2\right)^k$$

covers a neighborhood of $(\infty, \ldots, \infty)$.

Proof. We first describe the argument for $k = 1$ and then we sketch the general case.

Since we are dealing with the case $k = 1$ we omit the subscripts. We first re-write the equation defining $(p(z), q(z))$ in the following form:

$$p(z) + \tau(z) \cdot q(z) = \frac{2\pi i}{u(z)}.$$

Using this formula and recalling that τ and u are restrictions to $\mathrm{Def}(M)$ of holomorphic functions in a neighborhood of z^0 in $\left(\Pi^{2,+}\right)^n$ with $\tau(z^0) \notin \mathbb{R}$, the explicit formulae for $(p(z), q(z))$ can be exhibited and it is easily checked that they are restrictions to $\mathrm{Def}(M)$ of real analytic functions of z. Then the image N of the mapping $z \mapsto (p(z), q(z))$ is a "nice" set containing ∞ (namely, a subanalytic set, and moreover a closed one if we choose $\mathrm{Def}(M)$ to be compact). This implies that if ∞ does not belong to the interior of N we can find an analytic path in S^2 meeting N in ∞ only: we are going to prove now that this is absurd.

Remark first that we can also define another function $z \mapsto (p^0(z), q^0(z))$ by the equation

$$p^0(z) + \tau(z^0) \cdot q^0(z) = \frac{2\pi i}{u(z)}.$$

Moreover as u covers a neighborhood of $0 \in \mathbb{C}$ the mapping $z \mapsto (p^0(z), q^0(z))$ covers a neighborhood of $\infty \in S^2$. Let $R > 0$ be so large that

$$\gamma_R = \left\{u : |u| = 1/R\right\}$$

is contained in the image of $z \mapsto u(z)$; the set

$$\delta_R = \left\{ (P,Q) \in S^2 : P + \tau(z^0) \cdot Q = 2\pi i/u : \ u \in \gamma_R \right\}$$

is an analytic loop around ∞ in S^2, and it becomes arbitrarily close to ∞ as $R \to \infty$.

Let us choose a lifting α_R of γ_R to $\mathrm{Def}(M)$ which is a loop too (of course we may have to spin around γ_R several times), and denote by β_R the image of α_R under the mapping $z \mapsto (p(z), q(z))$. Since $\tau(z)$ converges to $\tau(z^0)$ as $z \to z^0$, we easily have that β_R is a loop in S^2 which is arbitrarily close to δ_R as $R \to \infty$. It follows that for $R \gg 0$ it meets any non-constant analytic curve starting at ∞; since δ_R is a loop in N this is absurd and our argument is complete.

We turn now to the general case. Remark first that the above argument can be repeated word-by-word for all k of the factors, so that we have that for $j = 1, ..., k$ the image of the mapping

$$\mathrm{Def}(M) \ni z \mapsto (p_j(z), q_j(z)) \in S^2$$

covers a neighborhood of ∞, but this is definitely not enough for concluding that a neighborhood of $(\infty, ..., \infty)$ is covered! Then we must consider all the factors simultaneously. We sketch the generalization of the above proof.

For large R the sphere

$$\left\{ u \in \mathbb{C}^k : \|u\| = R \right\}$$

represents the canonical generator of $H_{2k}(\mathbb{C}^k, \{0\})$. If we lift it to $\mathrm{Def}(M)$ and consider the image in $(S^2)^k$ we have a set which is arbitrarily close (as $R \to \infty$) to a multiple of a generator of

$$H_{2k}\left((S^2)^k, \{(\infty, ..., \infty)\} \right),$$

and then in particular it represents a non-trivial class in this homology group, and the conclusion follows from the same argument presented for the case $k = 1$. $\square$

E.6-iv Completion of the Deformed Hyperbolic Structures and Conclusion of the Proof

We are going to describe now the completion of M with respect to the non-complete hyperbolic structure associated to a point z of $\mathrm{Def}(M)$ different from z^0. The point z is kept fixed and we shall write u_j and v_j instead of $u_j(z)$ and $v_j(z)$.

We first remark that the completion is necessarily performed by adding something at each removed vertex of $\tilde{Q}$; in fact if we remove from $\tilde{Q}$ an arbitrarily small neighborhood of each vertex we get a compact space which is therefore complete with respect to the structure in question. Hence we only need to describe the completion of a neighborhood of the j-th vertex removed. We identify such a neighborhood with $L_j \times [0, \infty)$ and consider the hyperbolic structure induced by z on it (we recall that L_j is the link of the j-th vertex removed, diffeomorphic to a torus). We fix the index j too so that all the subscripts j can be omitted.

We define λ and μ to be the dilation components of the holonomy of the longitude and the meridian of the torus L with respect to the similarity structure induced on L by z; let us recall that by definition we have $u = \log \lambda$ and $v = \log \mu$ (where λ and μ are close to 1 and the logarithm function is the natural one extending $\log 1 = 0$ in the neighborhood of 1).

Lemma E.6.24. If $u = 0$ (which is equivalent to $v = 0$), then the structure on $L \times [0, \infty)$ is complete.

Proof. We know that $u = 0$ (or equivalently $v = 0$) means that the similarity structure on L is Euclidean, and we proved in E.6.5 that in this case $L \times [0, \infty)$ is complete. □

It follows from this lemma that we only need to discuss the case $u \neq 0$; we are actually going to check that if $u \neq 0$ the hyperbolic structure on $L \times [0, \infty)$ is not complete, so that the converse of the above lemma holds. The key point is to find a developing mapping for the non-complete structure on $L \times [0, \infty)$. We first describe a developing function for the similarity structure on the torus L in case it is not Euclidean (*i.e.* $\lambda \neq 1$ or $\mu \neq 1$).

As we remarked during the proof of E.6.22 we can cut L open along the longitude and meridian and realize the remaining "quadrilateral" P in $\mathbb{C}$ preserving the similarity structure; the vertices and edges of this quadrilateral are very close to those of a parallelogram and its opposite edges are identified by similarities having dilation components λ and μ. We fix now the universal covering $\mathbb{R}^2$ of the torus L, defined by the action of the automorphisms

$$\delta_1 : (x, y) \mapsto (x + 1, y) \quad \text{and} \quad \delta_2 : (x, y) \mapsto (x, y + 1),$$

corresponding to the longitude and the meridian respectively (we shall view this covering as a purely topological one, *i.e.* we allow changes of coordinates in $\mathbb{R}^2$ with the only condition that they commute with δ_1 and δ_2).

Lemma E.6.25. A developing mapping for the similarity structure of L is given by

$$D : \mathbb{R}^2 \ni (x, y) \mapsto \lambda^x \cdot \mu^y \in \mathbb{C}$$

where by definition $\lambda^x \cdot \mu^y = \exp(x \log \lambda + y \log \mu)$.

Proof. Let $P \subset \mathbb{C}$ be the "quadrilateral" described above. First of all we can assume that P is a true quadrilateral with straight edges. In fact straightening a pair of opposite edges corresponds to cutting some triangles and glueing them back in by the prescribed similarity, as represented in Fig. E.70. Remark that this argument strongly depends on the fact that P is "alomost" a parallelogram, as otherwise various pathologies could occur.

Now we can also assume (up to a similarity of $\mathbb{C}$) that P has vertices $1, \mu, \lambda \cdot \mu, \lambda$ (arranged in a positive order). Let P_0 be the square $[0, 1]^2$ in the universal covering $\mathbb{R}^2$ of L, and denote by γ_1 and γ_2 the dilations of ratio λ

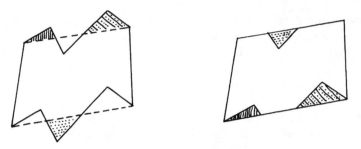

Fig. E.70. Straightening of the edges of a quadrilateral

and μ respectively. By the construction of P we can find a homeomorphism D_0 of P_0 onto P such that:

– the restriction to the interior is a similarity mapping with respect to the similarity structure $\mathbb{R}^2$ is endowed with as a covering of L (not with respect to the natural structure!).

– The action of δ_1 and δ_2 on ∂P_0 corresponds under D_0 to the action of γ_1 and γ_2 on ∂P. Denote by D_1 the developing mapping extending D_0; since D_1 must commute with the actions it must be given by

$$D_1(x,y) = (\gamma_1^{[x]} \circ \gamma_2^{[y]})(D_0(\{x\},\{y\}))$$

where $[t]$ denotes the integer part of a real number t and $\{t\} = t - [t]$.

Remark that D_1 agrees with D (defined in the statement of the lemma) when both x and y are integer numbers. Moreover P is simply connected and D is a covering mapping, so that we can find in $\mathbb{R}^2$ a "quadrilateral" P_0' with the same vertices as P_0 (whose edges are smooth lines but not straight) such that D maps P_0' bijectively onto P. The shape of P_0' is suggested in Fig. E.71.

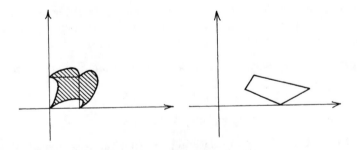

Fig. E.71. A "quadrilateral" with smooth edges mapped onto a straight quadrilateral by the developing function

Then we can find a homeomorphism $f : P_0' \to P_0$ such that $D = D_1 \circ f$; remark that $\delta_i(f(x,y)) = f(\delta_i(x,y))$ for $i = 1,2$ whenever one of these points is defined. Then we can extend f to an automorphism F of the plane commuting with δ_1 and δ_2, and of course we must still have $D = D_1 \circ F$. Since

D is obtained from D_1 by an allowed change of coordinates it is indeed a developing mapping for the similarity structure.   □

Figures E.72 and E.73 illustrate the mappings D_1 and D constructed in the above proof. Remark once again that they coincide on the vertices of the squares.

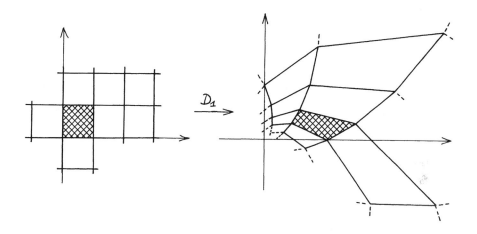

Fig. E.72. A developing mapping for the similarity structure on the torus

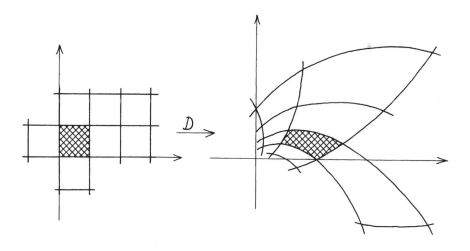

Fig. E.73. Another developing mapping for the similarity structure on the torus

We go back now to the determination of a developing mapping for the hyperbolic structure on $L \times [0, \infty)$. Let us fix the universal cover of $L \times [0, \infty)$ given by $\mathbb{R}^2 \times [0, \infty)$, from which $L \times [0, \infty)$ is obtained under the action of the following automorphisms corresponding to the longitude and the meridian of the torus:

$$\delta_1 : (x, y, t) \mapsto (x + 1, y, t) \qquad \delta_2 : (x, y, t) \mapsto (x, y + 1, t).$$

This covering is fixed up to (δ_1, δ_2)-equivariant changes of coordinates.

Lemma E.6.26. A developing mapping for the hyperbolic structure on $L \times [0, \infty)$ has the form

$$D(x, y, t) = \left(\lambda^x \cdot \mu^y, |\lambda|^x \cdot |\mu|^y \cdot (s_0 + t) \right)$$

where $\lambda^x \cdot \mu^y$ has the same meaning as above and s_0 is a positive number.

Proof. We first give a geometric construction and then we obtain an algebraic expression for the developing mapping. Let us recall that the space obtained by glueing the tetrahedra (without removing the vertices) was denoted by $\tilde{Q}$. Moreover the torus L we are dealing with corresponds to a vertex of $\tilde{Q}$.

We start by picking a certain tetrahedron in $\tilde{Q}$ a vertex of which corresponds to the vertex of $\tilde{Q}$ in question (we shall see it as a preferred vertex of the tetrahedron); we know its interior is isometric to the interior of an ideal tetrahedron in $\mathbb{H}^3$ with a certain modulus (one of the coordinates of our $z \in \mathrm{Def}(M)$, of course; as in Fig. E.74 we realize such an ideal tetrahedron in $\mathbb{H}^{3,+}$ in such a way that the preferred vertex is ∞.

Then we consider the three tetrahedra in $\tilde{Q}$ glueing to the first one along the faces containing the preferred vertex (of course they do not need to be different from each other and from the previous one); as in Fig. E.75 we realize them as ideal tetrahedra in $\mathbb{H}^{3,+}$ in such a way that the glueings with the previous ideal tetrahedron are represented by the identity.

Similarly we go on: remark that the condition that the structure is well-defined on the 1-skeleton (*i.e.* that $z \in \mathcal{H}(M)$) means precisely that no contradiction arises when iterating this procedure. Remark as well that of course all tetrahedra in $\tilde{Q}$ are represented infinitely many times, and moreover that we may have overlaps (we shall see that we actually have overlaps, and even more).

Now, what is the relation between this construction and the developing mapping of $L \times [0, \infty)$? The geometric construction is easily done: by our choice of ∞ as the vertex of the ideal tetrahedra corresponding to the vertex in question, we have that the subset of each of these ideal tetrahedra corresponding to the intersection with $L \times [0, \infty)$, is given by the intersection with a neighborhood of ∞. We can think of $[0, 1]^2 \times [0, \infty)$ (a fundamental domain in the universal cover of $L \times [0, \infty)$) as the union of some of these intersections of the ideal tetrahedra with a neighborhood of the vertex; then the developing mapping is geometrically represented as in Fig. E.76.

The opposite vertical faces of the "prism" in $\mathbb{H}^3$ are identified by two isometries of $\mathbb{H}^3$ (the holonomy of the longitude and the meridian); the action of δ_1 and δ_2 must correspond under the developing mapping to the action of

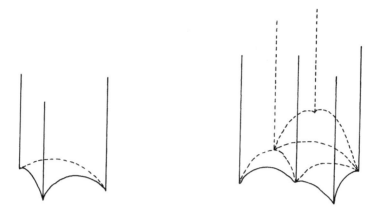

Fig. E.74. Realization of a tetrahedron in the upper half-space model

Fig. E.75. The three tetrahedra glueing to a fixed tetrahedron realized in the upper half-space

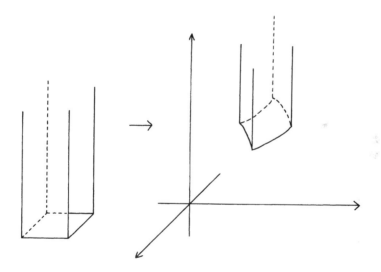

Fig. E.76. Geometric description of the developing mapping on the fundamental domain

these isometries, so that this developing mapping (which we denote by D) is essentially determined by the geometric construction.

But we can also give an explicit algebraic expression. By the geometric construction we have that D maps vertical half-lines in $\mathbb{R}^2 \times [0, \infty)$ to vertical half-lines in $\mathbb{H}^3$, so that we can set

$$D(x, y, t) = \big(D'(x, y), s(x, y, t)\big).$$

Moreover, by the very definition of the similarity structure on L, we can assume that D' is a developing mapping for the similarity structure of the torus, and then (by the previous lemma) that it has the form

$$D'(x,y) = \lambda^x \cdot \mu^y.$$

Now, the action of δ_1 corresponds in $\mathbb{H}^3$ via D to an isometry g_1 whose projection on $\mathbb{C}$ acts as $w \mapsto \lambda \cdot w$, and then we have

$$g_1(w,s) = (\lambda \cdot w, |\lambda| \cdot s)$$

and similarly δ_2 corresponds to

$$g_2(w,s) = (\mu \cdot w, |\mu| \cdot s).$$

The condition that D have the above form and that $D \circ \delta_i = g_i \circ D$ implies that up to an allowed change of coordinates for some $s_0 > 0$ we have:

$$D(x,y,t) = \left(\lambda^x \cdot \mu^y, |\lambda|^x \cdot |\mu|^y \cdot (s_0 + t)\right). \qquad \Box$$

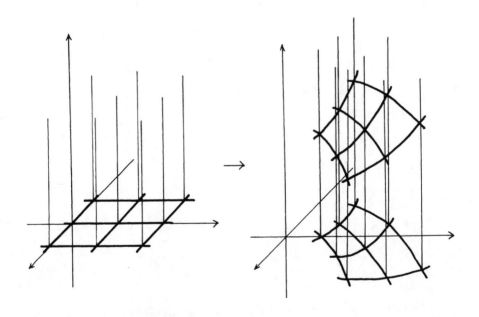

Fig. E.77. Global geometric description of the developing mapping

For a technical reason the expression we shall use for the developing mapping is not the one given in the above lemma: we divide it by $\sqrt{1 + s_0^2}$ (which, of course, does not alter the fact that it is a developing mapping):

$$D(x,y,t) = \frac{1}{\sqrt{1 + s_0^2}} \cdot \left(\lambda^x \cdot \mu^y, |\lambda|^x \cdot |\mu|^y \cdot (s_0 + t)\right).$$

Since the image of D is easily recognized to be a closed neighborhood of the vertical axis (even more: it contains a cone around this axis) with the vertical axis removed, the completion of the image of D is obtained by adding this axis. We are going to describe the way to associate to it a completion of the $L \times [0, \infty)$.

The first step is to describe a subset of $L \times [0, \infty)$ whose (unique) limit point not in $L \times [0, \infty)$ can be identified with a point of the vertical axis. Given $(x, y) \in \mathbb{R}^2$ we consider the geodesic arc in $\mathbb{H}^{3,+}$ passing through $D(x, y, 0)$ and $|\lambda|^x \cdot |\mu|^y$, which is a subset of a circle centred at the origin (this is the reason for the above technical change of the developing function, otherwise a constant factor would have appeared here). Then we "lift" such a half-open arc to $\mathbb{R}^2 \times [0, \infty)$ via D starting from $(x, y, 0)$ and we project the resulting curve to $L \times [0, \infty)$. The half-open arc we started from in $\mathbb{H}^3$ is geodesic and has finite length; since D and the projection onto $L \times [0, \infty)$ are local isometries, the half-open arc we get in $L \times [0, \infty)$ is still a geodesic of finite length. Then it has precisely one limit point (in the completion), and of course such a limit point cannot be in $L \times [0, \infty)$ (otherwise the arc in $\mathbb{H}^3$ would have its limit point in the image of D, and this is not the case). Hence to all $(x, y) \in \mathbb{R}^2$ we can associate a point (identified to $|\lambda|^x \cdot |\mu|^y \in \mathbb{R}_+$) of the completion of $L \times [0, \infty)$. Moreover it is easily checked (we leave it as an exercise to the reader) that all such points suffice to get the whole completion. Of course we still have to consider the equivalence relation giving the torus, *i.e.* we must identify these added points of $\mathbb{R}_+$ when they correspond to points identified in the torus.

We recall that we have denoted by p and q the only real solutions of the equation $p \cdot u + q \cdot v = 2\pi i$.

The points of the torus corresponding to which we must add the point $1 \in \mathbb{R}_+$ are the projections of the solutions of the equation

$$|\exp(x \cdot u + y \cdot v)| = 1 \quad \Leftrightarrow \quad \Re(x \cdot u + y \cdot v) = 0$$

which are given by $\mathbb{R} \cdot (p, q)$.

Of course p and q are not both 0; we define for $q = 0$ the number p/q to be 1 and separate two cases. A third case will be considered too as a subcase of the second one, and this, together with E.6.23 (and many other facts explained in this section) will provide the conclusion of the proof of Thurston's hyperbolic Dehn surgery Theorem E.5.1.

(1) $p/q \notin \mathbb{Q}$.

In the plane $\mathbb{R}^2$ (the universal cover of the torus, corresponding to a horizontal plane in the universal cover of $L \times [0, \infty)$) we draw the line corresponding to which we must add $1 \in \mathbb{R}_+$ in the completion. If we translate all the segments of the line intersecting the different squares to the same square, condition (1) implies that they cover a dense subset of the square.

We deduce from this that corresponding to a dense subset of the torus we must add one point only, and then the whole completion is obtained by adding

Fig. E.78. If p/q is not rational we get a dense subset of the square

one point only. Let us remark that this completion is not even a topological manifold.

(2) $P/q \in \mathbb{Q}$.

With the same argument as above we have now that the line translated to the square closes up after a finite time. Let p_1 and q_1 be co-prime integers such that $P/q = P_1/q_1$; these numbers determine the set of points of the square corresponding to which we must add $1 \in \mathbb{R}_+$ (see Fig. E.79). Of course if we look for the set corresponding to another $t \in \mathbb{R}_+$ different from 1 we get a line parallel to the previous one and then in the square the situation is essentially the same (see Fig. E.80).

Fig. E.79. Case of p/q rational: the points of the square corresponding to which we must add 1. The figure refers to the case $p_1 = 3$ and $q_1 = 2$

Fig. E.80. The same situation of Fig. E.79 with 1 replaced by some other t, which is supposed to be slighty bigger than 1

We can prove now that the completion is actually achieved by adding a loop. Figure E.81 describes a special proof of this in the example of Fig. E.79. In general, let r and s be integer numbers such that $r \cdot p_1 + s \cdot q_1 = 1$ and translate to the square the line $\mathbb{R} \cdot (-s, r)$. The resulting object represents a loop and it is easily seen that the points we must add to get the completion correspond bijectively to the points of such a loop (see Fig. E.82).

We have proved that the completion corresponds to adding a loop; moreover we have that the loop $p \cdot l + q \cdot m$ (we recall that l and m denote the longitude and the meridian of the torus) corresponds to one point only, *i.e.* it is homotopic to 0 in the completed space. It follows that the completion is topologically obtained by a Dehn surgery of coefficients (p_1, q_1). Remark that

Fig. E.81. Example of Fig. E.79 continued. It is easily seen that the points we must add correspond bijectively to the points of the loop obtained by gluing the endpoints of the segment thickened in the figure

Fig. E.82. How to apply the general proof working to the example of Fig. E.79. In this case we have $r = -1$ and $s = 2$

(p_1, q_1) is determined up to a change of sign, and indeed we have that opposite pairs correspond to the same surgery (E.4.2).

(3) We determine now the cases when the deformed hyperbolic structure extends to the completed manifold. We start by remarking that if we follow the loop $p_1 \cdot l + q_1 \cdot m$ we are actually spinning once around the added loop: this fact is quite evident if we think of the abstract operation of Dehn surgery, but it is convenient to see this using our developing mapping D, as in shown Fig. E. 83.

The condition that the hyperbolic structure extends to the completed manifold is then equivalent to the fact that while following $p_1 \cdot l + q_1 \cdot m$ we find on the added loop an angle of 2π with respect to the hyperbolic structure; this condition means that

$$p_1 \cdot u + q_1 \cdot v = \pm 2\pi i$$

which is equivalent to having $(p_1, q_1) = \pm(p, q)$, *i.e.* to the fact that (p, q) is a pair of co-prime integers. If we remark that in this case the loop we are adding is geodesic and the surgery torus is a tubular neighborhood of such a loop, we finally get the following result:

Proposition E.6.27. If the real solution $(p, q) \neq \infty$ of the equation

$$p \cdot u + q \cdot v = 2\pi i$$

is a pair of co-prime integers then the completion of $L \times [0, \infty)$ is obtained by performing a (weak) hyperbolic Dehn surgery of coefficients (p, q).

Remark E.6.28. If we define for $u = v = 0$ the solution of $p \cdot u + q \cdot v = 2\pi i$ to be ∞, according to E.6.24 the above proposition holds too, as by definition a surgery of coefficient ∞ corresponds to keeping the manifold unchanged.

According to the above proposition and to E.6.23, for the conclusion of the proof of Thurston's hyperbolic Dehn surgery Theorem E.5.1 we only have to establish the following (quite easy):

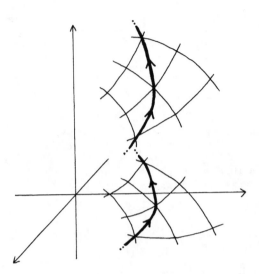

Fig. E.83. Proof that in the case of co-prime integers the structure extends to the completion

Proposition E.6.29. If $\{z_i\}$ is a sequence in $\mathrm{Def}(M)$ converging to z^0 such that $\forall i$ the completion M_i of the hyperbolic structure induced by z_i on M corresponds to a hyperbolic Dehn surgery, then the sequence M_i converges to M in $\mathcal{F}_3$.

Proof. Given $z \in \mathrm{Def}(M)$ different from z^0 and close to it, for $\nu = 1, ..., n$ we can consider a diffeomorphism ϕ_ν of the ideal tetrahedron $\Delta_{z_\nu^0}$ of modulus z_ν^0 onto the ideal tetrahedron Δ_{z_ν} of modulus z_ν. Moreover given any compact subset K of $\Delta_{z_\nu^0}$ as z_ν converges to z_ν^0 we can choose such a diffeomorphism to be arbitrarily close to an element of $\mathcal{I}^+(\mathbb{H}^3)$ (or to the identity, if the representatives of the tetrahedra are suitably chosen) with respect to the C^∞ topology on K.

Of course we can choose $\phi_1, ..., \phi_n$ in such a way that they respect the glueings, so that they define a mapping ϕ from M to M_z (the completion of M with respect to the hyperbolic structure induced by z). Such mapping can be lifted to (suitably chosen) universal covers and gives rise to an equivariant mapping which is arbitrarily C^∞-close to the identity on an arbitrarily large ball, and then the conclusion follows from the geometric characterization of the Chabauty topology. $\qquad\square$

E.7 Applications to the Study of the Volume Function and Complements about Three-dimensional Hyperbolic Geometry

We want to outline now some more advanced results in the theory of three-dimensional hyperbolic manifolds; we confine ourselves to sketches and we generally omit proofs.

The first questions we consider deal with the volume function. It has been proved in the previous section that non-trivial convergent sequences do exist in $\mathcal{F}_3$, but of course this does not imply that the image of the volume function on $\mathcal{F}_3$ is not a discrete subset of $\mathbb{R}_+$ (as happens for $n \neq 3$); in fact the volume may be constant along such a convergent sequences. Actually, it is possible to prove that this is not the case, *i.e.* that $\mathrm{vol}(\mathcal{F}_3)$ does have limit points. According to the results of Sect. C.2 we have that if a hyperbolic manifold is obtained by glueing the faces of ideal simplices in $\mathbb{H}^3$ of moduli $z_1, ..., z_n \in \Pi^{2,+}$ then its volume is given by the function

$$V(z_1, ..., z_n) = \Lambda(z_1) + ... + \Lambda(z_n)$$

(where Λ denotes the Lobachevsky function). This formula applies also to the situation we considered for the proof of the hyperbolic surgery theorem (whose notations we keep): since the completion is obtained by adding a set of volume 0, if $(z_1, ..., z_n) \in \mathrm{Def}(M)$ corresponds to a hyperbolic surgery then the volume of the surgered hyperbolic manifold is $V(z_1, ..., z_n)$. It is possible to show that Λ is a real analytic function on $\Pi^{2,+}$, which implies that V is real analytic on $(\Pi^{2,+})^n$; for instance it is checked in [Ne-Za] that

$$\Lambda(z) = \Im(\mathrm{Li}_2(z)) + \log|z| \cdot \arg(1 - z)$$

where, for $|z| \leq 1$,

$$\mathrm{Li}_2(z) = \sum_{j=1}^{\infty} \frac{z^j}{j^2}$$

and an analytic continuation on the whole $\Pi^{2,+}$ is considered. Moreover, using its explicit form, it is possible to show that V is not a constant function on $\mathrm{Def}(M)$ and it is not difficult to deduce from this that in the neighborhood of $(\infty, ..., \infty)$ corresponding to the coefficients giving hyperbolic surgeries we can find sequences convergent to $(\infty, ..., \infty)$ such that the volume of any element of the sequence is different from the volume of the limit manifold M. We obtain from this the following:

Corollary E.7.1. The volume of an element of $\mathcal{F}_3$ having $k \geq 1$ cusps is a limit point for $\mathrm{vol}(\mathcal{F}_3)$.

A careful study of the function V actually leads to a more precise estimate on the order of convergence of the volume function along a convergent sequence obtained by the hyperbolic surgery theorem: the following result was established in [Ne-Za].

Theorem E.7.2. Let $M \in \mathcal{F}_3$ have $k \geq 1$ cusps and let the notations of the previous section be fixed. If $\{z^i\}$ is a sequence in $\mathrm{Def}(M)$ convergent to z^0 with the property that z^i induces on M a hyperbolic structure whose completion M_ν is obtained by a hyperbolic Dehn surgery of coefficients $(p_1^i, q_1^i), ..., (p_k^i, q_k^i)$, then

$$\mathrm{vol}(M_\nu) = \mathrm{vol}(M) - \pi^2 \sum_{j=1}^k \frac{\mathcal{A}(L_j)}{r(p_j^i, q_j^i)} + O\left(\sum_{j=1}^k \frac{1}{(p_j^i)^4 + (q_j^i)^4} \right)$$

where $\mathcal{A}(L_j)$ denotes the area of the torus L_j with respect to the Euclidean structure induced by the complete hyperbolic structure of M and $r(p_j^i, q_j^i)$ is the length of the curve $p_j^i \cdot l_j + q_j^i \cdot m_j$ on the torus L_j with respect to this Euclidean structure; moreover by definition $r(p_j^i, q_j^i) = \infty$ if $(p_j^i, q_j^i) = \infty$.

The following important result deals with the volume function again and shades new light on the results of the previous sections; it can be found in [Th1, ch. 6].

Theorem E.7.3. If $M \in \mathcal{F}_3$ is obtained by hyperbolic Dehn surgery from $M_0 \in \mathcal{F}_3$ and $M \neq M_0$ then $\mathrm{vol}(M) < \mathrm{vol}(M_0)$.

The proof of this result requires the extension of the notion of Gromov norm to the case of manifolds with boundary and the generalization of the rigidity theorem stated in C.5.5. As straight-forward consequences we get the following very important corollaries which settle the study of the volume function in dimension three:

Corollary E.7.4. Given $v \in \mathbb{R}_+$ the set $\mathrm{vol}^{-1}(v) \subset \mathcal{F}_3$ is finite.

Corollary E.7.5. $\mathrm{vol}(\mathcal{F}_3)$ is a well-ordered subset of $\mathbb{R}_+$.

Remark that the above Theorem E.7.3 is stronger than E.7.2 in a sense (since it applies to all the elements of $\mathcal{F}_3$ obtained from M by hyperbolic surgery, not only to those coming from the hyperbolic surgery theorem) and weaker than E.7.2 in another sense (as it does not give a precise upper bound).

Before stating some further results we describe a problem naturally arising from the hyperbolic surgery theorem (*cf.* E.4.8 too). Let us denote by $\mathcal{I}_3$ the subset of $\mathcal{F}_3$ consisting of all the hyperbolic manifolds which can be obtained by hyperbolic Dehn surgery from a non-compact element of $\mathcal{F}_3$. Since we know that the volume function is proper and that if a sequence $\{M_i\}$ converges to M then M_i is obtained from M by hyperbolic surgery (for $i \gg 0$) it is easily verified that for any $c > 0$ the set

$$(\mathcal{F}_3 \setminus \mathcal{I}_3) \cap \mathcal{F}_3(c)$$

is finite. It follows that the complement of $\mathcal{I}_3$ in $\mathcal{F}_3$ is somewhat "small". The question is whether $\mathcal{I}_3$ is the whole $\mathcal{F}_3$ or not. It seems to be stated in [Ne-Za] that the answer is affirmative and that the proof can be found in [Th3]: we were not able to find such a result in [Th3] (though we tried to overcome the initial difficulty that it is not explicitly stated). The strongest fact we were able to obtain with the methods of [Th3] is the existence of generalized triangulations of an element of $\mathcal{F}_3$ by ideal tetrahedra (with a possibly null or negative volume) in the non-compact case (a result we exposed in E.5-ii with a slightly simpler proof) and, in the general case, a similar triangulation for subsets of full measure. Hence, in our opinion, the problem of whether $\mathcal{I}_3 = \mathcal{F}_3$ or not is open. Remark that at the end of the present section we shall prove in particular that it is possible to obtain the whole $\mathcal{F}_3$ by topological Dehn surgeries on non-compact elements of $\mathcal{F}_3$. Hence the above question may be related to the problem of the representation of the isotopy class of a simple closed curve in a compact hyperbolic manifold by a simple closed geodesic curve.

We obtained above as a corollary of E.7.3 that the fiber of any point in the image of the volume function on $\mathcal{F}_3$ is finite. Then it is a natural problem to ask whether it is possible or not to give a uniform upper bound for the number of elements of suh fiber. The answer is negative, since the following result to be found in [Wi] holds:

Theorem E.7.6. Given any $\nu \in \mathbb{N}$ there exist ν different elements of $\mathcal{F}_3$ having the same volume.

The quoted paper [Wi] exhibits explicit examples; the phenomenon originating these examples (deeply investigated in [Ad1]) is the following: let $M \in \mathcal{F}_3$ contain a thrice-punctured incompressible sphere S, cut M along S and glue the resulting pieces by any orientation-preserving homeomorphism; the result is still a hyperbolic manifold having the same volume as M. As was pointed out in [Ad1] this fact allows one to obtain an easily-applicable method for producing examples; we roughly describe applications of this method.

For the sake of brevity we shall say a link in S^3 is hyperbolic if its complement can be endowed with a complete finite-volume hyperbolic structure. Let L be a hyperbolic link and assume some part of the planar projection of L is in the situation represented in Fig. E.84 (remark that this projection is a forbidden one for the construction we described in Sect. E.5-iv). Let us replace such part with the projection shown in Fig. E.85 and let us denote by L' the new link obtained. Then L' is hyperbolic too and $\mathrm{vol}(S^3 \setminus L') = \mathrm{vol}(S^3 \setminus L)$.

Another example of Adam's results is described in Figg. E.86 and E.87.

We want to describe now in some detail the method for producing examples of elements of $\mathcal{F}_3$ implicit in our proof of the hyperbolic surgery theorem. Let us start with an element M of $\mathcal{T}_3$ (we recall that if M is the interior of a compact manifold $\overline{M}$ whose boundary consists of tori then such a presentation in $\mathcal{T}_3$ can be obtained starting from a triangulation of $\overline{M}$). It is quite natural to consider the following questions:

(1) does $\mathcal{H}(M)$ contain complete structures?
(2) does M belong to $\mathcal{F}_3$?

We saw that the first question is precisely equivalent to the existence in a product of n upper half-planes of a solution for a system of rational equations

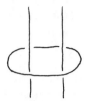

Fig. E.84. First example of Adam's methods to produce hyperbolic links. We can replace the projection represented here with the projection represented in Fig. E.85

Fig. E.85. First example of Adam's methods continued. The volume of the new link complement is the same as the volume of old one

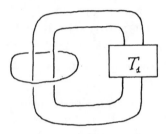

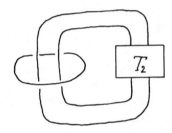

Fig. E.86. Second example of Adam's methods. If we have two hyperbolic links as represented here, the link represented in Fig. E.87 is hyperbolic

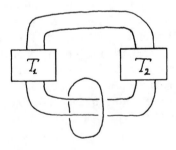

Fig. E.87. Second example of Adam's methods continued. The volume of the new link complement is the sum of the volumes of the old ones

(E.6.3 and E.6.8; n represents as usual the number of tetrahedra involved). It follows that this problem can be treated by numerical methods, so that it is possible to answer to (1) in an effective way. Moreover E.6.16 can be sharpened by the following result whose proof we outline:

Proposition E.7.7. If a solution exists then it is unique.

Proof. Given two solutions it is possible to construct a diffeomorphism of M onto itself identifying homologous tetrahedra and respecting the glueings. According to the rigidity theorem (in the non-compact case) this diffeomorphism is homotopic to an isometry, and its lifting to the universal cover $\mathbb{H}^3$ keeps fixed the points of the sphere at infinity. Since an ideal geodesic tetrahedron is determined by its points at infinity, homologous tetrahedra are actually isometric: in particular they have the same moduli, and then the conclusion follows. □

As for question (2), we go back to what we saw in E.5.7 and E.5.8. If $M \in \mathcal{F}_3$ we know that a solution of the same system of rational equations as above exists in Λ^n and not in $\left(\Pi^{2,+}\right)^n$, where $\Lambda = \mathbb{C} \setminus \{0, 1\}$. It follows that if such a solution does not exist we can state that M does not belong to $\mathcal{F}_3$. On the other hand the existence of a solution is not enough for concluding that $M \in \mathcal{F}_3$. Assume such a solution is given and think of the geometric situation: we can actually define a complete hyperbolic structure on M only if the algebraic sum of the oriented angles we "physically" find around each edge is 2π, and it is not easy to decide if this is true or not: remark for instance that it may happen that an edge lies inside a tetrahedron of negative volume, so that the tetrahedron contributes with -2π to the sum. The point we find not very satisfactory in this construction is that the condition on the angles is not purely algebraic and it does not depend only on the combinatorial data of the realization of M in T_3.

Going back to (1), we can describe some explicit examples. Using the algorithmic construction given in E.5-iv for the realization of the complement of a link in S^3 as an element of T_3, it is possible to prove by direct calculation that the eight knot, the Whitehead link, Borromeo's rings and the cyclic link with $k \geq 3$ components are hyperbolic. (We remark, however, that for these examples it is easier to consider directly the presentation of the complement as the union of two polyhedra with glued faces, as described in E.5-iv, and then use the particular symmetries of the situation to get the right angles: see [Th1].) These examples are quite relevant for the following reasons:

– the complement of the eight knot is fibered over S^1; if we recall what we saw in C.5.16 we obtain another feature of the flexibility of hyperbolic geometry in dimension 3. Remark that the existence of hyperbolic 3-manifolds fibering over S^1 is definitely not evident *a priori*; for instance if we consider the natural embeddings in $\mathbb{H}^3$ of the universal cover of the fibers (homeomorphic to $\mathbb{R}^2$) we get a foliation of $\mathbb{H}^3$ by a one-parameter family of these planes with the property that any two of them keep at a bounded distance, and it is not at all obvious that such a foliation exists;

– Borromeo's rings provide an example of what we remarked in E.5.2, *i.e.* that we may need to exclude infinitely many surgeries: if we fix a component of the link and we perform along it a surgery of coefficients $(1, 0)$, *i.e.* we close the cusp in the natural way, any other surgery on the other two components produces either a non-irreducible manifold (which is not hyperbolic) or a lens space (which is not hyperbolic too as its universal cover is the sphere);

– the cyclic links show that there exist hyperbolic manifolds with arbitrarily many cusps.

With the same method as above (algebraic description of the complete structures of $\mathcal{H}(M)$ for any element of $\mathcal{T}_3$) in [Ad-Hi-We] were determined (using automatic means of calculation) all hyperbolic links in S^3 having at most 4 components whose projection has at most 9 crossings (important invariants, and in particular the volume, were also calculated there).

The construction of examples of elements of $\mathcal{F}_3$ is also related to the problem of finding a hyperbolic three-manifold with minimal volume. In [Ma-Fo] hyperbolic manifolds with low "complexity" were considered (with the aid of a computer again) and a natural candidate for the minimal-volume element of $\mathcal{F}_3$ was found. (The notion of complexity is based on the minimal number of vertices of a standard spine, and the explicit bound considered in [Ma-Fo] for the complexity is 6.)

The (theoretical) determination of plenty of examples of elements of $\mathcal{F}_3$ is based on the great hyperbolization theorem of W. Thurston (the proof of which has not yet been entirely published: see [Th3] or [Mor]). We confine ourselves to the statement of this theorem (in the special case of non-compact finite-volume manifolds) without further discussion, as it would take us too far from the aims of the present book. Remark that the meaning of this theorem is that the necessary conditions stated at the end of Chapt. D for a non-compact manifold to be hyperbolic are "almost" sufficient:

Theorem E.7.8. Let $\overline{M}$ be a compact three-manifold with non-empty boundary made of tori. Assume the interior M of $\overline{M}$ is irreducible and that $\overline{M}$ is atoroidal. Then either M belongs to $\mathcal{F}_3$ or $\overline{M}$ is one of the following manifolds:
– the product of the 1-sphere and the closed 2-disc;
– the product of the unit interval and the torus;
– the oriented fiber bundle over the Klein bottle with the unit interval as fiber.

This theorem has the consequence that it is possible to describe a large class of links in S^3 whose complement is hyperbolic, the description being easily read on a generic planar projection of the link; see for instance [Ad2]. Using these results it is not so hard to prove the following:

Proposition E.7.9. Given any link L in S^3 (by a generic planar projection) there exists a (constructive) way to add some components to L in such a way that the resulting link is hyperbolic.

The Lickorish-Wallace theorem yields immediately the following:

Corollary E.7.10. All compact three-manifolds are obtained by (topological) Dehn surgeries along hyperbolic links.

This corollary, together with the hyperbolic surgery theorem, justifies the heuristical statement that "almost all three-manifolds are hyperbolic".

Chapter F.
Bounded Cohomology, a Rough Outline

In this chapter we point out some of the basic ideas of the theory of bounded cohomology we first met during the proof of the rigidity theorem (Sect. C.3: compare F.2.2 below). In particular we define the groups of singular bounded cohomology and we consider the natural class of cohomology arising from the problem of the existence of a global non-vanishing section on a flat fiber bundle (known as Euler class of the bundle). In connection with the notion of Euler class we introduce and develop the definition of amenable group.

Several proofs are omitted or incomplete, and only the first features of this theory are mentioned, but we hope this sketch may be useful as a first introduction to this theory. The chief reference for many of the results we will quote is [Gro3] once again; see also [Iv] for an approach closer to standard ideas of algebraic topology.

Though we will not be dealing in this chapter with hyperbolic geometry in the strict classical sense, we believe it is worth including this material as it introduces the reader to some examples of "hyperbolic" behaviour in a much more general sense. As a very important idea related to the subject of this chapter we mention the theory of hyperbolic groups recently developed by Gromov and other authors ([Gro4] is the original source; see [Gh-LH] for a more detailed exposition).

F.1 Singular Cohomology

Let $R \in \{\mathbb{Z}, \mathbb{R}\}$ and let X be a topological space; we shall refer to the notations of Sect. C.3 for singular cycles, chains, boundaries and homology R-modules of X. For $n \geq 0$ we define the set of n-cochains in X as the purely algebraic dual R-module $C^n(X; R)$ of $C_n(X; R)$. We shall denote by brackets the duality pairing:

$$C_n(X; R) \times C^n(X; R) \ni (z, c) \mapsto [z, c] \in R.$$

(Remark that by definition $[.,.]$ is bi-linear; we shall also use it as a symmetric function by setting $[c, z] = [z, c]$.)

The dual operator $\delta^n : C^n(X; R) \to C^{n+1}(X; R)$ of ∂_n, defined by the relation

$$[\partial_n z, c] = [z, \delta^n c] \qquad \forall z \in C_{n+1}(X; R), \ c \in C^n(X; R)$$

is called the n-th <u>coboundary operator</u>.

Identity $\partial_{n-1} \circ \partial_n = 0$ immediately implies $\delta^n \circ \delta^{n-1} = 0$ so that if we define $Z^n(X; R)$ as the kernel of δ^n (called the set of <u>cocycles</u>) and $B^n(X; R)$ as the range of δ^{n-1} (called the set of <u>coboundaries</u>), then we can introduce the <u>n-th singular cohomology R-module</u> of X as the quotient set

$$H^n(X; R) = {Z^n(X; R)}/{B^n(X; R)}$$

(we shall set $H^n(X; R) = \{0\}$ for $n < 0$).

If $A \subset X$, the same method applied to the modules $C_n(X, A; R)$ allows us to define the relative notions of cocycle, coboundary and singular cohomology.

Let us remark that by applying to the exact sequence

$$0 \to C_n(A; R) \hookrightarrow C_n(X; R) \xrightarrow{p_i} C_n(X, A; R) \to 0$$

the functor $\mathrm{Hom}(\,.\,, R)$ (*i.e.* by taking the dual of everything), another sequence is obtained

$$0 \to C^n(X, A; R) \to C^n(X; R) \to C^n(A; R) \to 0$$

and this sequence is exact too. In particular this method allows to identify $C^n(X, A; R)$ with the R-submodule of $C^n(X; R)$ of the annihilators of $C_n(A; R)$.

If we define $\delta^n : C^n(A; R) \to C^{n+1}(X, A; R)$ as the dual operator of ∂_n we obtain the exact diagram:

$$
\begin{array}{ccccccccc}
0 & \to & C^n(X, A; R) & \to & C^n(X; R) & \to & C^n(A; R) & \to & 0 \\
 & & \downarrow{\scriptstyle \delta^n} & & \downarrow{\scriptstyle \delta^n} & & \downarrow{\scriptstyle \delta^n} & & \\
0 & \to & C^{n+1}(X, A; R) & \to & C^{n+1}(X; R) & \to & C^{n+1}(A; R) & \to & 0
\end{array}
$$

from which we deduce that the <u>coboundary</u> operator is well-defined when passing to the quotient:

$$\delta^n : H^n(A; R) \to H^{n+1}(X, A; R).$$

We will list now the first essential properties of singular cohomology which follow quite easily from the definition and from the properties of singular homology. We shall often omit the specification of the underlying ring R.

Proposition F.1.1. (Naturality.) Given $f : (X, A) \to (Y, B)$ (which means that $f : X \to Y$ is continuous and $f(A) \subset B$) a homomorphism of R-modules

$$f^* : H^n(Y, B) \to H^n(X, A)$$

is associated to f for all $n \in \mathbb{Z}$; moreover $\mathrm{id}^* = \mathrm{id}$ and $(g \circ f)^* = f^* \circ g^*$.
(Coboundary morphism.) Given $f : (X, A) \to (Y, B)$ the following diagram is commutative:

$$H^n(B) \xrightarrow{\delta^n} H^{n+1}(Y, B)$$

$$\downarrow f^* \qquad\qquad \downarrow f^*$$

$$H^n(A) \xrightarrow{\delta^n} H^{n+1}(X, A).$$

(Cohomology exact sequence.) If $i : A \hookrightarrow X$ and $j : (X, \phi) \hookrightarrow (X, A)$ are the natural inclusions, the following sequence is exact:

$$0 \to H^0(X, A) \to ... \to H^n(X) \xrightarrow{i^*} H^n(A) \xrightarrow{\delta^n} H^{n+1}(X, A) \xrightarrow{j^*} (X) \to ... \ .$$

(Homotopy.) If $f, g : (X, A) \to (Y, B)$ are homotopic (under a homotopy F such that $F(A \times [0, 1]) \subset B$) then $f^* = g^*$.

(Excision.) If $U \subset A \subset X$, U is open and the interior of A contains the closure of U then the natural inclusion $\phi : (X \setminus U, A \setminus U) \hookrightarrow (X, A)$ induces a (surjective) isomorphism of R-modules

$$\phi^* : H^n(X, A) \xrightarrow{\sim} H^n(X \setminus U, A \setminus U) \ .$$

(Dimension.) if X consists of a single point then

$$H^n(X; R) \cong \begin{cases} R & \text{if } n = 0 \\ 0 & \text{otherwise.} \end{cases}$$

The properties listed above are analogous to (and proved in a similar way as) the Eilenberg-Steenrod axioms about homology. We have other facts making cohomology a much richer object than homology.

We start by remarking that a R-bi-linear mapping

$$[.,.] : H_n(X, A) \times H^n(X, A) \longrightarrow R$$

is naturally defined by $[\langle z \rangle, \langle c \rangle] = [z, c]$. This mapping induces a homomorphism

$$\alpha : H^n(X, A) \to H_n(X, A)^*$$

and it may be shown (see [Greenb1]) that the following holds:

Proposition F.1.2. If $R = \mathbb{R}$ then α is a (surjective) isomorphism, and if $R = \mathbb{Z}$ then α is onto.

According to this proposition one might think that (at least in the case $R = \mathbb{R}$) the passage from homology to cohomology has given nothing new: on the contrary we shall check now that $H^*(X)$ is naturally endowed with an R-algebra structure, and no such structure is defined on $H_*(X)$.

Remark F.1.3. Though we gave the definition of $H^n(X; R)$ with respect to the standard singular theory, we remark that the same method works for any algebraic complex $\{(C'_n, \partial'_n) : n \in \mathbb{Z}\}$ defining the homology of X. In particular if X is a smooth compact manifold (whence triangulated) then we can make use of the "simplicial theory" we briefly mentioned in C.3.

We recall that $H^*(X)$ is defined as the R-module generated by all the $H^n(X)$'s for $n \in \mathbb{Z}$; we want to define an operation

$$\cup : H^*(X) \times H^*(X) \to H^*(X)$$

called the cup product. We start by defining for all $p, q \geq 0$ the product

$$\cup : H^p(X) \times H^q(X) \to H^{p+q}(X).$$

(Since it will turn out that this product satisfies the distributive laws then for the global definition on $H^*(X)$ we will only need to extend by R-bi-linearity.)
Let us consider the following mappings:

$$\lambda_{p,q} : \Delta_p \to \Delta_{p+q} \qquad (t_0, ..., t_p) \mapsto (t_0, ..., t_p, 0, ..., 0)$$

$$\rho_{p,q} : \Delta_q \to \Delta_{p+q} \qquad (t_0, ..., t_q) \mapsto (0, ..., 0, t_0, ..., t_q).$$

Given $c \in C^p(X)$, $d \in C^q(X)$ we define $c \cup d \in C^{p+q}(X)$ by

$$[\sigma, c \cup d] = [(\sigma \circ \lambda_{p,q}), c] \cdot [(\sigma \circ \rho_{p,q}), d] \quad \text{for } \sigma : \Delta_{p+q} \to X$$

extended by linearity on $C_{p+q}(X)$).

Lemma F.1.4. $\cup$ is well-defined on the quotient spaces:

$$\cup : H^p(X) \times H^q(X) \to H^{p+q}(X).$$

Proof. we recall that if for $p \geq 0$ and $0 \leq i \leq p+1$ we set

$$j_{i,p} : \Delta_p \ni (t_0, ..., t_p) \mapsto (t_0, ..., 0_i, ..., t_p) \in \Delta_{p+1}$$

then for $\sigma : \Delta_{p+1} \to X$ we have

$$\partial \sigma = \sum_{i=0}^{p} (-1)^i (\sigma \circ j_{i,p}).$$

Then an easy calculation (based on the relations between the $\lambda_{p,q}$'s, the $\rho_{p,q}$'s and the $j_{i,p}$'s) proves that

$$\delta(c \cup d) = (\delta c) \cup d + (-1)^p \cdot c \cup (\delta d).$$

Then if $\delta c = \delta d = 0$ we have $\delta(c \cup d) = 0$ and moreover if $c = \delta c'$, $d = \delta d'$ then $c \cup d = \delta(c' \cup \delta d')$. $\square$

Since bi-linearity of $\cup$ is immediately checked an R-algebra structure is now defined on $H^*(X)$. We record the following interesting fact (to be found in [Greenb1]):

Proposition F.1.5. For $c \in H^p(X)$ and $d \in H^q(X)$ we have

$$d \cup c = (-1)^{p \cdot q} \cdot c \cup d.$$

Example F.1.6. Let M be an n-dimensional compact oriented manifold and let $F \subset M$ be a p-dimensional closed oriented submanifold (with $p < n$); set $q = n - p$. We can think of F as an element of $H^q(M, \mathbb{Z})$ in the following way: given $\langle z \rangle \in H_q(M, \mathbb{Z})$ we can choose the representative z to be the sum of simplices transversal to F, i.e. $z = \sum \lambda_i \cdot \sigma_i$ where each $\sigma_i : \Delta_q \to M$ is differentiable and if $t \in \Delta_q$ is such that $x = \sigma_i(t) \in F$ we have that $T_x M$ is the direct sum of $T_x F$ and the range of $d_t \sigma_i$. (It is easily verified that such a z exists, since given an arbitrary representative z' we can find a chain z satisfying the above properties and arbitrarily close to z', so that z and z' are homotopic and hence they represent the same homology class.) For $t \in \Delta_q$ we define $\epsilon_i(t)$ to be 0 if $\sigma_i(t) \notin F$, $+1$ if $d_t \sigma_i(u_1), ..., d_t \sigma_i(u_q), w_1, ..., w_p$ is a positive basis of $T_x M$ ($u_1, ..., u_q$ and $w_1, ..., w_p$ being positive bases of $\mathbb{R}^q$ and $T_x F$ respectively) and -1 otherwise. We set now

$$\alpha_F(z) = \sum_i \sum_{t \in \Delta_q} \lambda_i \cdot \epsilon_i(t).$$

(Remark that $F \cap \sigma_i(\Delta_q)$ is finite for all i, so that the sum is finite.) It may be shown that this number depends only on the homology class of z and that $\alpha_F \in H^q(M, \mathbb{Z})$. Moreover if F_1 and F_2 are oriented closed submanifolds of M then $\alpha_{F_1} \cup \alpha_{F_2}$ is given by α_F, where F denotes the tranversal intersection of F_1 and F_2 (i.e. the intersection of two manifolds arbitrarily close to F_1 and F_2 and transversal to each other, tranversality meaning as usual that the tangent spaces lie in generic position at the intersection points).

Similarly one could define the cup product with respect to the "simplicial theory" of the homology of a smooth compact manifold.

F.2 Bounded Singular Cohomology

In this section we shall always assume that the underlying ring is $\mathbb{R}$. For $c \in C^k(X)$ we set

$$\|c\|_\infty = \sup \left\{ |c(\sigma)| : \sigma : \Delta_k \to X \text{ continuous} \right\} \in [0, \infty]$$

and then we define

$$\hat{C}^k(X) = \left\{ c \in C^k(X) : \|c\|_\infty < \infty \right\}.$$

It is easily checked that $\delta^k\left(\hat{C}^k(X)\right) \subseteq \hat{C}^{k+1}(X)$ and hence it makes sense to define the k-th <u>bounded</u> <u>cohomology</u> <u>group</u> $\hat{H}^k(X)$ of X as the quotient space of the kernel of $\delta^k\big|_{\hat{C}^k(X)}$ by the range of $\delta^{k-1}\big|_{\hat{C}^{k-1}(X)}$. It easily follows from the definition that an $\mathbb{R}$-homomorphism (i.e. an homomorphism of real vector spaces) is defined:

$$\hat{H}^k(X) \longrightarrow H^k(X),$$

but in general it is neither one-to-one nor onto. Its range consists precisely of the cohomology classes of bounded chains.

Moreover a quotient pseudo-norm is defined on $\hat{H}^k(X)$ by

$$\|\beta\|_\infty = \inf\left\{\|c\|_\infty : c \in C^k(X),\ \langle c \rangle = \beta\right\} \qquad \beta \in \hat{H}^k(X).$$

It is not so hard check that the sequence $\hat{H}^*(X)$ satisfies the properties of naturality, homotopy and dimension, but in general not the other ones, so that its determination is much harder.

The following remark shades new light on the real meaning of our construction of bounded cohomology.

Remark F.2.1. If we set on $C_k(X)$ the norm

$$\left\|\sum_i a_i \cdot \sigma_i\right\|_1 = \sum_i |a_i|$$

(where $a_i \in \mathbb{R}$, $\sigma_i : \Delta_k \to X$ is continuous and we understand that the shortest possible expression is chosen, *i.e.* the σ_i's are different from each other), then the completion $\overline{C_k(X)}$ of $C_k(X)$ with respect to this norm is given by the set of all formal sums

$$\sum_{i=0}^{\infty} a_i \cdot \sigma_i$$

with $a_i \in \mathbb{R}$, $\sigma_i : \Delta_k \to X$ continuous and $\sum_i |a_i| < \infty$. Moreover well-known facts in functional analysis imply that the space $\hat{C}^k(X)$ defined above is isometrically isomorphic (as a normed space) to the (topological) dual space of $\overline{C_k(X)}$. In particular we have the following identities:

$$c \in \overline{C_k(X)} \ \Rightarrow\ \|c\|_1 = \sup\left\{|z(c)| : z \in \hat{C}^k(X),\ \|z\|_\infty = 1\right\}$$

$$z \in \hat{C}^k(X) \ \Rightarrow\ \|z\|_\infty = \sup\left\{|z(c)| : c \in \overline{C_k(X)},\ \|c\|_1 = 1\right\}.$$

We recall that in Sect. C.3 we introduced the Gromov norm $\|M\|$ of a manifold M.

Proposition F.2.2. (1) Given $z \in H_k(X)$ we have

$$\|z\|_1^{-1} = \inf\left\{\|\beta\|_\infty : \beta \in \hat{H}^k(X),\ [\beta, z] = 1\right\}$$

(this formula meaning that for $\|z\|_1 = 0$ the infimum is taken over the empty set.) In particular if M is an n-dimensional oriented compact manifold then

$$\|M\|^{-1} = \inf\left\{\|\beta\|_\infty : \beta \in \hat{H}^n(M),\ [\beta, [M]] = 1\right\}.$$

Proof. Let z be represented by a chain $\sum_i a_i \sigma_i$, let β be represented by a cochain γ and assume $[\beta, z] = 1$; then

$$1 = \sum_i a_i \gamma(\sigma_i) \le \sum_i |a_i| \cdot |\gamma(\sigma_i)| \le \left(\sum_i |a_i|\right) \cdot \|\gamma\|_\infty;$$

if we take the infimum over all representatives of β and z we obtain

$$1 \le \|z\|_1 \cdot \|\beta\|_\infty$$

which implies inequality $\ge$.

The case $\|z\|_1 = 0$ follows as well, so we prove inequality $\geq$ for $\|z\|_1 > 0$. Let c be a representative of z in $Z_k(X)$ and remark that

$$\|z\|_1 = \inf\left\{\|c'\|_1 : \langle c'\rangle = z\right\} = \inf\left\{\|c - d\|_1 : d \in B_k(X)\right\} = \mathrm{dist}(c, B_k(X)).$$

Then by the Hahn-Banach theorem we can find a continuous functional γ on $\overline{C_k(X)}$ such that:

(i) $\gamma(c) = 1$;

(ii) $\gamma\big|_{B_k(X)} = 0$;

(iii) $\|\gamma\|_\infty = \mathrm{dist}(c, B_k(X))^{-1} = \|z\|_1^{-1}$.

If we keep denoting by γ the restriction to $Z_k(X)$ we have by (ii)

$$[\delta(\gamma), d] = [\gamma, \partial(d)] = 0 \ \forall d \in C_{k+1}(X) \quad \Rightarrow \quad \delta(\gamma) = 0$$

hence γ is a cocycle and the element β of $H^k(X)$ it represents satisfies

$$[\beta, z] = 1 \qquad \|\beta\|_\infty \leq \|z\|_1^{-1}$$

which implies the conclusion. $\qquad\qquad\qquad\qquad\qquad\qquad\qquad\qquad\qquad\qquad$ □

Corollary F.2.3. If M is a compact oriented n-manifold then $\|M\| \neq 0$ if and only if there exists a bounded $\beta \in \hat{H}^n(M)$ which does not vanish on $[M]$.

Proposition F.2.4. For all X we have $\hat{H}^1(X) = \{0\}$.

Proof. Each 1-simplex σ can be represented as a mapping on S^1 (viewed as the unit sphere of C). Moreover if for $k \in \mathbb{N}$ we denote by $\sigma^{(k)}$ the composition of σ with the raising to the k-th power we have that σ is homologous to $1/k \cdot \sigma^{(k)}$, and the proof is easily completed. $\qquad\qquad\qquad\qquad\qquad\qquad\qquad\qquad$ □

As an application of the techniques developed in the present section we prove the following interesting result concerning the Gromov norm.

Theorem F.2.5. Let N and M be compact connected oriented manifolds; then:

(1) if the sum of the dimensions of N and M is k there exists a constant $c(k)$ depending on k only such that

$$\|N \times M\| \leq c(k) \cdot \|N\| \cdot \|M\|;$$

(2) $$\|N\| \cdot \|M\| \leq \|N \times M\|.$$

Proof. (1) Let n and m be the dimensions of N and M respectively, so that $n + m = k$, and remark that $\Delta_n \times \Delta_m$ can be canonically triangulated, *i.e.* it can be expressed as a sum of k-simplices, the number of k-simplices needed being bounded by an integer $c(k)$ depending on k only. Moreover if $\sigma : \Delta_n \to N$ and $\tau : \Delta_m \to M$ are singular simplices we can define a k-chain $\overline{\sigma \times \tau}$

in $N \times M$ by composing the product mapping $\sigma \times \tau$ with the expression of $\Delta_n \times \Delta_m$ as a sum of k-simplices. Now, let

$$\sum_i a_i \cdot \sigma_i \qquad \sum_j b_j \cdot \tau_j$$

be representatives of the fundamental classes of N and M respectively; then it is easily checked that

$$\sum_{i,j} a_i \cdot b_j \cdot \overline{\sigma_i \times \tau_j}$$

is a representative of the fundamental class of $N \times M$, and then

$$\|N \times M\| \leq c(k) \cdot \sum_{i,j} |a_i \cdot b_j| \leq c(k) \cdot \left(\sum_i |a_i|\right) \cdot \left(\sum_j |b_j|\right).$$

If we take the infimum over all representatives of $[N]$ and $[M]$ we obtain the required inequality.

(2) We start by remarking that for $\beta \in \hat{H}^n(X)$ and $\gamma \in \hat{H}^m(X)$ (where X is any topological space), $\beta \cup \gamma$ is naturally defined as an element of $\hat{H}^{n+m}(X)$ and moreover

$$\|\beta \cup \gamma\|_\infty \leq \|\beta\|_\infty \cdot \|\gamma\|_\infty.$$

For arbitrary $\varepsilon > 0$, it follows from F.2.2 that we can find $\beta \in \hat{H}^n(N)$ and $\gamma \in \hat{H}^m(M)$ such that

$$[\beta, [N]] = 1, \qquad \|\beta\|_\infty \leq \|N\|^{-1} + \varepsilon$$
$$[\gamma, [M]] = 1, \qquad \|\gamma\|_\infty \leq \|M\|^{-1} + \varepsilon.$$

Then we define $\overline{\beta} = \pi_N^*(\beta) \in \hat{H}^n(N \times M)$ (where π_N is the projection of $N \times M$ onto N), and similarly $\overline{\gamma} = \pi_M^*(\gamma)$. We easily have

$$[\overline{\beta} \cup \overline{\gamma}, [N \times M]] = [\beta, [N]] \cdot [\gamma, [M]] = 1$$

and

$$\|\overline{\beta} \cup \overline{\gamma}\|_\infty \leq \|\overline{\beta}\|_\infty \cdot \|\overline{\gamma}\|_\infty = \|\beta\|_\infty \cdot \|\gamma\|_\infty \leq (\|N\|^{-1} + \varepsilon)(\|M\|^{-1} + \varepsilon)$$

which implies (by F.2.2 again) that

$$\|N \times M\|^{-1} \leq (\|N\|^{-1} + \varepsilon)(\|M\|^{-1} + \varepsilon)$$

and arbitrariness of ε implies the conclusion

$$\|N\| \cdot \|M\| \leq \|N \times M\| \qquad\qquad \square$$

F.3 Flat Fiber Bundles

We recall that if M, E, F are topological spaces and if Γ is a topological group operating in a continuous way on F, a <u>fiber bundle</u> of total space E, fiber F,

basis M and structure group Γ is given by a continuous surjective mapping
$\pi : E \to M$ and an open covering $\{U_i\}$ of M (called a <u>trivializing</u> covering)
such that:

(1) $\forall i$ there exists a homeomorphism ϕ_i making the following diagram commutative:

$$\pi^{-1}(U_i) \overset{\phi_i}{\longrightarrow} U_i \times F$$

$$\searrow \pi \qquad \downarrow p_{U_i}$$

$$U_i$$

(2) if $U_i \cap U_j \neq \phi$ there exists a continuous mapping $h_{i,j} : U_i \cap U_j \to \Gamma$ such
that

$$(\phi_i \circ \phi_j^{-1})(u, f) = (u, h_{i,j}(u)(f)) \qquad \forall u \in U_i \cap U_j, \ f \in F.$$

We recall that these functions are called the <u>cocyles</u> of E and that they
determine E up to a suitable equivalence relation called <u>weak equivalence</u>: two
fiber bundles $\pi : E \to M$ and $\pi' : E' \to M$ with fiber F and group structure Γ
are said to be weakly equivalent if there exists a homeomorphism $W : E \to E'$
such that $\pi' \circ W = \pi$ and such that if $\{U_i\}$ is a trivializing covering for both
the bundles (with respective trivializations $\{\phi_i\}$ and $\{\phi'_i\}$), for all i's we have

$$W\big(\phi_i^{-1}(u, f)\big) = \phi_i'^{-1}\big(u, w_i(u)(f)\big) \qquad \forall u \in U_i, \ \forall f \in F$$

for a suitable continuous function $w_i : U_i \to \Gamma$.

The definition of a differentiable fiber bundle just requires all manifolds
and mappings to be smooth.

We shall say the fiber bundle is <u>flat</u> if in the above definition the open
covering and the functions ϕ_i can be chosen in such a way that all the $h_{i,j}$'s
are constant functions.

Remark F.3.1. Let $E \overset{\pi}{\longrightarrow} M$ be a fiber bundle with fiber F, let U_i be as in
the definition; then $\pi^{-1}(U_i)$ is identified with $U_i \times F$ (via the function ϕ_i) and
the condition that E be flat means that under two different identifications of
$p^{-1}(U_i \cap U_j)$ (via ϕ_i and ϕ_j) to $(U_i \cap U_j) \times F$, a horizontal leaf $(U_i \cap U_j) \times \{f\}$
corresponds to another horizontal leaf $(U_i \cap U_j) \times \{h_{i,j}(f)\}$.

We recall that an m-dimensional <u>foliation</u> of a topological n-manifold N
is given by a partition of N into m-submanifolds $\{M_\alpha\}$ such that there exists
an open covering $\{W_i\}$ of N and homeomorphisms $\gamma_i : W_i \to \mathbb{R}^n$ with the
following properties:

$-$if $x \in W_i \cap M_\alpha$ and $\gamma_i(x) = (u, v)$ with $u \in \mathbb{R}^{n-m}$, $v \in \mathbb{R}^m$, then

$$\gamma_i(W_i \cap M_\alpha) = \{u\} \times \mathbb{R}^m;$$

$-$if $W_i \cap W_j \neq \phi$ and $u \in \mathbb{R}^{n-m}$ then there exists $u' \in \mathbb{R}^{n-m}$ such that

$$(\gamma_i \circ \gamma_j^{-1})(\{u\} \times \mathbb{R}^m) = \{u'\} \times \mathbb{R}^m.$$

The M_α's will be called the <u>leaves</u> of the foliation.

Moreover we shall say two submanifolds M_1, M_2 of N (having dimension respectively m_1 and m_2) are <u>transversal</u> if for all $x \in M_1 \cap M_2$ the intersection of the tangent spaces $T_x M_1$ and $T_x M_2$ has dimension $m_1 + m_2 - n$. (Remark that this definition implies that if $m_1 + m_2 < n$ then M_1 and M_2 are transversal if and only if they do not meet.)

The following result is easily proved as an exercise and allows to understand better the meaning of the notion of flatness for a fiber bundle.

Proposition F.3.2. Let $E \xrightarrow{\pi} M$ be a flat fiber bundle with fiber F; then a foliation of dimension equal to the dimension of M is naturally defined on E, the leaves are transversal to the fibers and they project in a locally diffeomorphic way on M.

(Hint: given $x \in \pi^{-1}(U_i)$ let $\phi_i(x) = (p, f)$, consider $\phi_i^{-1}(U_i \cap \{f\})$ and define the leaf through x as the maximal connected subset of E obtained by glueing pieces of this type.)

Example F.3.3. Let $p : S^1 \times S^1 \to S^1$ be the projection on the first factor. Think of S^1 as the unit circle of $\mathbf{C}$ and let $\{U_1, U_2\}$ be the open covering of S^1:

$$U_1 = S^1 \setminus \{1\} \qquad U_2 = S^1 \setminus \{-1\}.$$

For $k \in \mathbf{R}$ consider the fiber bundle structure associated to the following homeomorphisms:

$$\phi_1 : p^{-1}(U_1) \ni (e^{i\theta}, e^{i\eta}) \mapsto (e^{i\theta}, e^{i(k\theta+\eta)}) \in U_1 \times S^1 \text{ for } -\pi < \theta < \pi$$

$$\phi_2 : p^{-1}(U_2) \ni (e^{i\theta}, e^{i\eta}) \mapsto (e^{i\theta}, e^{i(k\theta+\eta)}) \in U_2 \times S^1 \text{ for } 0 < \theta < 2\pi.$$

For all k's the structure is flat, but the shape of the leaves of the associated foliation is strikingly different, as shown in Figg. F.1 to F.3: if k is an integer multiple of 2π the leaves are plain circles, if k/π is rational the leaves spin around and close up after a finite number of laps, and if k/π is not rational the leaves are dense.

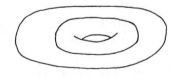

Fig. F.1. The case when k is an integer multiple of 2π

Let a flat fiber bundle $E \xrightarrow{\pi} M$ with fiber F and structure group Γ be fixed.

Proposition F.3.2 yields:

Corollary F.3.4. Given $x \in M$, $\overline{x} \in \pi^{-1}(x)$ and an arc $\sigma : [0,1] \to M$ starting at x there exists a unique arc $\overline{\sigma}_{\overline{x}}$ in E with the following properties:

Fig. F.2. The case when k is a rational multiple of π

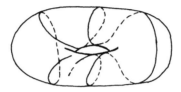

Fig. F.3. The case when k has irrational ratio with π

(i) $\overline{\sigma}_{\overline{x}}$ starts at $\overline{x}$;

(ii) $\pi \circ \overline{\sigma}_{\overline{x}} = \sigma$;

(iii) the range of $\overline{\sigma}_{\overline{x}}$ is contained in the leaf passing through $\overline{x}$.

Proof. According to the fact that leaves project on M in a locally diffeomorphic way, the lift exists in the neighborhood of all points, and a standard maximality argument allows us to conclude.   □

Proposition F.3.5. Given x and σ as above, if we identify $\pi^{-1}(x)$ and $\pi^{-1}(\sigma(1))$ to F (via two of the trivializations), then the mapping

$$\psi(\sigma) : \pi^{-1}(x) \to \pi^{-1}(\sigma(1))$$
$$\overline{x} \mapsto \overline{\sigma}_{\overline{x}}(1)$$

is an element of Γ and it depends only on the homotopy class of σ.

In case σ is a loop and only one trivialization is used for the identification of $\pi^{-1}(x)$ with F, ψ defines a representation

$$\Psi : \Pi_1(M, x) \to \Gamma.$$

Moreover if the trivialization of $\pi^{-1}(x)$ is changed, the resulting representation is conjugate to the previous one (*i.e.* they are obtained from each other by composition with an inner automorphism of Γ).

If $A : \Pi_1(M, x) \to \Pi_1(M, x')$ is the isomorphism associated to any path joining x to x', and if $\Psi' : \Pi_1(M, x') \to \Gamma$ is built as above, then the representations $\Psi' \circ A$ and Ψ of $\Pi_1(M, x)$ into Γ are conjugate to each other.

Proof. Since $[0, 1]$ is compact we can find numbers

$$0 = t_0 < t_1 < ... < t_N = 1$$

such that $\sigma([t_{i-1}, t_i])$ is contained in a trivializing open set U_i with trivialization

$$\phi_i : \pi^{-1}(U_i) \to U_i \times F.$$

If γ_i is such that $\phi_{i-1}^{-1}(u, f) = \phi_i^{-1}(u, \gamma_i(f))$ $\forall u \in U_i \cap U_{i-1}$, $f \in F$, then the definition of $\overline{\sigma}_{\overline{x}}$ easily implies that for $\overline{x} = \phi_0^{-1}(x, f)$ we have

$$\overline{\sigma}_{\overline{x}}(1) = \phi_N^{-1}(\sigma(1), (\gamma_N \circ \dots \circ \gamma_1)(f)),$$

whence $\psi(\sigma) = \gamma_N \circ \dots \circ \gamma_1$ and the first part is proved.

Assume now $\Sigma : [0,1] \times [0,1] \to M$ is a homotopy, $i.e.$ it is continuous and $\Sigma(0, s) = x$, $\Sigma(1, s) = x'$ $\forall s$. For $s \in [0, 1]$ we set $\sigma^{(s)}(t) = \Sigma(t, s)$; given $\overline{x} \in \pi^{-1}(x)$ we must check that the function

$$s \mapsto \overline{\sigma^{(s)}}_{\overline{x}}(1)$$

is constant. Connectedness of $[0, 1]$ implies that it suffices to prove that this mapping is locally constant, and then by the compactness of $[0, 1]$ we can assume that there exist numbers $0 = t_0 < t_1 < \dots < t_N = 1$ such that $\Sigma([t_{i-1}, t_i] \times [0, 1]) \subset U_i$, where U_i is as above. Then by the above argument we still have that for $\overline{x} = \phi_0^{-1}(x, f)$

$$\overline{\sigma^{(s)}}_{\overline{x}}(1) = \phi_N^{-1}(x', (\gamma_N \circ \dots \circ \gamma_1)(f)) \qquad \forall s$$

which implies our assertion.

Homotopy invariance implies that if a trivialization of $\pi^{-1}(x)$ is fixed, then the above construction defines a homomorphism

$$\Psi : \Pi_1(M, x) \to \Gamma.$$

Now let α be a path joining x to another point x' in M, fix a trivialization of $\pi^{-1}(x')$ and let $\Psi' : \Pi_1(M, x') \to \Gamma$ be the associated homomorphism (remark that for $x' = x$ we are only considering a change of trivialization on $\pi^{-1}(x)$, so that $\psi(\alpha)$ may be non-trivial even if α is constant). Let us pick a path $\alpha : [0, 1] \to M$ joining x to x'. Then it is very easily checked (via a construction analogous to the previous ones) that

$$\Psi(\langle\sigma\rangle) = \psi(\alpha)^{-1} \cdot \Psi'(\langle\alpha\sigma\alpha^{-1}\rangle) \cdot \psi(\alpha) = \psi(\alpha)^{-1} \cdot \Psi'(A(\langle\sigma\rangle)) \cdot \psi(\alpha)$$

and then the proof is over.   □

From now on we shall always assume M is (arcwise) connected. We shall call the conjugacy class of homomorphisms $\Pi_1(M) \to \Gamma$ whose existence is established in the above proposition the <u>holonomy</u> of the flat fiber bundle. (Compare to B.1.14; remark as well that the above proof is somewhat similar to the arguments presented in B.1.3; see also Remark F.3.8 for another relationship between the concepts of flat fiber bundle and (X, G)-structure.)

Theorem F.3.6. Let M be a connected topological manifold and let Γ be a topological group operating continuously on another topological manifold

F. To each conjugacy class of homomorphisms $\Pi_1(M) \to \Gamma$ there corresponds a flat fiber bundle on M with fiber F and structure group Γ having it as holonomy; such a fiber bundle is unique up to weak equivalence.

Proof. Let $\rho : \Pi_1(M, x) \to \Gamma$ be a fixed representative of the conjugacy class in question. Consider the universal cover $\tilde{M} \xrightarrow{p} M$ and define the action of $\Pi_1(M, x)$ of $\tilde{M} \times F$ in the following way:

$$\langle \sigma \rangle : (y, f) \to \left(\langle \sigma \rangle(y), \rho(\langle \sigma \rangle)^{-1}(f) \right)$$

(recall that $\Pi_1(M, x)$ operates as a group of homeomorphisms on $\tilde{M}$). Define E as the quotient space on $\tilde{M} \times F$ under such an action and let π be the natural projection of E onto M.

If U is a connected and simply connected open subset of M and U' is a connected component of $p^{-1}(U)$ we define

$$\theta_{(U, U')} : \pi^{-1}(U) \to U \times F$$

in the following way: if $\overline{u} \in \pi^{-1}(U)$ then it has one and only one representative of the form (y, f) with $y \in U'$; then we set

$$\theta_{(U, U')}(\overline{u}) = (p(y), f).$$

If $(U_1, U_1)'$ and $(U_2, U_2)'$ are pairs as above and $U_1 \cap U_2 \neq \phi$ then we can find $\langle \sigma_{1,2} \rangle \in \Pi_1(M, x)$ which is a homeomorphism of $U_1' \cap p^{-1}(U_1 \cap U_2)$ onto $U_2' \cap p^{-1}(U_1 \cap U_2)$. Then it is easily checked that

$$\theta_{(U_1, U_1')} \circ \theta_{(U_2, U_2')}^{-1}(u, f) = \left(u, \rho(\langle \sigma_{1,2} \rangle)^{-1}(f) \right)$$

and hence E has the required structure of a flat fiber bundle.

In order to compute the holonomy of E we fix $\tilde{x} \in p^{-1}(x)$ and $\overline{x} \in \pi^{-1}(x)$; moreover we assume $\overline{x}$ is represented by a pair $(\tilde{x}, f)$. Given a loop σ at x, the lift $\overline{\sigma}_{\overline{x}}$ of σ to the leaf passing through $\overline{x}$ is obtained by projecting to E the lift $\tilde{\sigma}$ of σ at $\tilde{M}$ starting at $\tilde{x}$ times $\{f\}$:

$$\overline{\sigma}_{\overline{x}}(t) = \tau(\tilde{\sigma}(t), f)$$

(where $\tau : \tilde{M} \times F \to E$ is the natural projection). In particular we have:

$$\overline{\sigma}_{\overline{x}}(1) = \tau(\tilde{\sigma}(1), t) = \tau(\langle \sigma \rangle(\tilde{x}), f) = \tau(\tilde{x}, \rho(\langle \sigma \rangle)(f))$$

and hence the holonomy of E is the conjugacy class of ρ.

Assume now $E' \xrightarrow{\pi'} M$ is another flat fiber bundle with the same holonomy. We can find a trivialization

$$\phi : {\pi'}^{-1}(U) \to U \times F$$

in such a way that the concrete homomorphism $\Pi_1(M, x) \to \Gamma$ associated to ϕ (by F.3.5) is ρ itself. Let us define a mapping $J : \tilde{M} \times F \to E'$ in the following way: fix $\tilde{x} \in p^{-1}(x)$; for $(\tilde{y}, f) \in \tilde{M} \times F$ let $\tilde{\alpha}$ be a path joining $\tilde{x}$ to $\tilde{y}$ and set

$$\alpha = p \circ \tilde{\alpha} \qquad \overline{x} = \phi^{-1}(x, f) \qquad J(\tilde{y}, f) = \overline{\alpha}_{\overline{x}}(1).$$

It is quite evident that the definition is independent of α.

We claim that J is onto: given $\overline{y} \in E'$ let $y = \pi'(\overline{y})$ and find a path α joining x to y. If we lift α (in E') to the leaf passing through $\overline{y}$ we obtain a path whose starting point is $\overline{x} \in {\pi'}^{-1}(x)$; let $\overline{x} = \phi^{-1}(x, f)$. Now, the lift of α to $\tilde{M}$ starting at $\tilde{x}$ stops at some point $\tilde{y} \in p^{-1}(y)$. Then by construction $J(\tilde{y}, f) = \overline{y}$ and our claim is proved.

Now, it easily follows from the definition of J and from our choice of ρ as a concrete representative of the holonomy of E' that

$$J(\tilde{y}_1, f_1) = J(\tilde{y}_2, f_2) \quad \Leftrightarrow \quad \exists g \in \Pi_1(M, x) \ s.t. \ \tilde{y}_2 = g(\tilde{y}_1), \ f_1 = \rho(g)(f_2).$$

and hence J induces a bijective mapping of E onto E'. We leave as an exercise to prove that such a bijection actually defines a weak equivalence of flat fiber bundles. □

According to the above result the theory of flat fiber bundles with structure group Γ on a connected manifold M reduces to the theory of representations of the fundamental group of M in Γ.

Example F.3.7. A differentiable n-manifold is called <u>affine</u> if there exists a differentiable atlas with respect to which the changes of chart are affine (*i.e.* they are restrictions of mappings of the form $\mathbb{R}^n \ni x \mapsto Ax + a$, with $A \in \mathrm{Gl}(n, \mathbb{R})$ and $a \in \mathbb{R}^n$). Of course the tangent bundle to an affine manifold is a flat fiber bundle with group structure $\mathrm{Gl}(n, \mathbb{R})$ and fiber $\mathbb{R}^n$. A special case of affine manifolds is represented by the flat (or Euclidean) manifolds we defined in Sect. B.1. We recall that as a corollary of the Gauss-Bonnet formula we proved in B.3.4 that the only oriented compact surface supporting a flat structure is the torus; in the sequel we shall see an important generalization of this fact (F.5.8).

Remark F.3.8. We close this section by mentioning a relationship between the concept of flat fiber bundle introduced above and the notion of (X, G)-structure defined in Sect. B.1: see [Su-Th] for details. Let M be a (not necessarily complete) hyperbolic manifold (or, more generaly, a (X, G)-manifold, according to the definition given in B.1). We can associate to a representative $\rho : \Pi_1(M) \to G$ of the holonomy of M a flat fiber bundle $\pi : E \to M$ with fiber X and structure group G. Let us consider the natural foliation of E with leaves tranversal to the fibers: then the (X, G)-structure on M can be identified with a section of E being tranversal to the leaves of the foliation. (We recall that a section of E is a smooth mapping $s : M \to E$ such that $\pi \circ s = $ id.) Conversely, given a flat fiber bundle $\pi : E \to M$ with fiber X and structure group G and a section of E being transversal to the leaves, an (X, G)-structure is naturally defined on M and its holonomy is the same as that of E.

F.4 Euler Class of a Flat Vector Bundle

We shall consider in this section a fixed connected oriented compact differentiable m-manifold M. The objects we will be dealing with are smooth oriented flat vector bundles of rank n, $i.e.$ flat differentiable fiber bundles $\xi = (E \xrightarrow{\pi} M)$ with fiber $\mathbb{R}^n$ and group structure $\mathrm{Gl}^+(n, \mathbb{R})$. Our aim is to define an element $\mathcal{E}(\xi) \in H^n(M, \mathbb{Z})$ representing the obstruction to the possibility of finding a global non-vanishing section of ξ.

CONSTRUCTION OF $\mathcal{E}(\xi)$.

It will be proved in a short time that if $n > m$ then ξ does have a non-vanishing section, so we confine ourselves to the case $n \leq m$.

We define $\mathcal{E}(\xi)$ with respect to the "simplicial theory" of homology on M, $i.e.$ the value of $\mathcal{E}(\xi)$ is defined on the algebraic sums of the n-simplices appearing in a fixed triangulation of M and then it is checked that this value is zero on the "simplicial boundaries".

Assume a triangulation is fixed and denote by $M^{(j)}$ the j-th skeleton of the triangulation. The general definition is based on the fact that it is possible to find smooth sections on $M^{(n)}$ which do not vanish on $M^{(n-1)}$ and whose range is transversal to M (viewed as the range of zero section in E); if s is such a section and σ is a n-simplex of the triangulation then the intersection of σ with the range of s is a finite number of points to each of which we can give a sign $+1$ or -1 (in the usual way, using orientations); then $\mathcal{E}(\xi)(\sigma)$ is the sum of such signs. $\mathcal{E}(\xi)$ is extended by linearity on all the algebraic sums of the n-simplices appearing in the fixed triangulation. In order to prove that this procedure defines an element of $H^n(M, \mathbb{Z})$ we should check that:
1) if c is a simplicial n-cycle then $\mathcal{E}(\xi)(c)$ is independent of s;
2) if c is a simplicial $(n+1)$-chain then $\mathcal{E}(\xi)(\partial c) = 0$.

We will not do this. On the contrary we shall describe another definition making use of a special class of sections on the n-skeleton (namely the simplicial sections) and we shall prove the definition is well-given in this case only. The use of simplicial sections allows to prove quite easily a further result due to Sullivan (F.4.10): in fact our method was borrowed from Sullivan's original proof.

We first remark that, since all the simplices σ are simply connected and E is flat, it follows from F.3.5 that E is the trivial bundle over σ, and of course the same must hold for some neighborhood of σ; it follows that all the simplices lie inside some trivializing open set. Let $v_1, ..., v_l$ be the vertices of the triangulation.

Lemma F.4.1. If $s(v_i) \in \pi^{-1}(v_i)$ is arbitrarily chosen for all i's then by affine extension on all charts a section s of ξ is well-defined.

Proof. Let τ be a simplex and assume by simplicity that it has vertices $v_1, ..., v_p$; let U be a simply connected neighborhood of τ and consider a trivialization $\phi : \pi^{-1}(U) \to U \times \mathbb{R}^n$ and let $\phi(s(v_i)) = (v_i, x_i)$ for $i = 1, ..., p$. We define s on τ by

$$s\left(\sum_{i=1}^{p} t_i \cdot v_i\right) = \phi^{-1}\left(\sum_{i=1}^{p} t_i \cdot v_i, \sum_{i=1}^{p} t_i \cdot x_i\right) \qquad \left(\sum t_i = 1\right).$$

Since the changes of chart have the form $(v, x) \mapsto (v, Ax)$ for $A \in Gl^+(n, \mathbb{R})$, convex combinations are preserved and then s is well-defined. $\qquad \square$

Lemma F.4.2. For a generic choice of the $s(v_i)$'s the section s built in the above lemma:

(A) does not vanish on $M^{(n-1)}$;

(B) on each n-simplex σ vanishes in at most one point p belonging to the interior of σ, and at this point

$$T_p(E) = T_p M \oplus \left(d_p s\big|_{M^{(n)}}\right)(T_p \sigma)$$

(which means that the range of the restriction of s to the n-skeleton is transversal to the 0-section).

Proof. We recursively choose $s(v_1), ..., s(v_l)$ in such a way that the section obtained from them satisfies the following properties:

(1) if τ is a p-simplex, with $p \leq n-1$, with vertices $v_{i_0}, ..., v_{i_p}$, if U is a trivializing open set containing τ with trivialization $\phi : \pi^{-1}(U) \to U \times \mathbb{R}^n$, and if $\phi(s(v_{i_j})) = (v_{i_j}, x_j)$ then $x_0, ..., x_p$ are linearly independent.

(2) if τ is an n-simplex with vertices $v_{i_0}, ..., v_{i_n}$, if U, ϕ and the x_j's are as in (1), then $x_1 - x_0, ..., x_n - x_0$ are linearly independent.

We can check now that at each step the choices we must exclude have zero Lebesgue measure in the fiber (identified with $\mathbb{R}^n$). (The forbidden choices are actually described in a very explicit way, but we only need to know that their complement is not empty, so that it suffices to say that they have zero measure.)

For $s(v_1)$ we only have to exclude 0.

Assume $s(v_1), ..., s(v_r)$ are chosen in such a way that properties (1) and (2) hold for all simplices having vertices contained in $\{v_1, ..., v_r\}$. Let h be the (finite) number of different simplices of dimension at most $n-1$ having one vertex in v_{r+1} and the others in $\{v_1, ..., v_r\}$, and let k be the number of n-dimensional simplices with the same condition on the vertices. For the choice of $s(v_{r+1})$ we must exclude from $\mathbb{R}^n = \pi^{-1}(v_{r+1})$ h linear subspaces of dimension at most $n-1$ and k affine subspaces of dimension $n-1$, that is, a set of zero measure.

Property (A) follows at once from (1).

As for (B), for all n-simplices s is expressed in suitable coordinates as a function

$$f : \Delta_n \to \Delta_n \times \mathbb{R}^n \qquad t \mapsto \left(t, \sum_{i=0}^{n} t_i \cdot x_i\right)$$

where $\{x_0, ..., x_n\} \subset \mathbb{R}^n$ is such that each subset of n elements is linearly independent, and the same holds for $x_1 - x_0, ..., x_n - x_0$. We only have to prove that the second component of f vanishes in at most one point and it has surjective differential in such a point: both facts immediately follow from the properties of $\{x_0, ..., x_n\}$. □

We are now ready for the definition of $\mathcal{E}(\xi)$ on a n-simplex σ of the triangulation in question: consider a section s built as in the above lemma and set:

$$\mathcal{E}(\xi)(\sigma) = \begin{cases} 0 & \text{if } s \text{ does not vanish on } \sigma; \\ 1 & \begin{cases} \text{if } s(p) = 0 \text{ for } p \in \sigma \text{ and given positive bases } t_1, ..., t_m \\ \text{and } x_1, ..., x_n \text{ of } T_p M \text{ and } T_p \sigma \text{ respectively, then} \\ t_1, ..., t_m, d_p(s|_\sigma)(x_1), ..., d_p(s|_\sigma)(x_n) \text{ is positive;} \end{cases} \\ -1 & \text{otherwise;} \end{cases}$$

$\mathcal{E}(\xi)$ is then extended by linearity.

Proposition F.4.3. If c is a cycle then $\mathcal{E}(\xi)(c)$ does not depend on the section s.

Proof. Let c be such that $\partial c = 0$ and let s and s' be simplicial sections satisfying the conditions of F.4.2. Let $\mathcal{E}(\xi)(c)$ and $\mathcal{E}(\xi)'(c)$ be obtained by the above construction respectively from s and s'. We want to prove that $\mathcal{E}(\xi)(c) = \mathcal{E}(\xi)'(c)$. Let us assume first that this is true when s and s' differ only on one vertex, and let us conclude the proof. By F.4.2 we can find a vector $u_1 \in \pi^{-1}(v_1)$ arbitrarily close to $s'(v_1)$ and such that the section s_1 extending

$$s_1(v_1) = u_1, s_1(v_2) = s(v_2), ..., s_1(v_l) = s(v_l)$$

satisfies F.4.2. Similarly we can go on and find at last a section s'' such that $\mathcal{E}(\xi)''(c) = \mathcal{E}(\xi)(c)$ and $s''(v_j)$ is arbitrarily close to $s'(v_j)$ for all j. This implies immediately that $\mathcal{E}(\xi)''(c) = \mathcal{E}(\xi)'(c)$ and hence the proof is over.

We are left to prove that $\mathcal{E}(\xi)'(c) = \mathcal{E}(\xi)(c)$ when s and s' differ only on one vertex. Since we do not want to get involved into complicated combinatorial technicalities we confine ourselves to the case $n = 1$.

We assume s and s' differ only on v_1, and we re-write c as

$$c = \sum_i \alpha_i \sigma_i + \sum_j b_j \tau_j + \sum_k c_k \rho_k$$

where the $\sigma_i's$ have v_1 as first vertex, the τ_j's have v_1 as second vertex and the ρ_k's do not contain v_1. Remark that $\partial c = 0$ implies

$$\sum_i a_i = \sum_j b_j$$

and moreover $\mathcal{E}(\xi)(\rho_k) = \mathcal{E}(\xi)'(\rho_k)$ for all k.

The only forbidden value for the section at v_1 is 0, and it is easily checked that if $s(v_1)$ and $s'(v_1)$ lie in the same component of $\pi^{-1}(x) \setminus \{0\}$ then

$$\mathcal{E}(\xi)(\sigma_i) = \mathcal{E}(\xi)'(\sigma_i) \quad \mathcal{E}(\xi)(\tau_j) = \mathcal{E}(\xi)'(\tau_j) \qquad \forall i, j.$$

Conversely if $s'(v_1) = -s(v_1)$ we can assume (by a change of the order of the σ_i's and an interchange of s and s', if necessary) that $\mathcal{E}(\xi)(\sigma_1) = 1$. This implies (see Fig. F.4) that $\mathcal{E}(\xi)(\sigma_i) \in \{0, 1\}$ for all i and $\mathcal{E}(\xi)(\tau_j) \in \{-1, 0\}$ for all j.

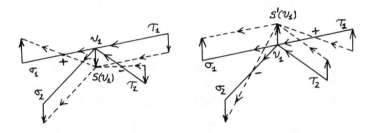

Fig. F.4. Independence of the definition of $\mathcal{E}(\xi)$ on the section

Moreover we have $\mathcal{E}(\xi)(\sigma_i) = 0$ if and only if $\mathcal{E}(\xi)'(\sigma_i) \neq 0$, and the same holds for the τ_j's, so that up to a suitable change of order we have

$$
\begin{aligned}
\mathcal{E}(\xi)(\sigma_i) = 1, \quad \mathcal{E}(\xi)'(\sigma_i) = 0 && \text{for } i \leq p \\
\mathcal{E}(\xi)(\sigma_i) = 0, \quad \mathcal{E}(\xi)'(\sigma_i) = -1 && \text{for } i > p \\
\mathcal{E}(\xi)(\tau_j) = 0, \quad \mathcal{E}(\xi)'(\tau_j) = 1 && \text{for } i \leq q \\
\mathcal{E}(\xi)(\tau_j) = -1, \quad \mathcal{E}(\xi)'(\tau_j) = 0 && \text{for } i > q
\end{aligned}
$$

so that

$$\mathcal{E}(\xi)(c) = \sum_{i \leq p} a_i - \sum_{j > q} b_j + \sum_k c_k \cdot \mathcal{E}(\xi)(\rho_k)$$

$$\mathcal{E}(\xi)'(c) = -\sum_{i > p} a_i + \sum_{j \leq q} b_j + \sum_k c_k \cdot \mathcal{E}(\xi)(\rho_k)$$

and hence

$$\mathcal{E}(\xi)(c) - \mathcal{E}(\xi)'(c) = \sum_i a_i - \sum_j b_j = 0. \qquad \square$$

Proposition F.4.4. If c is a sum of $(n+1)$-simplices then

$$\mathcal{E}(\xi)(\partial c) = 0.$$

Proof. It suffices to prove this when c is a simplex. If $v_0, ..., v_{n+1}$ are the vertices of c and they are positively ordered, then if we call σ_i the face of c opposite to v_i we have that

$$\partial c = \sum_{i=0}^{n+1} (-1)^i \cdot \sigma_i.$$

Let s be represented in a suitable trivialization by a simplicial function

$$f : c \to \mathbb{R}^n$$

and set $x_i = f(v_i)$. If 0 does not belong to the convex hull of any subset of $n+1$ elements of $\{x_0, ..., x_{n+1}\}$ then of course $\mathcal{E}(\xi)(\sigma_i) = 0$ for all i and we are done. On the contrary, assume by simplicity that 0 belongs to the convex hull of $x_0, ..., x_n$ (remark that our choice of s implies that it belongs to the interior of such a convex hull).

For $i = 0, ..., n$ consider the infinite cone C_i with vertex at 0 and symmetric to the convex hull of the points $x_0, ..., \hat{x}_i, ..., x_n$, as represented in Fig. F.5.

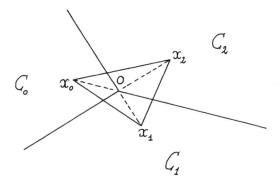

Fig. F.5. The Euler class takes value 0 on boundaries

Our choice of s implies that $x_{n+1} \notin \cup_i \partial C_i$; moreover it is easily checked that 0 belongs to the convex hull of $x_0, ..., \hat{x}_i, ..., x_{n+1}$ (which means that f vanishes on σ_i) if and only if $x_{n+1} \in C_i$. Since $\mathbb{R}^n$ is covered by the C_i's and $\overset{\circ}{C}_i \cap \overset{\circ}{C}_j = \phi$ for $i \neq j$, we have $x_{n+1} \in C_i$ for exactly one index $i = i_0$. It is immediately deduced from the picture that

$$\mathcal{E}(\xi)(\sigma_{i_0}) = (-1)^{n-i_0} \cdot \mathcal{E}(\xi)(\sigma_{n+1})$$

and hence

$$\mathcal{E}(\xi)(\partial c) = (-1)^{i_0} \cdot \mathcal{E}(\xi)(\sigma_{i_0}) + (-1)^{n+1} \cdot \mathcal{E}(\xi)(\sigma_{n+1}) = 0. \qquad \square$$

The above results imply that $\mathcal{E}(\xi)$ is perfectly well-defined as an element of $H^n(M)$. In the special case $m = n$ we can introduce the number $\mathcal{E}(\xi)([M])$ denoted by $\chi(\xi)$ and called the <u>Euler number</u> of the flat fiber bundle.

Remark F.4.5. $\chi(\xi)$ represents the obstruction to the existence of a non-vanishing global section of ξ in the sense that $\chi(\xi) \neq 0$ implies that no such section exists, but the converse is not true in general.

Proposition F.4.6. If n is odd then $\chi(\xi) = 0$.

Proof. Let s be a simplicial section as in F.4.2, and remark that $-s$ satisfies the same hypothesis. Moreover the sign of the intersection of s and $-s$ with the simplices are opposite to each other (as $-I$ reverses the orientation on $\mathbb{R}^n$) and hence

$$\mathcal{E}(\xi) = -\mathcal{E}(\xi). \qquad \square$$

The following result is due to Hopf (see [Mi2]).

Theorem F.4.7. If TM denotes the tangent bundle to an affine compact connected oriented manifold M then $\chi(TM)$ equals the Euler-Poincaré characteristic $\chi(M)$ of M.

We shall state now a result generally known as "abstract version of the Milnor-Sullivan theorem". We shall provide a sketch of the proof at the end of the next section; we address for a complete proof to [Gro3].

Theorem F.4.8. Let ξ be a flat vector bundle of rank n on a m-manifold M. Then $\mathcal{E}(\xi)$ as an element of $H^n(M,\mathbb{R}) = H^n(M,\mathbb{Z}) \otimes \mathbb{R}$ is bounded by 1, that is

$$\|\mathcal{E}(\xi)\|_\infty \le 1.$$

Remark F.4.9. The above result together with F.2.2 implies the inequality $\|M\| \ge |\chi(\xi)|$. This fact provides a sufficient condition for a manifold to have non-zero Gromov norm, that is existence of a flat vector bundle based on the manifold with non-vanishing Euler number.

We shall prove now some geometric results leading to the proof of F.4.8. The first fact is a straight-forward corollary of the particular construction we gave of the Euler class, as we mentioned at the beginning of the section. As above we shall keep fixed a flat oriented vector bundle ξ of rank n on M.

Proposition F.4.10 (Sullivan). If an n-cycle c in M is expressed as a sum $\sum_i a_i \cdot \sigma_i$, where the σ_i's are simplices of a triangulation of M, then

$$|\mathcal{E}(\xi)(c)| \le \sum_i |a_i|.$$

In particular, if $n = m$ and there exists a triangulation of M containing k n-simplices, then $|\chi(\xi)| \le k$.

Proof. It follows from our definition of $\mathcal{E}(\xi)$ that $\mathcal{E}(\xi)(\sigma_i) \in \{-1, 0, +1\}$, and the conclusion follows immediately. The second fact is now straight-forward.
$$\square$$

Let us remark that the above result can be considered as a special case of F.4.8, *i.e.* as a version of the Milnor-Sullivan theorem with respect to the "simplicial theory" of homology.

We consider now the set $\mathcal{F}_M$ of all weak-equivalence classes of oriented flat vector n-bundles on M and we set

$$\mathcal{E}_M = \{|\chi(\xi)| : \xi \in \mathcal{F}_M\}.$$

The above proposition implies that $\sup \mathcal{E}_M$ depends only on something finer than M (that is, the number k of simplices appearing in a triangulation). Remark as well that $\#\mathcal{F}_M$ is bounded by a constant depending on k too; we leave it to the reader as an exercise to give explicit bounds on this constant.

The following result is apparently due to Lusztig.

Proposition F.4.11. $\#\mathcal{F}_M$ actually depends only on $\Pi_1(M)$.

Proof. According to F.3.6 a flat vector n-bundle on M is determined by a (conjugacy class of) homomorphism $\Pi_1(M) \to \mathrm{Gl}^+(n, \mathbb{R})$. Since M is compact $\Pi_1(M)$ has a finite number j of generators and a finite number h of independent relations on such generators. Then the set of all representations $\Pi_1(M) \to \mathrm{Gl}^+(n, \mathbb{R})$ can be identified with a subset of $\mathrm{Gl}^+(n, \mathbb{R})^j$ defined by h algebraic equations: in particular it is naturally endowed with the structure of a real algebraic manifold. Then the Tarski-Seidenberg theorem implies that it has a finite number of connected components bounded by a constant depending only on j and h; moreover it is easily checked that two representations lying in the same connected component define equivalent bundles, and the proof is over. □

The next result, to be found in [Sm], sharpens Proposition F.4.10.

Proposition F.4.12 (Smillie). If an n-cycle c in M is expressed as a sum $\sum_i a_i \cdot \sigma_i$, where the σ_i's are simplices of a triangulation of M, then

$$|\mathcal{E}(\xi)(c)| \le 2^{-n} \cdot \sum_i |a_i|.$$

In particular, if $n = m$ and there exists a triangulation of M containing k n-simplices, then $|\chi(\xi)| \le 2^{-n} \cdot k$.

Proof. Let $v_1, ..., v_p$ be the vertices of the triangulation and assume $v_1, ..., v_l$ are the vertices involved in the cycle c (*i.e.* v_q is a vertex of one of the σ_i's if and only if $q \le l$). We checked in F.4.2 that a generic choice of a section on the vertices provides a "good" section, that is a section not vanishing on the $(n-1)$-skeleton and having transversal zeros on the n-skeleton. Hence we can choose a section s such that all simplicial sections s' such that

$$s'(v_q) = \pm s(v_q) \text{ for } q \le l \qquad s'(v_q) = s(v_q) \text{ for } q > l$$

are good. 2^l different sections are defined. Since all the σ_i's have $n+1$ vertices, 2^{n+1} different sections are defined over each σ_i. We claim that exactly two of them (opposite to each other) vanish on σ_i. In order to check this we must prove that:

if $\{v_0, ..., v_n\} \subset \mathbb{R}^n$ is such that the convex hull of each set of the form

$$\{\pm v_0, ..., \hat{v}_i, ..., \pm v_n\}$$

does not contain zero, then there exist precisely two choices of $\varepsilon_0, ..., \varepsilon_n \in \{\pm 1\}$ (opposite to each other) such that the convex hull of $\varepsilon_0 v_0, ..., \varepsilon_n v_n$ contains 0.

The choice exists. Since $v_0, ..., v_n$ are surely linearly dependent, we can find a non-trivial linear combination $\sum_{i=0}^{n} \alpha_i v_i = 0$. Then we set

$$t_i = \frac{|\alpha_i|}{\sum_j |\alpha_j|} \qquad \varepsilon_i = \text{sgn}(\alpha_i)$$

and we have that $\sum_{i=0}^{n} t_i \cdot (\varepsilon_i v_i) = 0$ is the required convex combination.

The choices are exactly two. Assume by contradiction 0 is in the convex hull of both $v_0, \varepsilon_1 v_1, ..., \varepsilon_n v_n$ and $v_0, \varepsilon_1 v_1, ..., \varepsilon_r v_r, -\varepsilon_{r+1} v_{r+1}, ..., -\varepsilon_n v_n$ (we can assume this, up to a suitable change of order): then it is easily checked that 0 is in the convex hull of $v_0, \varepsilon_1 v_1, ..., \varepsilon_r v_r$, and this is absurd if $r < n$.

Our claim is proved.

Now, let $s^1, ..., s^{2^l}$ be all the different sections in question; since each of them can be used for the definition of the Euler class we have that

$$2^l |\mathcal{E}(\xi)(c)| \le \sum_{j=1}^{2^l} \sum_i |a_i| \cdot z(s^j, \sigma_i) = \sum_i |a_i| \sum_{j=1}^{2^l} z(s^j, \sigma_i)$$

where $z(s^j, \sigma_i)$ denotes the number of zeros of s^j on σ_i. On σ_i the 2^l different sections divide into 2^{n+1} groups of 2^{l-n-1} sections equal to each other. Exactly two of these groups contain zeros, so that

$$\sum_{j=1}^{2^l} z(s^j, \sigma^i) = 2 \cdot 2^{l-n-1} = 2^{l-n}$$

$$\Rightarrow 2^l |\mathcal{E}(\xi)(c)| \le \sum_i |a_i| \cdot 2^{l-n} \Rightarrow |\mathcal{E}(\xi)(c)| \le 2^{-n} \cdot \sum_i |a_i|.$$

The second assertion is now straight-forward. $\qquad\qquad\qquad\qquad\qquad\square$

F.5 Flat Vector Bundles on Surfaces and the Milnor-Sullivan Theorem

We consider in this section the case when M is a compact connected oriented surface of genus $g \ge 1$ (remark that for $g = 0$ we have the sphere which is simply connected, so that there exists no non-trivial flat fiber bundle). Let $\xi = (E \xrightarrow{\pi} M)$ be a flat oriented vector bundle of rank 2. We can associate to ξ a fiber bundle in circles in the following way: if E_x denotes the fiber over x consider on $E_x \setminus \{0\}$ the equivalence relation $\sim$

$$v \sim w \Leftrightarrow \exists \lambda > 0 \text{ such that } v = \lambda \cdot w;$$

on the set

$$S(E) = \bigcup_{x \in M} \left(E_x \setminus \{0\} \big/ \sim \right)$$

a natural projection $\bar{\pi}$ onto M is defined; moreover if U is a trivializing open set for ξ then $\bar{\pi}^{-1}(U)$ is canonically identified with $U \times S^1$: if we define this identification to be a diffeomorphism we easily obtain that a flat fiber bundle $\bar{\xi} = (S(E) \overset{\bar{\pi}}{\longrightarrow} M)$ is defined, the fiber being S^1 and the structure group being the group of all orientation preserving diffeomorphisms of S^1.

If ρ is the holonomy of ξ we shall denote by $\bar{\rho}$ the holonomy of the associated bundle $\bar{\xi}$.

Remark F.5.1. If a metric is defined on ξ then $\bar{\xi}$ is equivalent as a fiber bundle to the bundle of the unit circles in ξ. Remark however that such a bundle is not endowed with a natural flat structure, while our construction provides this structure. A case in which the flat structure is preserved is when the group structure of ξ is not the whole $Gl^+(2, \mathbb{R})$ but only $SO(2)$.

Remark F.5.2. It is easily checked (and this is the reason we introduced $\bar{\xi}$) that there exists a non-trivial section of ξ if and only if there exists a global section of $\bar{\xi}$.

We shall discuss now the way to obtain the Euler number of the bundle ξ from the bundle $\bar{\xi}$. Actually, we shall introduce the definition of the Euler number of an arbitrary oriented flat bundle in circles on M as the number representing the obstruction to the existence of a global section. Hence we fix a flat circle bundle $\eta = (S \overset{\pi}{\longrightarrow} M)$ on M which need not be obtained from a rank 2 oriented vector bundle.

CONSTRUCTION OF $\chi(\eta)$.

Let us consider a small closed disc D in M and let us denote by M_1 the closure of $M \setminus D$; remark that both D and M_1 are bounded by the same circle: this circle will be denoted by ∂D or ∂M_1 in order to emphasize the manifold of which we are considering it a boundary. We shall consider the circle bundles

$$\eta_0 = \left(\pi^{-1}(D) \overset{\pi_0}{\longrightarrow} D \right) \qquad \eta_1 = \left(\pi^{-1}(M_1) \overset{\pi_1}{\longrightarrow} M_1 \right)$$

naturally obtained from η.

Lemma F.5.3. η_1 is equivalent (as a fiber bundle, but not as a flat fiber bundle) to the product $M_1 \times S^1$. Hence there exists a trivialization $\phi_1 : \pi^{-1}(M_1) \to M_1 \times S^1$, and given $y_1 \in S^1$ a section on ∂M_1 of the form

$$\partial M_1 \ni u \mapsto \phi_1^{-1}(u, y_1)$$

extends to the whole M_1.

Proof. M_1 retracts on the bouquet of $2g$ circles (as M is obtained by glueing pairs of sides in a polygon with $4g$ sides). An oriented flat bundle in circles on a bouquet of circles is necessarily trivial (hint: each orientation-preserving diffeomorphism of S^1 is isotopic to the identity) and hence our bundle is trivial too. As for the fact that η_1 may actually be non-trivial as a flat bundle, recall example F.3.3. The conclusion is now obvious.   □

Now, D is simply connected and hence η_0 is certainly isomorphic to the product bundle, so we fix a trivialization $\phi_0 : \pi^{-1}(D) \to D \times S^1$.

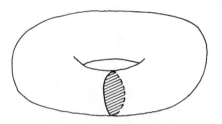

Fig. F.6. A fixed trivialization of $\pi^{-1}(D)$

In particular, ∂D is identified with a meridian $\partial D \times \{y_0\}$ in the torus $\partial D \times S^1$. Then a section of η_0 on ∂D is represented by a loop on the torus meeting each longitude in exactly one point. We omit the proof of the following result and leave it as an exercise.

Lemma F.5.4. A section on ∂D extends to D if and only if the loop it is represented by on the torus is isotopic to a loop which does not meet ∂D (and hence to a loop of the form $\partial D \times \{y\}$).

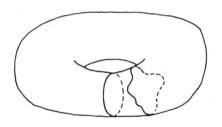
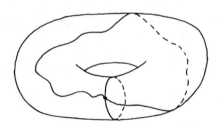

Fig. F.7. Extension exists   **Fig. F.8.** Extension does not exist

We define now
$$f : \partial M_1 \times S^1 \to \partial D \times S^1$$
as the restriction of $\phi_0 \circ \phi_1^{-1}$. According to the above lemmas in order to investigate the existence of a global section of η we only have to check whether a meridian $\partial M_1 \times \{y_1\}$ in $\pi_1^{-1}(\partial M_1)$ corresponds (up to isotopy) to a meridian $\partial D \times \{y\}$ in $\pi_0^{-1}(\partial D)$. This fact justifies heuristically the definition of $\chi(\eta)$ as the algebraic sum of the transversal intersections of $f(\partial M_1 \times \{y_1\})$ with $\partial D \times \{y_0\}$. The phrase "algebraic sum of the tranversal intersections" means as usual that we first perturb a little the loops in order to make them tranversal, then we consider orientations in order to give intersections a sign, and we finally take the sum of these signs. An example is shown in Fig. F.9.

Fig. F.9. Computation of the transversal intersection

We shall prove neither that $\chi(\eta)$ is well-defined nor that if $\overline{\xi}$ is obtained from a vector bundle ξ then $\chi(\overline{\xi}) = \chi(\xi)$; however, both are true and obtained by standard arguments of differential topology.

In the remainder of this section we will be discussing of the possibility of giving a bound to $\chi(\eta)$ depending only on the genus g of the surface we are considering. We shall denote by $\rho : \Pi_1(M) \to \mathrm{Diff}^+(S^1)$ the holonomy of the fixed bundle η. Firstly we will be concerned with bundles satisfying the following further property: ρ is said to <u>preserve antipodality</u> if

$$\rho(\gamma)(-y) = -\rho(\gamma)(y)$$

for all $\gamma \in \Pi_1(M)$ and $y \in S^1$ (viewed as the unit circle of C).

Remark F.5.5. If η is obtained from a rank 2 vector bundle then its holonomy preserves antipodality.

The following inequality, due to J. Milnor [Mi1], has a very important consequence we will discuss before proving the result.

Theorem F.5.6. If η is a flat oriented circle bundle on M and its holonomy preserves antipodality then $|\chi(\eta)| \leq g - 1$.

Corollary F.5.7. If $g \geq 2$ then M supports no affine structure.

Proof. If M supports an affine structure then its tangent bundle TM is endowed with a flat structure. Then, if η is the associated circle bundle, by F.4.9 and F.5.6 we have

$$2(g - 1) = |\chi(M)| = |\chi(TM)| = |\chi(\eta)| \leq g - 1 \implies g \leq 1. \qquad \square$$

Proposition F.5.8. The torus is the only surface supporting an affine structure.

Proof. It suffices to show that the sphere S^2 does not support such a structure. Assume the converse; since there exists a triangulation of S^2 with four 2-simplices, F.4.9 and Smillie's Proposition F.4.14 imply

$$2 = |\chi(TS^2)| \le 2^{-2} \cdot 4 = 1 \Rightarrow \text{contradiction.} \qquad \square$$

As we did in Chapt. B for the space of all hyperbolic metrics on a surface of genus $g \ge 2$, it is possible to study the space of all affine structures on the torus (up to a natural equivalence relation): this study was carried out in [Na-Ya], leading to the fact that such a space has dimension 4.

As suggested by F.5.8, Sullivan formulated in [Su1] the following still open conjecture.

SULLIVAN'S CONJECTURE. If M is an n-dimensional connected compact oriented manifold supporting an affine structure, then its Euler-Poincaré characteristic $\chi(M)$ is 0.

Proof of Theorem F.5.6. We confine ourselves to a sketch and address the reader to [Su1] for a complete proof in a more general setting.

As usual we represent M as the quotient of a polygon with $4g$ edges, and we consider a disc D as suggested by the picture here: ∂D is the union of $4g$ arcs "parallel" to the edges of the polygon.

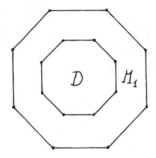

Fig. F.10. Representation of M as a polygon with glued edges and definition of D and M_1

The bundle η_0 induced by η on D is isomorphic to the product bundle, so that we can represent it as a solid torus as depicted; Fig. F.11 also contains a canonical section on ∂D extending to D.

Let us consider a global section of η_1, the bundle induced by η on the closure M_1 of $M \setminus D$ (as we saw above, such a section exists as η_1 is a product bundle). If we consider its restriction to ∂M_1 and read the restriction on ∂D using the above trivialization of η_0, we get the situation of Fig. F.12 (in which only a piece of the section is depicted).

We must compute how many laps are made by this section around the hole of the torus.

Now, let us order the arcs of ∂D as $a_1, ..., a_{4g}$, and let us assume a_i and a_{i+2} correspond to identified edges of the polygon for $i \equiv 1, 2$ modulo 4. If we isolate the contribution to the total number of laps of the arc a_i we get thesituation in $a_i \times S^1$ described in Fig. F. 13. By the definition we gave of

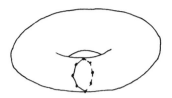

Fig. F.11. The bundle on D and the section on the boundary extending to D

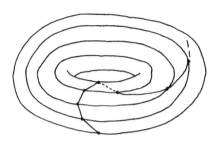

Fig. F.12. A global section on M_1 restricted to the boundary and represented in the trivialization of the bundle on D

the Euler number of the bundle we have

$$\chi(\eta) = \frac{1}{2\pi} \cdot \sum_{i=1}^{4g} \theta_i.$$

We first remark that there is no loss in assuming $|\theta_i| < 2\pi$; assume for instance $\theta_i \geq 2\pi$ and replace the section by the one represented in Fig. F.14, which corresponds to the "shorter" arc connecting the two points on the circumferences and having the required orientation.

Let us remark now that by definition we have that q_i is the image of p_i under the holonomy of the loop corresponding to the arc a_i; it follows from the assumption that the holonomy preserves antipodality that we actually have $|\theta_i| \leq \pi$. Moreover, for $i \equiv 1, 2$ modulo 4 we have that the arcs a_i and a_{i+2} correspond to the same loop but the orientation is opposite. Then we have that $\theta_i \cdot \theta_{i+2} \leq 0$, and if one of them is 0 then the other one is 0 too; this implies that

$$|\theta_i + \theta_{i+2}| < \pi.$$

We are now ready for the required estimate on the Euler number:

$$\chi(\eta) \leq \frac{1}{2\pi} \cdot \sum_{i \equiv 1,2 \bmod 4} |\theta_i + \theta_{i+2}| < \frac{1}{2\pi} \cdot 2g \cdot \pi = g. \qquad \square$$

Remark F.5.9. Consider the subdivision of M into $4g - 2$ "triangles" as suggested by Fig. F.15. Unfortunately this is not a triangulation (all the tri-

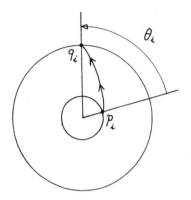
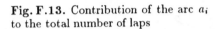

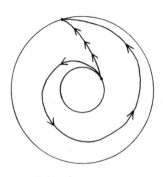

Fig. F.13. Contribution of the arc a_i to the total number of laps

Fig. F.14. What to do if the angle is more than 2π

Fig. F.15. A subdivision of M into triangles

angles have the same vertex), but, if it were, Milnor's inequality could have been proved as a corollary of Smillie's Proposition F.4.14:

$$|\chi(\eta)| \leq 1/4 \cdot (4g - 2) = g - 1/2 \;\Rightarrow\; |\chi(\eta)| \leq g - 1.$$

As for the general case (without the assumptions on the holonomy) we have the following result due to Wood [Woo].

Theorem F.5.10. Let η be an oriented flat fiber bundle in circles over a compact connected surface M of genus g; then

$$|\chi(\eta)| \leq 2(g - 1) = -\chi(M).$$

Proof. The method presented above for the proof of Milnor's inequality allows one to check that $|\chi(\eta)| \leq 2g-1$ (since we have no assumption on the holonomy we must replace π by 2π). Hence:

$$|\chi(\eta)| \leq -\chi(M) + 1.$$

In order to obtain the stronger inequality we use the same trick of C.4.7: we consider a connected d-fold covering M_d of M and we remark that η allows us to define an oriented flat fiber bundle in circles η_d on M_d (the holonomy of η_d being given by the composition of the holonomy of η with the natural inclusion of the fundamental groups). Moreover it is easily checked that $\chi(\eta_d) = d \cdot \chi(\eta)$ and hence by the inequality established above

$$d \cdot |\chi(\eta)| \leq -d \cdot \chi(M) + 1.$$

If we divide by d and consider the limit as d goes to infinity we obtain the required inequality.   □

Proposition F.5.11. Wood's inequality is sharp, *i.e.* $\forall g \geq 1$ there exist cases when equality holds.

Proof. The case $g = 1$ is obvious. Let $g \geq 2$, consider a hyperbolic structure on M and let $\rho : \Pi_1(M) \to \mathcal{I}^+(\mathbb{H}^2)$ be the associated holonomy. Since $\mathcal{I}^+(\mathbb{H}^2)$ can be viewed as a subgroup of $\text{Diff}^+(S^1)$ (the action being given by the extension to the sphere at infinity), an oriented flat fiber bundle in circles η is defined. We claim that η is equivalent to the unit tangent bundle $T_1 M$ with respect to the fixed hyperbolic metric. In fact, let us consider the bijective mapping

$$T_1 \mathbb{H}^2 \to \mathbb{H}^2 \times S^1$$
$$(x, v) \mapsto (x, p_\infty(x, v))$$

(where $p_\infty(x, v)$ denotes the endpoint of the geodesic half-line starting at x with velocity v).

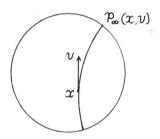

Fig. F.16. The bundle η is equivalent to the unit tangent bundle with respect to the hyperbolic metric

The natural action of the holonomy on $\mathbb{H}^2 \times S^1$ defining the flat fiber bundle (compare to the proof of F.3.6) corresponds under such a bijection to the natural action of the fundamental group of M on $T_1 \mathbb{H}^2$, giving the unit tangent bundle $T_1 M$. Our claim is proved.

Now, we have that

$$|\chi(T_1 M)| = -\chi(M)$$

and the proof is over.   □

We conclude this section with a rough sketch of the proof the Milnor-Sullivan theorem: several aspects are just outlined, but we think it is enough to understand the idea of the proof.

We first remark that by F.4.10 the theorem is true with respect to the "simplicial theory" (as we remarked after the proof). The key idea for the general case is to associate to M a topological space $K(M)$ with the following properties:

(1) $K(M)$ has an almost-polyhedral structure (namely, it can be expressed as the limit of a growing sequence of polyhedra);

(2) $K(M)$ is homotopically equivalent to M (which implies that the modules of bounded cohomology $\hat{H}^*(K(M))$ and $\hat{H}^*(M)$ are canonically isometric);

(3) in $K(M)$ the simplicial theory of homology and cohomology equals the singular one.

Then the conclusion follows from the same argument presented for Sullivan's Proposition F.4.10.

For the definition of $K(M)$ we first introduce the set Σ of all singular simplices in M (*i.e.* the union over $h \in \mathbb{N}$ of the set of all continuous mappings $\sigma : \Delta_h \to M$) which are one-to-one on the vertices. For all $\sigma \in \Sigma$ we choose a fixed copy Δ_σ of the standard simplex (the dimension being such that σ can be considered to be defined on Δ_σ). Then we set

$$K(M) = \left(\bigcup_{\sigma \in \Sigma} \Delta_\sigma \right)\Big/_\sim$$

where the equivalence relation $\sim$ is defined in the following way: if $x_1 \in \Delta_{\sigma_1}$ and $x_2 \in \Delta_{\sigma_2}$ then $x_1 \sim x_2$ if and only if there exist faces $F_1 \subseteq \Delta_{\sigma_1}$ and $F_2 \subseteq \Delta_{\sigma_2}$ and an orientation-preserving simplicial isomorphism $f : F_1 \to F_2$ such that

$$\sigma_2\big|_{F_2} \circ f = \sigma_1\big|_{F_1} \quad \text{and} \quad x_2 = f(x_1)$$

(we just mean that common faces must be glued together). We define on $K(M)$ the weakest topology with respect to which all the Δ_σ's are closed.

Remark that $K(M)$ has a natural structure of "generalized infinite polyhedron": when writing $K(M)$ we shall understand this further structure too, while $|K(M)|$ will be used for the mere topological space.

A continuous function $S : |K(M)| \to M$ is naturally defined by the requirement that for all $\sigma \in \Sigma$ the restriction of S to Δ_σ equals σ itself. The following technical result may be found in [Moo]:

Proposition F.5.12. S is indeed a homotopy equivalence.

We just remark that our construction of $K(M)$ works for any topological space M, and the function S turns out to be a homotopy equivalence as soon as M is "good enough"; this is true in particular in our situation of a smooth compact manifold.

Proposition F.5.12 implies that M and $|K(M)|$ have "the same" homology, cohomology and bounded $\mathbb{R}$-cohomology. Moreover we can consider on $K(M)$

the natural "simplicial" homology, cohomology and bounded $\mathbb{R}$-cohomology theory: the latter will be denoted by $\hat{H}^*_{(s)}(K(M))$. Gromov proved the following quite technical result.

Theorem F.5.13. $\hat{H}^*(K(M))$ is canonically isometric to $\hat{H}^*_{(s)}(K(M))$.

We are now ready for the conclusion of the proof of Theorem F.4.8 by the same argument presented for F.4.10. We consider the pull-back bundle of ξ with respect to S (denoted by $S^*(\xi)$) and we remark the following:

– for any flat vector bundle η of rank n on a finite oriented polyhedron Q the Euler class $\mathcal{E}(\eta)$ is defined just as for a compact manifold as an element of $H^n(Q, \mathbb{Z})$, and the same proof as in F.4.10 allows one to check that the norm of $\mathcal{E}(\eta)$ is bounded by 1 with respect to the $\mathbb{Z}$-theory;

– though $K(M)$ is not a finite polyhedron it can be expressed as the union of a growing sequence of finite polyhedra $\{Q_k\}$, and its bounded $\mathbb{Z}$-cohomology is obtained as a "limit" of the bounded $\mathbb{Z}$-cohomologies of these finite polyhedra;

– the sequence $\mathcal{E}(S^*(\xi)|_{Q_k})$ converges to an element

$$\mathcal{E}(S^*(\xi)) \in H^n_{(s)}(K(M), \mathbb{Z})$$

corresponding, under the natural isomorphism given by the homotopy equivalence, to $\mathcal{E}(\xi) \in H^*(M, \mathbb{Z})$. Since we have

$$\|\mathcal{E}(S^*(\xi))|_{Q_k}\|_\infty \leq 1 \quad \forall k$$

we also have $\|\mathcal{E}(s^{(}\xi))\|_\infty \leq 1$ (with respect to the $\mathbb{Z}$-theory) and Propositions F.6.1 and F.6.2 allow us to conclude immediately.

If we replace Sullivan's proposition (F.4.10) by Smillie's one (F.4.12), the same method allows us to prove the following sharper version of the Milnor-Sullivan theorem:

Corollary F.5.14. $\|\mathcal{E}(\xi)\|_\infty \leq 2^{-n}$; in particluar, if $n = m$, we have

$$\|M\| \geq 2^n \cdot |\chi(\xi)|.$$

F.6 Sullivan's Conjecture and Amenable Groups

We recall that Sullivan conjectured that if a connected compact oriented n-manifold M supports an affine structure then its Euler-Poincaré characteristic must be zero. We will discuss in this section some facts concerning this conjecture.

The first idea is to introduce the fiber bundle in spheres $T_1 M$ associated to the tangent bundle to M (the definition works just like the one given at the beginning of section 4 for the two-dimensional case). $T_1 M$ is a bundle on M with fiber S^{n-1} and projection p (while the projection of TM will be denoted by π); moreover if M supports an affine structure then $T_1 M$ is naturally endowed with a flat structure.

Proposition F.6.1. Assume $p^* : H^n(M) \to H^n(T_1 M)$ is one-to-one; if M supports an affine structure then $\chi(M) = 0$.

Proof. Let $p^*(TM)$ denote the pull-back fiber bundle of TM on $T_1 M$, so that we have the following diagram:

$$
\begin{array}{ccc}
p^*(TM) & & TM \\
\downarrow{\scriptstyle \pi_1} & & \downarrow{\scriptstyle \pi} \\
T_1 M & \xrightarrow{\ p\ } & M
\end{array}
$$

We claim that $p^*(TM)$ has a non-vanishing section: in fact if we consider any differential metric on M we can identify $T_1 M$ with the unit tangent bundle

$$\{(x, v) : x \in M, \ v \in T_x M, \ \|v\|_x = 1\}.$$

Then

$$T_1 M \ni (x, v) \mapsto v \in \pi_1^{-1}(x, v)$$

is the desired non-vanishing section.

Now, if M is affine then TM and $p^*(TM)$ are flat, so that Euler classes can be defined, and moreover it can be shown that $\mathcal{E}(p^*(TM)) = p^*(\mathcal{E}(TM))$. Existence of a non-vanishing section implies that $\mathcal{E}(p^*(TM)) = 0$ and injectivity of p^* yields $\mathcal{E}(TM) = 0$, which implies $\chi(M) = \chi(TM) = 0$. $\quad\square$

According to F.6.1 Sullivan's conjecture leads quite naturally to the following problem: let $\xi = (E \xrightarrow{p} M)$ be a flat fiber bundle with structure group Γ, fiber F and holonomy ρ; under what conditions is the homomorphism $p^* : H^n(M) \to H^n(E)$ one-to-one?

Let us discuss first the case when ξ is a covering. In the following we shall keep assuming the basis of the bundle is the manifold M, though a topological space is often enough.

Lemma F.6.2. If ξ is a covering with automorphisms group G then the homomorphism $p^* : C^n(M) \to C^n(E)$ is one-to-one and its range is given by

$$\{\beta \in C^n(E) : g^*(\beta) = \beta \ \forall g \in G\}.$$

Proof. Assume $p^*(\gamma) = 0$ and let σ be a singular n-simplex in M. Since Δ_n is simply connected there exists a lift $\tilde{\sigma}$ of σ. Then

$$0 = p^*(\gamma)(\tilde{\sigma}) = \gamma(p \circ \tilde{\sigma}) = \gamma(\sigma) \ \Rightarrow \ \gamma = 0.$$

As for the second fact, inclusion $\subseteq$ is quite obvious:

$$g^*(p^*(\gamma)) = (p \circ g)^*(\gamma) = p^*(\gamma).$$

For the opposite inclusion, assume $g^*(\beta) = \beta \ \forall g \in G$ and define $\gamma \in C^n(M)$ in the following way: for a singular simplex σ in M consider any lift $\tilde{\sigma}$ and set

$\gamma(\sigma) = \beta(\tilde{\sigma})$. The property of β implies that γ is well-defined and moreover $p^*(\gamma) = \beta$. $\square$

The following result is established quite easily, but its proof provides an idea which turns out to be very fruitful since it can be generalized to much more interesting situations.

Proposition F.6.3. Let ξ be a finite covering: then

$$p^* : H^n(M) \to H^n(E)$$

is one-to-one.

Proof. Remark that the automorphisms group G is finite and for $\beta \in C^n(E)$ set

$$r(\beta) = \frac{1}{\#G} \cdot \sum_{g \in G} g^*(\beta).$$

Assume $\gamma \in C^n(M)$ is such that $p^*(\gamma) = \delta\beta_1$ for some $\beta_1 \in C^{n-1}(E)$ (which means that $p^*(\langle\gamma\rangle) = 0$); then, using the above lemma, we have

$$p^*(\gamma) = r(p^*(\gamma)) = r(\delta\beta_1) = \delta(r(\beta_1));$$

moreover $r(\beta_1) = p^*(\gamma_1)$ and then for some γ_1

$$p^*(\gamma) = \delta(p^*(\gamma_1)) = p^*(\delta\gamma_1) \;\Rightarrow\; \gamma = \delta\gamma_1$$

which means that $\langle\gamma\rangle = 0$ and hence the conclusion. $\square$

We are now going to describe the most general situation in which the argument presented above works. We first remark that Proposition F.6.1 holds also with the assumption that p^* is one-to-one as a homomorphism of $\hat{H}^n(M)$ into $\hat{H}^n(T_1 M)$ (in fact by the Milnor-Sullivan theorem Euler classes are bounded). So the general problem we will consider is whether

$$p^* : \hat{H}^n(M) \to \hat{H}^n(E)$$

is one-to-one for a fixed smooth flat bundle $\xi = (E \xrightarrow{p} M)$ with fiber F, structure group Γ and holonomy ρ; n is now an arbitrary fixed integer, and not necessarily the dimension of M. The idea for the generalization of the proof of F.6.3 is to consider ξ as a "covering with group Γ" and take a "mean over Γ" in order to retract $\hat{H}^n(E)$ onto $\hat{H}^n(M)$.

This is the reason for introducing the class of <u>amenable</u> groups, defined as those groups Γ for which there exists a linear functional $\int d\gamma$ on the space $B(\Gamma)$ of bounded real functions on Γ, satisfying the following properties:
(1) $f \in B(\Gamma)$, $f \geq 0 \;\Rightarrow\; \int f(\gamma)d\gamma \geq 0$;
(2) $\int 1 d\gamma = 1$;
(3) $\gamma_0 \in \Gamma, f \in B(\Gamma) \;\Rightarrow\; \int f(\gamma \cdot \gamma_0)d\gamma = \int f(\gamma)d\gamma$.
For a general treatise about amenable groups see [Greenl].

We shall prove later that if such a functional exists then there exists another one (denoted by $\int d\gamma$ as well) satisfying (1), (2), (3) and the additional property:

$(3)'\ \gamma_0 \in \Gamma, f \in B(\Gamma)\ \Rightarrow\ \int f(\gamma_0 \cdot \gamma) d\gamma = \int f(\gamma) d\gamma.$

The next result shows the way the argument presented above for the proof of F.6.3 can be generalized to the case of amenable structure group.

Theorem F.6.4 (Trauber). If in the above situation the group Γ is amenable then $p^* : \hat{H}^n(M) \to \hat{H}^n(E)$ is one-to-one.

Proof. We first remark that homology, cohomology and bounded cohomology of M and E can be defined starting from the following special classes of chains:

$$C_{*k}(M) = \left\{ \sum_i a_i \sigma_i :\ \sigma_i : \Delta_k \to M,\ \sigma_i(\Delta_k) \subset U \text{ (trivializing set)} \right\}$$
$$C_{*k}(E) = \left\{ \sum_i a_i \sigma_i :\ \sigma_i : \Delta_k \to E,\ p(\sigma_i(\Delta_k)) \subset U \text{ (trivializing set)} \right\}.$$

Given $\sigma : \Delta_k \to E$ with $p(\sigma(\Delta_k)) \subset U$ and a trivialization $\phi = (p, \lambda) : p^{-1}(U) \xrightarrow{\sim} U \times F$, we define for $\gamma \in \Gamma$ another simplex $\gamma\sigma$ by

$$(\gamma\sigma)(t) = \phi^{-1}\Big(p(\sigma(t)), \gamma\big(\lambda(\sigma(t))\big) \Big).$$

Another trivialization $\phi_1 = (p, \lambda_1)$ is related to ϕ by the relation

$$\phi_1^{-1}(u, f) = \phi^{-1}(u, \gamma_1(f))$$

where $\gamma_1 \in \Gamma$ is fixed; it follows that

$$\phi_1^{-1}\Big(p(\sigma(t)), \gamma\big(\lambda_1(\sigma(t))\big) \Big) = \phi^{-1}\Big(p(\sigma(t)), (\gamma_1 \cdot \gamma \cdot \gamma_1^{-1})\big(\lambda(\sigma(t))\big) \Big)$$

and the construction of $\gamma\sigma$ with respect to ϕ_1 produces $(\gamma_1 \cdot \gamma \cdot \gamma_1^{-1})\sigma$ with respect to ϕ.

We claim now that $p^* : \hat{C}_*^k(M) \to \hat{C}_*^k(E)$ is one-to-one and its range consists of

$$\big\{ c \in \hat{C}_*^k(E) : c(\gamma\sigma) = c(\sigma)\ \forall \text{ trivialization } \forall \gamma \in \Gamma \big\}.$$

Assume $p^*(c_1) = 0$; for a simplex $\sigma \in C_{*k}(M)$ and a trivialization ϕ around $\sigma(\Delta_k)$, fix $f_0 \in F$ and set $\tilde{\sigma}(t) = \phi^{-1}(\sigma(t), f_0)$; then $\tilde{\sigma} \in C_{*k}(E)$ and hence:

$$0 = p^*(c)(\tilde{\sigma}) = c(p \circ \tilde{\sigma}) = c(\sigma)$$

and injectivity of p^* is proved. If $c \in \hat{C}_*^k$ of course we have

$$p^*(c)(\gamma\sigma) = c((p \circ (\gamma\sigma)) = c(p \circ \sigma) = p^*(c)(\sigma)$$

whenever $\gamma\sigma$ is defined. Conversely, given $c \in \hat{C}_*^k(E)$ with the required invariance property, for a simplex $\sigma \in C_{*k}(M)$ we define $\tilde{\sigma}$ as above and set

$$c_1(\sigma) = c(\tilde{\sigma}).$$

Invariance of c implies that $c_1(\sigma)$ is well-defined, and moreover we have that $|c_1(\sigma)| \le \|c_1\|_\infty$, so that $c_1 \in \hat{C}_*^k(M)$, and obviously $p^*(c_1) = c$. Our claim is proved.

Now we can define the retraction of $\hat{C}_\star^k(E)$ onto the range of $p^\ast$ as in F.6.3; given $c \in C_\star^k(E)$, for a simplex $\sigma \in C_{\star k}(E)$ we set

$$r(c)(\sigma) = \int c(\gamma\sigma)d\gamma$$

where all the $\gamma\sigma$'s are performed with respect to the same trivialization. We have that:

(i) $|c(\gamma\sigma)| \leq \|c\|_\infty \ \forall \gamma$ so that the mean makes sense;

(ii) bi-invariance of the mean implies that if we change trivialization we obtain the same result;

(iii) $|r(c)(\sigma)| \leq \|c\|_\infty$ so that $r(c)$ is bounded;

(iv) left-invariance of the mean implies that for any choice of $\gamma_1\sigma$ we have $r(c)(\gamma_1\sigma) = r(c)(\sigma)$. Now the retraction is constructed, and the conclusion of the proof works in the same way as in F.6.3. $\qquad\square$

The above theorem, together with the version of Lemma F.6.1 for bounded cohomology, yields the following:

Theorem F.6.5 (Hirsch-Thurston). Let M be an affine compact connected oriented n-manifold and assume the range of the holonomy of the associated sphere fiber bundle T_1M is amenable; then $\chi(M) = 0$.

The original proof of the above theorem (to be found in [Hi-Th]) made use of a partially different machinery: the mean was taken over the complexes of differential forms (via the theorem of De Rham) and bounded cohomology was not even mentioned. However the present approch to Sullivan's conjecture, and the proof presented above of the Hirsch-Thurston theorem can be viewed as a starting point of the theory of bounded cohomology.

We shall discuss now some interesting properties of amenable groups. In the following we shall denote means over a group G not by $\int dg$ but simply by μ.

For $g_0 \in G$ we shall denote by ρ_{g_0} and by λ_{g_0} respectively the right and the left translations by g_0:

$$\rho_{g_0}(g) = g \cdot g_0 \qquad \lambda_{g_0}(g) = g_0 \cdot g.$$

Property (3) of the definition of amenable groups means that the mean is invariant under right translations. In the following result we recall the definition and prove an elementary fact we have already used above; we recall that $B(G)$ denotes the vector space of all bounded real functions on G.

Lemma F.6.6. If G admits a (right-invariant) mean then it admits a bi-invariant mean, i.e. a linear functional μ on $B(G)$ satisfying the following properties:

(1) $f \in B(G)$, $f \geq 0 \ \Rightarrow \ \mu(f) \geq 0$;

(2) $\mu(1) = 1$;

(3) $g_0 \in G$, $f \in B(G) \ \Rightarrow \ \mu(f \circ \rho_{g_0}) = \mu(f)$;

(3)' $g_0 \in G$, $f \in B(G) \ \Rightarrow \ \mu(f \circ \lambda_{g_0}) = \mu(f)$.

Proof. Let μ be any mean, *i.e.* a functional satisfying (1), (2) and (3). We define a fuction $\Phi : B(G) \to B(G)$ by:

$$(\Phi(f))(g) = \mu(f \circ \lambda_{g^{-1}}).$$

Then we set $\overline{\mu}(f) = \mu(\Phi(f))$, and it is quite easily verified that $\overline{\mu}$ satisfies all the properties (1),...,(3)'. $\square$

We have already used the following elementary fact.

Remark F.6.7. Finite groups are amenable.

In order to give other characterizations of amenable groups we need to introduce a new concept. We shall consider a fixed group G with identity e.

A subset S of G is called <u>symmetric</u> if $x \in S \Rightarrow x^{-1} \in S$. Given a finite symmetric part S of G we define the <u>S-boundary</u> of any other subset A of G as

$$\partial_S(A) = \{a \in A : \exists s \in S \text{ s.t. } a \cdot s \notin A\} = A \setminus \left(\bigcap_{s \in S} A \cdot s\right).$$

We recall moreover that a group is said to be <u>finitely presented</u> if it can be expressed as the quotient of a finitely generated free group under a finite number of relations.

We state now the main theorem about amenable groups; its proof will be deferred for a while.

Theorem F.6.8. The following conditions are pairwise equivalent:
(i) G is amenable;
(ii) $\forall k \in \mathbb{N}, \forall f_1, ..., f_k \in B(G), \forall g_1, ..., g_k \in G,$

$$\inf_{g \in G} \sum_i (f_i - f_i \circ \rho_{g_i})(g) \leq 0;$$

(iii) for any finite symmetric part S of G and for any $\varepsilon > 0$ there exists a finite non-empty subset A of G such that

$$\#(\partial_S(A)) \leq \varepsilon \cdot \#A.$$

These conditions imply the next one, and they are equivalent to it for finitely presented groups:
(iv) for any $n \in \mathbb{N}$, for any connected Riemannian n-manifold V such that $\Pi_1(V) \cong G$ and for any $\varepsilon > 0$ there exists a domain $\Omega \subset \tilde{V}$ (the universal covering of V, endowed with the natural Riemannian structure) with $(n-1)$-measurable boundary such that the $(n-1)$-measure of $\partial\Omega$ does not exceed ε times the n-measure of Ω.

Before sketching the proof of F.6.8 we discuss some interesting consequences.

Lemma F.6.9. If G has a finite number of generators it suffices to check condition (iii) for a fixed arbitrary finite symmetric part generating G.

Proof. Assume S_1 is a fixed symmetric part of G generating it, with $\#S_1 = h < \infty$. Any other finite part of G is contained in S_1^k for some $k \in \mathbb{N}$; hence we can refer to $S = S_1^k$; remark that $\forall A$ finite

$$\partial_{S_1^k}(A) = \partial_{S_1^{k-1}}\left(\partial_{S_1}(A)\right).$$

Then we easily have

$$\#\partial_{S_1^k}(A) \le h^{k-1} \cdot \#\partial_{S_1}(A)$$

and the conclusion follows at once. □

Proposition F.6.10. For all $k \in \mathbb{N}$ the group $\mathbb{Z}^k$ is amenable.

Proof. Let $e_1, ..., e_k$ be the canonical generators and let

$$S = \left\{ \pm e_1, ..., \pm e_k \right\}.$$

If $n \in \mathbb{N}$ and A_n denotes a cube of edge n in $\mathbb{Z}^k$, *i.e.* for instance:

$$A_n = \left\{ (z_1, ..., z_k) : 1 \le z_1, ..., z_k \le n \right\}$$

then we have that $\#A_n = n^k$. Moreover the S-boundary of A_n is the real boundary of A_n, *i.e.*

$$\partial_S(A_n) = \left\{ (z_1, ..., z_k) : 1 \le z_1, ..., z_k \le n, \ z_i = 1 \text{ or } z_i = n \text{ for some } i \right\}.$$

Then $\#A_n = c_k \cdot n^{k-1}$ where c_k is a constant depending only on k; hence

$$\lim_{n \to \infty} \frac{\#\partial_S(A_n)}{\#A_n} = 0$$

and the conclusion follows from Remark F.6.9. □

Proposition F.6.11. (a) If H is a subgroup of G and G is amenable then H is amenable;

(b) if H is a normal subgroup of G and both H and G/H are amenable then G is amenable;

(c) if G is the union of an increasing sequence of amenable groups then G is amenable;

(d) if G is such that all finitely generated subgroups of G are amenable, then G is amenable;

(e) all Abelian groups are amenable;

(f) if G is solvable (*i.e.* there exists a tower of subgroups

$$G = G_0 \supset G_1 \supset ... \supset G_m = \{e\}$$

such that G_{i+1} is normal in G_i and G_i/G_{i+1} is Abelian) then G is amenable.

Proof. (a) Let μ be a mean on G; since G is partitioned in the right lateral classes of H we can find a set S of G such that G is the disjoint union of $\{H \cdot g : g \in S\}$ (we are actually taking a representative of all the points of

G/H). For $f \in B(H)$ we define $\hat{f} \in B(G)$ such that for $h \in H$ and $g \in S$ we have $\hat{f}(h \cdot g) = f(h)$; of course $\hat{f}$ is well-defined and bounded. We set now $\mu_H(f) = \mu(\hat{f})$ and claim that μ_H is a mean on H. Since the mapping $f \to \hat{f}$ is linear μ_H is linear; we have $\hat{1} = 1$ so that $\mu_H(1) = 1$; of course if $f \geq 0$ we have $\hat{f} \geq 0$ and then $\mu_H(f) \geq 0$. Now, let $h_0 \in H$ and define $f_1(h) = f(h_0 \cdot h)$; since for $h \in H$ we have $h_0 \cdot h \in H$ we have for $g \in S$

$$\hat{f}_1(h \cdot g) = f_1(h) = f(h_0 \cdot h) = \hat{f}(h_0 \cdot (h \cdot g))$$

and then $\hat{f}_1$ is obtained from $\hat{f}$ by a left translation, which implies $\mu(\hat{f}_1) = \mu(\hat{f})$ and then $\mu_H(f_1) = \mu_H(f)$.

(b) Let μ_1 and μ_2 be left-invariant means on H and G/H respectively. For all $\alpha \in G/H$ let us fix $g_\alpha \in G$ such that $\alpha = g_\alpha \cdot H$. Given $f \in B(G)$ let us define

$$f_{g_\alpha}(h) = f(g_\alpha \cdot h);$$

of course we have $f_{g_\alpha} \in B(H)$. Then we set

$$\hat{f}(\alpha) = \mu_1(f_{g_\alpha});$$

we have again $\hat{f} \in B\left(G/H\right)$ and hence we can define

$$\mu(f) = \mu_2(\hat{f}).$$

The functional μ defined on $B(G)$ is easily checked to be linear, to have value 1 on the function 1 and to have non-negative values on non-negative functions. We are left to check left-invariance; since all $g \in G$ can be written as $g = h_0 \cdot g_{\alpha_0}$ for some $h_0 \in H$ and $\alpha_0 \in G/H$ it suffices to check that for $f_1 = f \circ \lambda_{h_0}$ and for $f_1 = f \circ \lambda_{g_{\alpha_0}}$ we have $\mu(f_1) = \mu(f)$.

– normality of H implies that for some $h_1 \in H$ we have $h_0 \cdot g_\alpha = g_\alpha \cdot h_1$ and then

$$(f_1)_{g_\alpha}(h) = f(h_0 \cdot g_\alpha \cdot h) = f(g_\alpha \cdot h_1 \cdot h) = (f_{g_\alpha} \circ \lambda_{h_1})(h)$$

which implies $\hat{f}_1 = \hat{f}$ and then the conclusion;

– again, normality of H implies that $g_{\alpha_0} \cdot g_\alpha = g_{\alpha_0 \cdot \alpha} \cdot h_1$ for some $h_1 \in H$; then

$$(f_1)_{g_\alpha}(h) = f(g_{\alpha_0} \cdot g_\alpha \cdot h) = f(g_{\alpha_0 \cdot \alpha} \cdot h_1 \cdot h) = (f_{g_{\alpha_0 \cdot \alpha}} \circ \lambda_{h_1})(h)$$

and then $\hat{f}_1 = f_1 \circ \lambda_{\alpha_0}$ and the conclusion follows.

(c) We use condition (iii) of F.6.8: if G is the union of the growing sequence $\{G_k\}$ and S is a finite symmetric part of G them $S \subset G_k$ for k large enough, and the conclusion follows at once.

(d) We use condition (iii) of F.6.8 again: the subgroup generated by a finite symmetric part of G is finitely generated, and the proof is over.

(e) According to point (d) it suffices to show that finitely generated Abelian groups are amenable. This fact is deduced from point (b), using the fact that a finitely generated Abelian group is the direct sum of $\mathbb{Z}^k$ and a finite Abelian group, and both these are normal and amenable (by F.6.7 and F.6.10).

(f) This is deduced immediately from points (b) and (e).   □

The above result implies that several pleasant groups are amenable (remark however that we are far from being able to give explicit means: we will discuss this later). The next proposition provides the first interesting examples of non-amenable groups.

Proposition F.6.12. (a) The free group with two generators is not amenable; (b) the fundamental group of a compact hyperbolic manifold is not amenable.

Proof. (a) Let G be the free group generated by the symbols X and Y. Let us remark first that each element of G has a unique expression of minimal length in the generators (we only need to get rid of expressions $X \cdot X^{-1}$, $X^{-1} \cdot X$, $Y \cdot Y^{-1}$ and $Y^{-1} \cdot Y$). We define now two functions f_1 and f_2 in the following way:

$$f_1(w) = \begin{cases} 1 & \text{if the shortest expression of } w \text{ ends with } X \\ 0 & \text{otherwise;} \end{cases}$$

$$f_2(w) = \begin{cases} 1 & \text{if the shortest expression of } w \text{ ends with } Y \\ 0 & \text{otherwise.} \end{cases}$$

It is very easily checked that

$$f_1(w \cdot X) - f_1(w) + f_2(w \cdot Y) - f_2(w) \geq 1 \quad \forall w \in G$$

and hence point (ii) of F.6.8 implies that G is not amenable.

(b). We shall use point (a) of F.6.11 and check that if M is hyperbolic and compact then $\Pi_1(M)$ contains a subgroup isomorphic to the free group with two generators. We proved in B.4.4 that the elements of $\Pi_1(M)$ viewed as elements of $\mathcal{I}(\mathbf{H}^n)$ are isometries of hyperbolic type. Of course $\Pi_1(M)$ cannot be infinite cyclic, and hence we can find elements X and Y of $\Pi_1(M)$ which are not multiples of the same isometry; D.3.9 implies that the axes of X and Y have no common endpoint.

We claim that there exist positive integers p and q such that the group generated by X^p and Y^q is free. Let us choose horoballs at the endpoints as suggested by Fig. F.17.

We can find p and q such that $X^p(\mathbf{H}^n \setminus A) \subseteq A'$ and $Y^p(\mathbf{H}^n \setminus B) \subseteq B'$, which implies that no relation can hold between X^p and Y^q, and the proof is over.   □

Remark F.6.13. In the case of surfaces point (b) of the above proposition can be proved in another interesting way: if $\Pi_1(M)$ were amenable then the unit tangent bundle to M (a flat bundle: compare to the proof of F.5.11) would have zero Euler class (by F.6.5), which implies $\chi(M) = 0$ and this is absurd.

Point (b) of the above proposition admits the following (quite hard) generalization to be found in [Av]:

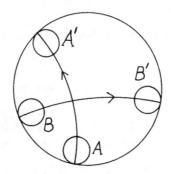

Fig. F.17. Disjoint horoballs at the endpoints of the axes of the isometries

Theorem F.6.14. Let M be a connected compact Riemannian manifold with non-positive curvature everywhere, which is not a flat manifold; then the fundamental group of M is not amenable.

We turn now to a sketch of the proof of F.6.8.

Proposition F.6.15. (i) $\Leftrightarrow$ (ii).

Proof. (i) $\Rightarrow$ (ii): assume by contradiction

$$\inf_{g \in G} \sum_i (f_i - f_i \circ \rho_{g_i})(g) = \varepsilon > 0;$$

by property (3) of the definition of the mean μ we have

$$\mu\Big(\sum_i (f_i - f_i \circ \rho_{g_i})\Big) = 0$$

while by properties (1) and (2) of the definition and the assumption we have

$$\mu\Big(\sum_i (f_i - f_i \circ \rho_{g_i})\Big) \geq \varepsilon > 0$$

and this is a contradiction.

(ii) $\Rightarrow$ (i): let $E \subset B(G)$ be the linear subspace generated by all the functions of the form $f - f \circ \rho_g$ for $f \in B(G)$ and $g \in G$. (ii) implies that the distance (with respect to the natural sup norm in $B(G)$) from 1 to E is 1; then it follows from the Hahn-Banach theorem that there exists a continuous linear functional μ on $B(G)$ such that $\|\mu\| = \mu(1) = 1$ and μ vanishes on E. Properties (2) and (3) of the definition of the mean are obviously verified, while (1) is checked in the following way: if $f \in B(G)$ and $f \geq 0$ we have for k large enough (namely, for $k \geq \|f\|_\infty$), $\|k \cdot 1 - f\|_\infty = k$, and then

$$k \geq |\mu(k \cdot 1 - f)| = |k - \mu(f)|$$

which entails $\mu(f) \geq 0$. $\qquad\qquad\square$

Proposition F.6.16. (iii) $\Rightarrow$ (ii).

Proof. Given $f_1, ..., f_k \in B(G)$ and $g_1, ..., g_k \in G$ let us consider the finite symmetric set $S = \{g_i^{\pm 1} : i = 1, ..., k\}$. Let $\varepsilon > 0$ be arbitrary and let A be a finite set such that $\#(\partial_S(A)) \leq \varepsilon \cdot \#A$.

For a fixed i we easily have

$$\sum_{a \in A} (f_i - f_i \circ \rho_{g_i})(a) = \sum_{a \in \partial_S A} (f_i - f_i \circ \rho_{g_i})(a)$$

(the other terms kill each other). If we call F the function

$$g \mapsto \sum_i (f_i - f_i \circ \rho_{g_i})(g)$$

we deduce from above that

$$\#A \cdot \inf(F) \leq \sum_{a \in A} F(a) = \sum_{a \in \partial_S A} F(a) \leq \#\partial_S A \cdot \sum_i 2\|f_i\| \leq 2\varepsilon \cdot \#A \cdot \sum_i \|f_i\|$$

whence $\inf(F) \leq 2\varepsilon \sum_i \|f_i\|$ and arbitrariness of ε implies the conclusion. $\square$

The following implication we shall prove, (ii) $\Rightarrow$ (iii), requires a little more work. The proof, by contradiction, is deduced from the following result due to Folner:

Theorem F.6.17. let G be a group, S a finite part of G and $c > 0$ such that for any finite subset A of G

$$\#A \leq c \cdot \#(\partial_S A).$$

Then there exist functions $\{f_s\}_{s \in S} \subset B(G)$ such that:
(a) $\|f_s\| \leq c \ \forall s \in S$;
(b) $\sum_{s \in S} (f_s - f_s \circ \rho_s)(g) \geq 1 \ \forall g \in G$.

Proof. We claim that it suffices to prove that for any finite subset A of G the f_s's can be found in such a way that (a) holds and (b) holds for $g \in A$: for any finite $A \subset G$ denote by $\mathcal{S}_A$ the set of all S-tuples of functions satisfying properties (a) and (b) on A; $\mathcal{S}_A$ is a closed subset of

$$\prod_{s \in S} [-c, c]^G$$

which is compact (by the Tychonov theorem), and hence $\mathcal{S}_A$ is compact. If $\mathcal{S}_A \neq \phi \ \forall A$, the family of compact spaces $\{\mathcal{S}_A : A \subset G \text{ finite}\}$ has the property that the intersection of a finite subfamily is non-empty, which implies that the intersection of the whole family is non-empty; an S-tuple in this intersection satisfies (a) and (b), and our claim is proved.

Given a finite subset A of G we look for the f_s's between the functions supported in A. If we think of the unknown $y_{(s,a)}$ as the value of f_s in a, the conditions on the f_s's are re-written in the following system of inequalities:

$$\begin{cases} |y_{(s,a)}| \leq c & \forall (s,a) \in S \times A \\ \sum_{s \in S} \left(y_{(s,a)} - y_{(s,a \cdot s)}\right) \geq 1 & \forall a \in A \end{cases}$$

(where it is understood that for $a \cdot s \notin A$ we have $y_{(s,a \cdot s)} = 0$).

It can be shown (see [Ci]) that a system of linear inequalities in $y \in \mathbb{R}^k$

$$\begin{cases} L_1 y \geq c_1 \\ \cdots \\ L_n y \geq c_n \end{cases}$$

(where $L_1, ..., L_n$ are linear functionals on $\mathbb{R}^k$) has a solution if and only if

$$\alpha_1, ..., \alpha_n \geq 0, \ \sum_i \alpha_i L_i = 0 \ \Rightarrow \ \sum_i \alpha_i c_i \leq 0.$$

Let us use this general criterion in our situation; we must show that if $\alpha_a \geq 0$ (for $a \in A$) and $\beta_{(s,a)}, \gamma_{(s,a)} \geq 0$ (for $(s,a) \in S \times A$) are constants such that

$$\sum_{(s,a)} \left(\beta_{(s,a)} \cdot y_{(s,a)} - \gamma_{(s,a)} \cdot y_{(s,a)}\right) + \sum_{(s,a)} \alpha_a \left(y_{(s,a)} - y_{(s,a \cdot s)}\right) = 0 \ \forall y$$

then we have

$$\sum_{(s,a)} \beta_{(s,a)} \cdot (-c) + \sum_{(s,a)} \gamma_{(s,a)} \cdot (-c) + \sum_a \alpha_a \leq 0.$$

The first relation is re-written as

$$\sum_{(s,a)} \left(\beta_{(s,a)} - \gamma_{(s,a)}\right) \cdot y_{(s,a)} + \sum_{(s,a)} \alpha_a \left(y_{(s,a)} - y_{(s,a \cdot s)}\right) = 0 \ \forall y$$

and the second one as

$$\sum_a \alpha_a \leq c \cdot \sum_{(s,a)} \left(\beta_{(s,a)} + \gamma_{(s,a)}\right).$$

If we set $\lambda_{(s,a)} = \beta_{(s,a)} - \gamma_{(s,a)}$ we have $|\lambda_{(s,a)}| \leq \beta_{(s,a)} + \gamma_{(s,a)}$; it follows that it suffices to prove the following implication:
given $\alpha_a \geq 0$ and $\lambda_{(s,a)} \in \mathbb{R}$ such that

$$(\star) \qquad \sum_{(s,a)} \lambda_{(s,a)} \cdot y_{(s,a)} + \sum_{(s,a)} \alpha_a (y_{(s,a)} - y_{(s,a \cdot s)}) = 0 \ \forall y$$

then

$$\sum_a \alpha_a \leq c \cdot \sum_{(s,a)} |\lambda_{(s,a)}|.$$

Assume $(\star)$ is verified; then the coefficients of all the $y_{(s,a)}$'s vanish:

$$\alpha_a - \alpha_{a \cdot s^{-1}} + \lambda_{(s,a)} = 0$$

(where we implicitly mean that $\alpha_{a \cdot s^{-1}} = 0$ if $a \cdot s^{-1} \notin A$); then

$$\lambda_{(s,a)} = \alpha_{a \cdot s^{-1}} - \alpha_a.$$

Let us number the elements of A by numbers $1, ..., p$. We simplify the notation and write i instead of a_i (no confusion will arise, as we shall use only the order relation between numbers); moreover we assume the numbering is such that $\alpha_1 \leq \alpha_2 \leq ... \leq \alpha_p$. We define for a technical reason α_0 to be zero. Moreover for $0 \leq r \leq p$ we set

$$A_r = \{r+1, ..., p\}.$$

Given a pair (s, a) we define the following subset of $\{0, 1, ..., p-1\}$:

$$R_{(s,a)} = \begin{cases} \text{(i)} & \{0, 1, ..., a-1\} & \text{if } a \cdot s^{-1} \notin A \\ \text{(ii)} & \{a, a+1, ..., a \cdot s^{-1} - 1\} & \text{if } a < a \cdot s^{-1} \in A \\ \text{(iii)} & \{a \cdot s^{-1}, a \cdot s^{-1} + 1, ..., a-1\} & \text{if } A \ni a \cdot s^{-1} < a. \end{cases}$$

We claim that

$$|\lambda_{(s,a)}| = \sum_{r \in R_{(s,a)}} (\alpha_{r+1} - \alpha_r).$$

In fact it is easily checked that in the same three cases as in the definition of $R_{(s,a)}$

$$\sum_{r \in R_{(s,a)}} (\alpha_{r+1} - \alpha_r) = \begin{cases} \text{(i)} & \alpha_a - \alpha_0 = \alpha_a = -\lambda_{(s,a)} \\ \text{(ii)} & \alpha_{a \cdot s^{-1}} - \alpha_a = \lambda_{(s,a)} \\ \text{(iii)} & \alpha_a - \alpha_{a \cdot s^{-1}} = -\lambda_{(s,a)}. \end{cases}$$

We can prove now that for $r \in \{0, 1, ..., p-1\}$ we have

$$r \in R_{(s,a)} \Leftrightarrow a \in (A_r \cdot s) \triangle A_r = (A_r \cdot s \setminus A_r) \bigcup (A_r \setminus A_r \cdot s).$$

In fact it is easily checked that

$$a \in A_r \cdot s \setminus A_r \Leftrightarrow \begin{pmatrix} a \notin A_r \\ a \cdot s^{-1} \in A_r \end{pmatrix} \Leftrightarrow \begin{pmatrix} a \leq r \\ a \cdot s^{-1} > r \end{pmatrix}$$

$$a \in A_r \setminus A_r \cdot s \Leftrightarrow \begin{pmatrix} a \in A_r \\ a \cdot s^{-1} \notin A_r \end{pmatrix} \Leftrightarrow \begin{pmatrix} a > r \\ \text{either } a \cdot s^{-1} \notin A \text{ or } a \cdot s^{-1} \leq r \end{pmatrix}$$

and the assertion follows at once.

We are now ready for the required estimate:

$$\sum_s \sum_a |\lambda_{(s,a)}| = \sum_s \sum_a \sum_{r \in R_{(s,a)}} (\alpha_{r+1} - \alpha_r) =$$

$$= \sum_s \sum_{r=0}^{p-1} \sum_{a \in (A_r \cdot s) \triangle A_r} (\alpha_{r+1} - \alpha_r) =$$

$$= \sum_{r=0}^{p-1} |\alpha_{r+1} - \alpha_r| \cdot \sum_s \#((A_r \cdot s) \triangle A_r) \geq \sum_{r=0}^{p-1} (\alpha_{r+1} - \alpha_r) \cdot \#(\partial_S(A_r)) \geq$$

$$\geq \sum_{r=0}^{p-1} (\alpha_{r+1} - \alpha_r) \cdot \frac{1}{c} \cdot \#A_r = \frac{1}{c} \cdot \sum_{r=0}^{p-1} (p - r) \cdot (\alpha_{r+1} - \alpha_r) =$$

$$= \frac{1}{c} \cdot \left\{ \sum_{j=1}^{p} (p - (j-1)) \cdot \alpha_j - \sum_{j=0}^{p-1} (p - j) \cdot \alpha_j \right\} =$$

$$= \frac{1}{c} \cdot \left\{ \sum_{j=1}^{p} (p - (j-1)) \cdot \alpha_j - \sum_{j=1}^{p} (p - j) \cdot \alpha_j \right\} = \frac{1}{c} \cdot \sum_{j=1}^{p} \alpha_j.$$

$\square$

Proposition F.6.18. (iii) $\Rightarrow$ (iv).

Proof. Let D be a compact fundamental domain of $\tilde{V}$ (with respect to the action of the fundamental group) such that ∂D has finite $(n-1)$-measure. Let us set:

$$S = \{ g \in \Pi_1(V) : g(D) \cap D \neq \phi \}.$$

S is a finite symmetric part of $\Pi_1(V)$.

For $\varepsilon > 0$ let A be given by (iii) for ε and S, and let us set $\Omega = A \cdot D$ (the orbit of D under A). We assert that $\partial\Omega \subset (\partial_S A)(\partial D)$; in fact if $a \in A \setminus \partial_S A$ then $\Omega \supset a \cdot s \cdot D$ $\forall s \in S$, which implies that $a \cdot D$ is contained in the interior of Ω; moreover the orbit under A of the interior of D is contained in the interior of Ω, and our assertion is proved.

If $\mathcal{M}_n$ and $\mathcal{M}_{n-1}$ denote respectively the n-measure and the $(n-1)$-measure in $\tilde{V}$, then we have:

$$\mathcal{M}_n(\Omega) = \#A \cdot \mathcal{M}_n(D)$$

$$\mathcal{M}_{n-1}(\partial\Omega) \leq \#(\partial_S A) \cdot \mathcal{M}_{n-1}(\partial D)$$

whence

$$\mathcal{M}_{n-1}(\partial\Omega) \leq \varepsilon \cdot \frac{\mathcal{M}_{n-1}(\partial D)}{\mathcal{M}_n(D)} \cdot \mathcal{M}_n(\Omega)$$

and the proof is easily completed. $\square$

For the proof of Theorem F.6.8 only implication (iv) $\Rightarrow$ (iii) for finitely presented groups is missing, but we shall not prove it, since it is quite hard: we refer to [Gr-La-Pa].

Remark F.6.19. We recall that in F.6.10 we used condition (ii) of F.6.8 to prove that $\mathbb{Z}^k$ is amenable. We checked in F.6.15 that the implication (ii) $\Rightarrow$ (i) makes use of the theorem of Hahn-Banach: since this theorem is "purely existential" this method provides no way to construct explicit means. In fact, as we remarked before, no explicit mean is known for any infinite amenable group, not even for $\mathbb{Z}$!

We conclude our review of the various characterizations of amenable groups by mentioning the relationship (namely, the equivalence) between non-amenability and existence of paradoxal partitions (see [LH-Sk]).

Fix a group G and consider its action onto itself by multiplication on the left. A pa̲ra̲do̲xal partition of G is given by:
– two finite partitions $\{S_i\}_{i=1,...,k}$ and $\{S'_i\}_{i=1,...,l}$ of G (a partition being a collection of pairwise disjoint subsets whose union covers the whole space);
– elements $g_1, ..., g_k$ and $g'_1, ..., g'_l$ of G such that the collection of sets

$$\{g_i(S_i): \ i = 1, ..., k\} \bigcup \{g'_j(S'_j): \ j = 1, ..., l\}$$

is still a partition of G.

(The paradox comes from the fact that one could expect that both

$$\{g'_j(S'_j): \ j = 1, ..., l\} \quad \text{and} \quad \{g_i(S_i): \ i = 1, ..., k\}$$

cover G, as they are obtained by translating the elements of a partition, and intersections are avoided.)

The following holds:

Theorem F.6.20. G admits paradoxal partitions if and only if it is not amenable.

Remark that the implication that if G is amenable it cannot have para-doxal partitions is quite evident, while the converse is harder: its proof requires in particular the Hahn-Banach theorem we have already needed in F.6.8.

It is possible to show that SO(3) is not amenable (in fact it contains a free group with two generators). By a suitable generalization of the above notions to the situation of a group acting on a set, it is possible to obtain from this as a corollary of the above theorem the classical Banach-Tarski paradox:

Theorem F.6.21. Let U and V be compact subsets of $\mathbb{R}^3$. Then there exist finite partitions $\{U_i\}_{i=1,...,n}$ and $\{V_i\}_{i=1,...,n}$ of U and V respectively, and Euclidean isometries $g_1, ..., g_n$ of $\mathbb{R}^3$ such that $g_i(U_i) = V_i$.

We shall discuss now another notion related to amenability. If S is a finite set of generators of a group G and n is a natural number we denote by $B_S(n)$ the set of all elements of G which can be written as a word in $S \cup S^{-1} \cup \{e\}$ with length not greater than n; moreover we set $b_S(n) = \#B_S(n)$. We shall say G has a:
– po̲ly̲nomi̲al gr̲owt̲h of degree not greater than d if there exists $c > 0$ such that

$$b_S(n) \le c \cdot n^d + 1 \qquad \forall n \in \mathbb{N};$$

– exponential growth if there exist $c > 0$ and $\alpha > 1$ such that

$$b_S(n) \geq c \cdot \alpha^n \qquad \forall\, n \in \mathbb{N};$$

– sub-exponential growth in all other cases (*i.e.* if it has neither polynomial nor exponential growth).

In the above definition we understood that these notions are independent of the set S of generators; in fact this is the case, as it is easily deduced from the next result.

Lemma F.6.22. If S_1 and S_2 are finite sets of generators of G then there exist positive integers k_1 and k_2 such that

$$b_{S_1}(n) \leq b_{S_2}(k_1 \cdot n) \quad b_{S_2}(n) \leq b_{S_1}(k_2 \cdot n) \qquad \forall\, n \in \mathbb{N}.$$

Proof. It suffices to choose k_1 and k_2 in such a way that $S_1 \subseteq B_{S_2}(k_1)$ and $S_2 \subseteq B_{S_1}(k_2)$, and of course this is possible because of finiteness.   □

If G has polynomial growth we define the degree of the growth as the least possible integer d for which the definition works; remark that the above lemma implies that the degree is well-defined.

Example F.6.23. (A) As a set S of generators of $\mathbb{Z}^k$ choose the canonical basis: then it is easily checked that

$$b_S(n) = (2n + 1)^k$$

which implies that $\mathbb{Z}^k$ has polynomial growth of degree k.
(B) If G is the free group with $k \geq 2$ generators and S consists of such generators, then it is possible to show (we leave it as an exercise) that

$$b_S(n) = \frac{k(2k - 1)^n - 1}{k - 1}$$

and then G has exponential growth.
(C) Let V be a compact connected Riemannian manifold, fix $x_0 \in \tilde{V}$ (the Riemannian universal cover of V) and consider the function

$$\mathbb{R}_+ \ni r \mapsto \mathcal{M}_n\big(B(x_0, r)\big)$$

(where $\mathcal{M}_n$ denotes the n-measure associated to the Riemannian structure, and $B(x_0, r)$ denotes the ball of center x_0 and radius r. It may be shown (see [Gh-LH]) that this function has the same type of growth as $\Pi_1(V)$.

The first important result relating amenability and type of growth is the following:

Proposition F.6.24. If a finitely generated group G is not amenable then it has exponential growth.

Proof. If G is not amenable and S is a finite set of generators, according to F.6.8 we can find $c > 0$ such that for any finite subset A of G

$$\#A \le c \cdot \#(\partial_{\hat{S}} A)$$

(where $\hat{S} = S \cup S^{-1}$, the symmetrized of S).

Let us use this for $A = B_S(n)$, and remark that

$$\partial_{\hat{S}} B_S(n) = B_S(n) \setminus B_S(n-1);$$

then

$$b_S(n) \le c \cdot \big(b_S(n) - b_S(n-1)\big).$$

Since we can obviously assume $c > 1$ we have

$$b_S(n) \ge \frac{c}{c-1} \cdot b_S(n-1) \quad \Rightarrow \quad b_S(n) \ge \left(\frac{c}{c-1}\right)^n;$$

as $\frac{c}{c-1} > 1$ this implies that G has exponential growth.   □

The above result implies the following criterion of amenability:

Corollary F.6.25. *If a group is finitely generated and it has polynomial growth then it is amenable.*

It may be conjectured that for finitely generated groups non-amenability is equivalent to having exponential growth (*i.e.* the converse of F.6.24 holds). Unfortunately this is not the case:

Proposition F.6.26. *There do exist finitely generated amenable groups having exponential growth.*

Proof. Consider the group G of affine automorphisms of $\mathbb{R}$ generated by the following mappings g and h:

$$g(x) = 2x \qquad h(x) = x + 1.$$

The group of the commutators is a group of translations, and hence it is Abelian; it follows from point (f) of F.6.11 that G is amenable.

On the other hand, if $S = \{g, h\}$, then $B_S((n+1)^2)$ contains all the translations

$$x \mapsto x + 2^{-n} \cdot p \quad \text{for } 0 \le p \le 2^n$$

and this implies quite easily that G has exponential growth.   □

We state now a very important result providing a useful characterization of groups having polynomial growth.

Theorem F.6.27. *For a finitely generated group G the following facts are equivalent:*

(A) G *has polynomial growth;*

(B) G *is almost-nilpotent.*

(We address to [Gro2] for the proof: implication (A) $\Rightarrow$ (B) is due to Gromov, and implication (B) $\Rightarrow$ (A) is due to Wolf.)

We state now one of the key results of the abstract theory of bounded cohomology, *i.e.* the fact that the bounded cohomology of a topological space depends essentially only on its fundamental group. The following holds:

Theorem F.6.28. (1) Let X_1 and X_2 be topological spaces and let $f : X_1 \to X_2$ be a continuous mapping such that $f_* : \Pi_1(X_1) \to \Pi_1(X_2)$ is surjective and has amenable kernel. Then for all n's the mapping:

$$\hat{f}^* : \hat{H}^n(X_2) \to \hat{H}^n(X_1)$$

is a bijective isometry (with respect to the norms $\| \cdot \|_\infty$).
(2) If $\Pi_1(X)$ is amenable then $\hat{H}^n(X) = \{0\}$ for all n's.
(3) If M is a connected oriented compact manifold and $\Pi_1(M)$ is amenable then $\|M\| = 0$.

The proof of point (1) is too complicated to be even sketched: we address the reader to [Gro3]. We just mention the key idea for the proof of point (2) in case X has a special form: if G is a finitely presented group it is possible to show (by the Eilenberg-McLane construction) that there exists, unique up to homotopy equivalence, a finite connected CW-complex $K(G,1)$ such that $\Pi_1(K(G,1)) \cong G$ and $\Pi_n(K(G,1)) = \{0\}$ for $n \geq 2$. The universal cover $\tilde{K}(G,1)$ of $K(G,1)$ is contractible and has automorphisms group G, an amenable group. Then Trauber's argument works more or less as for F.6.4, leading to the fact that the natural homomorphism

$$\hat{H}^*(K(G,1)) \to \hat{H}^*(\tilde{K}(G,1))$$

is one-to-one, which implies that the groups $\hat{H}^*(K(G,1))$ are trivial.

Point (3) is readily deduced ("by duality") from point (2) and F.2.2.

The above theorem shows that for several pleasant spaces bounded cohomology is trivial. On the other hand it is a frequent phenomenon that when bounded cohomology is not trivial then it is actually extremely large. We conclude with a sketch of the description of this fact for surfaces.

Fix a (compact, connected, oriented) surface of genus $g \geq 2$; consider a hyperbolic structure on M and associate to each differential two-form $\omega \in \Lambda^2(M)$ on M a 2-cochain c_ω by setting:

$$c_\omega(\sigma) = \int_{\Delta_2} \bar{\sigma}^*(\omega)$$

where $\bar{\sigma}$ is the straightening of the simplex σ (with respect to the fixed hyperbolic structure); c_ω is recognized to be a bounded cocycle. It follows that a mapping

$$\Phi : \Lambda^2(M) \to \hat{H}^2(M) \qquad \omega \mapsto [c_\omega]$$

is naturally defined. The next result, to be found in [Ba-Gh], gives an idea of how large $\hat{H}^2(M)$ is:

Theorem F.6.29. The above mapping Φ is one-to-one.

Subject Index

The following index does not contain the terms which we presume are well-known to the reader, even if they appear as definitions in the book (*e.g.* the 'signature' of a bi-linear form). For the terms appearing most frequently in the book (*e.g.* 'hyperbolic manifold') we indicate the first occurrence only (*i.e.* the page where they are defined); these terms are preceded by an *asterisk. In general, the boldface numeral indicates the page where the term is defined, and the normal numeral indicates further occurrences. A few terms are defined more than once; of course the definitions are always consistent: they usually refer to slightly different cases or generalizations.

Notation Index

For the reader's convenience we list the non-standard notations most frequently used in the book, and explain their meaning; we omit the symbols whose meaning is self-evident (such as 'Conf(M)' for the group of conformal automorphisms of a manifold M).

$\partial\mathbb{H}^n$	Boundary of hyperbolic n-space
$\mathcal{D}_*(G)$	Set of all discrete torsion-free subgroups of a topological group G
$\mathbb{D}^n$	Disc model of hyperbolic n-space
$\mathcal{E}(\xi)$	Euler class of a flat vector bundle ξ
ε_n	The n-th Margulis constant
$\mathcal{F}_n$	Family of all finite-volume complete oriented hyperbolic n-manifolds
$\mathcal{F}_n(c)$	Family of elements of $\mathcal{F}_n$ having volume at most c
$\mathcal{H}_n$	Family af all n-dimensional complete oriented hyperbolic manifolds
$\mathcal{H}(M)$	(Not necessarily complete) hyperbolic structures supported by a manifold M of $\mathcal{T}_3$
$\mathbb{H}^n$	Hyperbolic n-space (as an abstract Riemannian manifold)
$\overline{\mathbb{H}^n}$	Hyperbolic n-space together with its boundary
$i_{x_0,\alpha}$	Inversion with respect to the sphere of centre x_0 and radius $\sqrt{\alpha}$
$\mathbb{I}^n$	Hyperboloid model of hyperbolic n-space
$\mathcal{I}(M)$	Group of isometries of a Riemannian manifold M
$\mathcal{I}^+(M)$	Group of orientation-preserving isometries of an oriented Riemannian manifold M
Λ	Lobachevsky function
$M_{d_1,\dots,d_k}$	Manifold obtained by Dehn surgery of coefficients $d_1,\dots,d_k$ from a manifold M whose boundary consists of tori
$M_{(0,\varepsilon]}$	The ε-thin part of a hyperbolic manifold M
$M_{[\varepsilon,\infty)}$	The ε-thick part of a hyperbolic manifold M
$\|M\|$	Gromov norm of a manifold M
$P_{x,y}$	Parallel transport along the unique geodesic line joining two points x and y of $\mathbb{H}^n$

$\mathbf{H}^{n,+}$ Half-space model of hyperbolic n-space

$\mathcal{S}_n$ Set of geodesic simplices in $\overline{\mathbf{H}^n}$

T_g Compact oriented surface of genus g

$\mathcal{T}_3$ Family of oriented three-manifolds bounded by tori and obtained by glueing tetrahedra along faces and removing vertices

τ_g Teichmüller space (hyperbolic structures on a surface of genus g, up to isometries isotopic to the identity)

v_n Maximal volume of a geodesic simplex in $\overline{\mathbf{H}^n}$

$\sphericalangle (v, w)$ Angle between two vectors v and w

References

[Ab]   W. Abikoff. *The real analytic theory of Teichmüller space.* Lecture Notes in Math. 820, Springer-Verlag, Berlin-Heidelberg-New York (1980).

[Ad1]   C. C. Adams. *Thrice-punctured spheres in hyperbolic 3-manifolds.* Trans. Am. Math. Soc., **287** (1985), pp. 645-656.

[Ad2]   C. C. Adams. *Augmented alternated link complements are hyperbolic.* In London Math. Soc. Lecture Notes Series, no. 112 (D. B. A. Epstein editor), pp. 115-130.

[Ad-Hi-We] C. C. Adams, M. Hildebrand, J. Weeks. *Hyperbolic invariants of knots and links.* To appear in Trans. Amer. Math. Soc.

[Av]   A. Avez. *Variétés riemanniennes sans points focaux.* C. R. Acad. Sc. Paris, **270** (1970), pp. 188-191.

[Ba-Gr-Sc] W. Ballman, M. Gromov, V. Schroeder. *Manifolds of nonpositive curvature.* Birkhäuser, Boston (1985).

[Ba-Gh]   J. Barge, E. Ghys. *Surfaces et cohomologie bornée.* Inv. Math., **92** (1988), pp. 509-526.

[Bea]   A. F. Beardon. *The geometry of discrete groups.* Graduate Texts in Math. 91, Springer-Verlag, Berlin-Heidelberg-New York (1983).

[Ber]   M. Berger. *Geometry*, part I. Universitext, Springer-Verlag, Berlin-Heidelberg-New York (1987).

[Boo]   W. M. Boothby. *An introduction to differentiable manifolds and Riemannian geometry*, (Second edition). Academic Press Inc., New York (1986).

[Bor1]   A. Borel. *Compact Clifford-Klein forms of symmetric spaces.* Topology, **2** (1963), pp. 111-122.

[Bor2]   A. Borel. *Introduction aux groupes arithmétiques.* Hermann, Paris (1969).

[Bu-Ka]   P. Buser, H. Karcher. *Gromov's almost flat manifolds.* Astérisque 81, Soc. Math. de France (1981).

[Cas]   B. G. Casler. *An embedding theorem for connected 3-manifolds with boundary.* Proc. Amer. Math. Soc., **16** (1965), pp. 559-566.

[Ca-Ep-Gr] R. D. Canary, D. B. A. Epstein, P. Green. *Notes on notes of Thurston.* London Math. Soc. Lecture Notes Series 111 (D. B. A. Epstein editor), (1984).

[Chab]   C. Chabauty. *Limites d'ensemble et géométrie des nombres.* Bull. Soc. Math. France, **78** (1950), pp. 143-151.

[Char]   L. S. Charlap. *Bieberbach groups and flat manifolds.* Universitext, Springer-Verlag, Berlin-Heidelberg-New York (1986)

[Ci]   P. G. Ciarlet. *Introduction à l'optimization.* Masson, Paris (1982).

[Co]   H. S. M. Coxeter. *Non-Euclidean geometry.* Toronto Univ. Press (1942).

[DC]   M. P. Do Carmo. *Differential geometry of curves and surfaces.* Prentice-Hall Inc., London (1976).

[Ep1]   D. B. A. Epstein. *Curves on 2-manifolds and isotopies.* Acta. Math., **115** (1966), pp. 83-107.

[Ep2]   D. B. A. Epstein. *Isomorfismi conformi e geometria iperbolica.* Quaderno C.N.R., Università di Cagliari (1983).

[Ep-Pe]   D. B. A. Epstein, R. C. Penner. *Euclidean decomposition of non-compact hyperbolic manifolds.* J. Diff. Geom., **27**, n.1 (1988), pp. 67-80.

[Fa-La-Po] A. Fathi, F. Laudenbach, V. Poenaru et al. *Travaux de Thurston sur les surfaces.* Astérisque 66-67 (1979) pp. 1-284.

[Fe]   W. Fenchel. *Elementary geometry in hyperbolic spaces.* de Gruyter Studies in Math., no. 11, Berlin-New York (1989).

[Fo]   L. R. Ford. *Automorphic functions.* Chelsea, New York (1951).

[Ga-Hu-La] S. Gallot, D. Hulin, J. Lafontaine. *Riemannian geometry.* Universitext, Springer-Verlag, Berlin-Heidelberg-New York (1987).

[Gh-LH]   E. Ghys, P. de La Harpe (editors). *Sur les groupes hyperboliques d'après Mikhael Gromov.* Progress in Math. 77, Birkhäuser, Boston (1990).

[Greenb1]   M. J. Greenberg. *Lectures on algebraic topology.* Benjamin, London (1967).

[Greenb2]   M. J. Greenberg. *Euclidean and non-Euclidean geometries.* W. H. Freeman and Co., S. Francisco (1974).

[Greenl]   F. P. Greenleaf. *Invariant means in topological groups and their applications.* Van Nostrand, New York-London (1969).

[Gro1]   M. Gromov. *Hyperbolic manifolds according to Thurston and Jorgensen.* Sém. Bourbaki 32ᵉ année (1979-80) n. 546, pp. 1-14.

[Gro2]   M. Gromov. *Groups of polynomial growth and expanding maps.* Publ. Math. I.H.E.S. **53** (1981), pp. 53-78.

[Gro3]   M. Gromov. *Volume and bounded cohomology.* Publ. Math. I.H.E.S. **56** (1982), pp. 5-99.

[Gro4]   M. Gromov. *Hyperbolic Groups.* In *Essays in Group Theory*, S. M. Gersten (editor), M.S.R.I. Publ. 8, Springer-Verlag, Berlin-Heidelberg-New York (1987), pp. 75-263.

[Gr-La-Pa] M. Gromov, J. Lafontaine, P. Pansu. *Structures métriques pour les variétés riemanniennes.* Cedic F. Nathan, Paris (1981).

[Gr-PS]   M. Gromov, I. Piatetski-Shapiro. *Non-arithmetic groups in Lobachevskji space.* Publ. Math. I.H.E.S. **66** (1988), pp. 93-103.

328   References

[Ha-Mu]   U. Haagerup, H. J. Munkholm. *Simplices of maximal volume in hyperbolic n-space*. Acta Math., **147** (1981) n.1-2, pp. 1-11.

[He]   S. Helgason. *Differential geometry, Lie groups and symmetric spaces.* Academic Press Inc., New York (1978).

[Hic]   N. J. Hicks. *Notes on differential geometry.* Van Nostrand, New York-London (1965).

[Hir]   M. W. Hirsch. *Differential topology.* Graduate Texts in Math. 33, Springer-Verlag, Berlin-Heidelberg-New York (1976).

[Hi-Th]   M. W. Hirsch, W. P. Thurston. *Foliated bundles, invariant measures and flat manifolds.* Ann. of Math., **101** (1975), pp. 369-390.

[Iv]   N. V. Ivanov. *Foundations of the theory of bounded cohomology.* Journal of Soviet Math., **37** (1987) pp. 1090-1114.

[Jo]   T. Jorgensen. *Compact 3-manifolds of constant negative curvature fibering over the circle.* Ann. of Math., **106** (1977), pp. 61-72.

[Ka-Ma]   D. Kazhdan, G. Margulis. *A proof of Selberg's hypothesis.* Math. Sb. (117) **75** (1968), pp. 163-168.

[Ko-No]   S. Kobayashi, K. Nomizu. *Foundations of differential geometry.* Interscience, New York (1963).

[La]   S. Lang. *Complex analysis.* Graduate Texts in Math. 103, Springer-Verlag, Berlin-Heidelberg-New York (1985).

[LH-Sk]   P. de La Harpe, G. Skandalis. *Un résultat de Tarski sur les actions moyennables de groupes et les partitions paradoxales.* L'Enseignement Math., **32** (1986), pp. 121-138.

[Mag]   W. Magnus. *Noneuclidean tesselations and their groups.* Academic Press, New York (1974).

[Mar]   G. A. Margulis. *Isometry of closed manifolds of constant negative curvature with the same fundamental group.* Soviet. Math. Dokl., **11** (1970), pp. 722-723.

[Mask1]   B. Maskit. *On Poincaré theorems for fundamental polygons.* Adv. in Math., **7** (1971), pp. 219-230.

[Mask2]   B. Maskit. *Kleinian groups.* Grund. der Math. Wissenschaften 287, Springer-Verlag, Berlin-Heidelberg-New York (1988).

[Mass1]   W. S. Massey. *Algebraic topology: an introduction.* Harcourt Brace and World, New York (1967).

[Mass2]   W. S. Massey. *Singular homology theory.* Springer-Verlag, Berlin-Heidelberg-New York (1980).

[Ma-Fo]   S. V. Matveev, A. T. Fomenko. *Constant energy surfaces of Hamiltonian systems, enumerating of three-dimensional manifolds in increasing order of complexity, and computation of volumes of closed hyperbolic manifolds.* Uspekhi Mat. Nauk 43:1 (1988), pp. 5-22. Russian Math. Surveys **43:1** (1988), pp. 3-24.

[Me]   W. W. Menasco. *Polyhedra representation of links complements.* Contemporary Math., **20** (1983), pp. 303-325.

[Mi1]   J. W. Milnor. *On the existence of a connection with curvature zero.* Comment. Math. Helv. **32** (1958), pp. 215-233.

[Mi2]   J. W. Milnor. *Topology from the differentiable viewpoint.* Univ. Press of Virginia, Charlottesville (1965).

[Mi3]   J. W. Milnor. *Hyperbolic geometry: the first 150 years.* Bull. Amer. Math. Soc., **6**, n. 1 (1982), pp. 9-23.

[Moo]   J. C. Moore. *Semi-simplicial complexes and Postnikov systems.* Symp. de Top. Alg. Mexico (1958), pp. 232-248.

[Mor]   J. W. Morgan *On Thurston's uniformization theorem for three-dimensional manifolds.* In *The Smith conjecture".* (Morgan-Bass editors) Academic Press Inc., New York (1984).

[Mos]   G. D. Mostow. *Strong rigidity of locally symmetric spaces.* Ann. of Math. Studies 78, Princeton (1973).

[Mu]   H. J. Munkholm. *Simplices of maximal volume in hyperbolic space, Gromov's norm and Gromov's proof of Mostow's rigidity theorem (following Thurston).* Top. Symp. Siegen 1979, Lecture Notes in Math. 788, Springer-Verlag, Berlin-Heidelberg-New York (1980) pp. 109-124.

[Na-Ya]   T. Nagano, K. Yagi. *The affine structures on the real two-torus.* Osaka J. Math., **11** (1974), pp. 181-210.

[Na]   R. Narasimhan. *Complex analysis in one variable.* Birkhäuser, Boston (1984).

[Ne-Za]   W. D. Neumann, D. Zagier. *Volumes of hyperbolic 3-manifolds.* Topology, **24**, n. 3 (1985) pp. 307-332.

[Pe]   C. Petronio. *An algorithm producing hyperbolicity equations for a link complement in S^3.* Preprint Scuola Normale Superiore, Pisa (1991). To appear in Geom. Ded.

[Pr]   G. Prasad. *Strong rigidity of $\mathbb{Q}$-rank 1 lattices.* Inv. Math., **21** (1973), pp. 255-286.

[Ra]   M. S. Raghunathan. *Discrete subgroups of Lie groups.* Ergebn. der Math. 58, Springer-Verlag, Berlin-Heidelberg-New York (1972).

[Rol]   D. Rolfsen. *Knots and links.* Lect. Series 7, Publish or perish, Berkeley (1976).

[Roy]   H. L. Royden. *Real analysis*, Second edition. McMillan, New York (1968).

[Ro-Sa]   C. P. Rourke, B. J. Sanderson. *Introduction to piecewise-linear topology.* Ergebn. der Math. 69, Springer-Verlag, Berlin-Heidelberg-New York (1972).

[Si]   C. L. Siegel. *Topics in complex function theory*, vol. II. Wiley-Interscience, New York (1971).

[Sm]   J. Smillie. *An obstruction to the existence of affine structures.* Inv. Math., **64** (1981), pp. 411-415.

[Sp]   M. Spivak. *A comprehensive introduction to differential geometry*, vol. II. Publish or Perish, Berkeley (1970).

330   References

[Su1]   D. Sullivan. *A generalization of Milnor's inequality concerning affine foliations and affine manifolds.* Comm. Math. Helv., **51** (1976), pp. 183-189.

[Su2]   D. Sullivan. *Seminar on conformal and hyperbolic geometry.* Publ. Math. I.H.E.S., March 1982.

[Su-Th]   D. Sullivan, W. P. Thurston. *Manifolds with canonical coordinate charts: some examples.* L'Enseignement Math., **29** (1983), pp. 15-25.

[Th1]   W. P. Thurston. *The geometry and topology of three-manifolds.* Princeton University Mathematics Department (1979).

[Th2]   W. P. Thurston. *Three dimensional manifolds, kleinian groups and hyperbolic geometry.* Bull. Amer. Math. Soc., **6**, n. 3 (1982), pp. 357-381.

[Th3]   W. P .Thurston. *Hyperbolic structures on 3-manifolds.* Ann. of Math., **124** (1986), pp. 203-246.

[Ve]   E. Vesentini. *Capitoli scelti della teoria delle funzioni olomorfe.* Gubbio, UMI (1984).

[Wa]   H. C. Wang. *Topics in totally discontinuous groups.* In *Symmetric spaces*, ed. Boothby-Weiss, New York (1972), pp. 460-485.

[Wi]   N. J. Wielenberg. *Hyperbolic 3-manifolds which share a fundamental polyhedron.* In: *Riemannian surfaces and related topics*, proceedings of the 1978 Stony Brook Conference, Princeton University Press, Princeton N.J. (1981), pp. 505-513.

[Wol]   J. A. Wolf. *Spaces of constant curvature.* McGraw-Hill, New York (1967).

[Woo]   J. Wood. *Bundles with totally disconnected structure group.* Comment. Math. Helv., **46** (1971), pp. 257-273.

Universitext

Printing and Binding: Strauss GmbH, Mörlenbach